The Biologic Role of Dehydroepiandrosterone (DHEA)

The Biologic Role of Dehydroepiandrosterone

(DHEA)

Editors
M. Kalimi · W. Regelson

W DE G

Walter de Gruyter · Berlin · New York 1990

Editors

Mohammed Kalimi, Ph. D.
Professor of Physiology
Medical College of Virginia
Virginia Commonwealth University
Richmond, Virginia 23298
U.S.A.

William Regelson, M. D.
Professor of Medicine and Microbiology
Medical College of Virginia
Virginia Commonwealth University
Richmond, Virginia 23298
U.S.A.

∞ Printed on acid-free paper which falls within the guidelines of the ANSI to ensure permanence and durability

Library of Congress Cataloging-in-Publication Data

The Biologic role of dehydroepiandrosterone (DHEA) / editors, M. Kalimi, W. Regelson.
Includes bibligraphical references.
Includes indexes.
ISBN 3-11-012243-X
1. Dehydroepiandrosterone--Therapeutic use--Testing.
2. Dehydroepiandrosterone--Physiological effect.
I. Kalimi, M. (Mohammed), 1939- . II. Regelson, William
[DNLM: 1. Dehydroepiandrosterone--pharmacology.
2. Dehydroepiandrosterone--therapeutic use. WK 150 B615]
RM296.5.D45B56 1990
615'.36--dc20 90-13958

CIP-Titelaufnahme der Deutschen Bibliothek

The biologic role of dehydroepiandrosterone : (DHEA) / ed. M. Kalimi ; W. Regelson. - Berlin ; New York : de Gruyter, 1990
ISBN 3-11-012243-X
NE: Kalimi, Mohammed [Hrsg.]

Printed in Germany · Printing: Gerike GmbH, Berlin. - Binding: Dieter Mikolai, Berlin.

PREFACE

Dehydroepiandrosterone (DHEA) is a native steroid that declines with progressive age, and is found in the brain at concentrations equal to that in the adrenal cortex. For many years, the role of DHEA or its sulfate has focussed on its place as an intermediate in sex steroid synthesis. More recently, DHEA or related analogs have been found to prevent carcinogenesis and to show anti-diabetogenic action. There is a growing interest in DHEA's clinical place in atherosclerosis, hypertension, memory disorders, fat mobilization and cancer prevention and treatment. Importantly, DHEA or its metabolites up regulate host resistance to viral or bacterial infection. DHEA's biologic role is protean, as it has collagenolytic activity, and acts as a naturally occurring digitalis. The clinical experience with DHEA suggest it may be useful in fatigue syndromes and in lowering blood cholesterol. Clinical trials are in progress in cancer, diabetes, obesity, hypercholesterolemia, Alzheimer's disease and multiple sclerosis.

DHEA acts as a precursor steroid and/or a "buffer hormone" that alters "state dependency" by interacting with other hormones. Our text explores DHEA's broad biologic action and its potential relevance to clinical disease.

We wish to thank Eva Gross for her efforts in assembling and preparing manuscripts, and we wish to complement Arthur Schwartz and Etienne-Emile Baulieu who have helped to develop appropriate attention to this interesting native steroid.

Richmond

W. Regelson
M. Kalimi

Contents

DEHYDROEPIANDROSTERONE (DHEA): THE PRECURSOR STEROID: INTRODUCTORY REMARKS.

William Regelson, Mohammed Kalimi and Roger Loria

Departments of Medicine, Physiology and Microbiology, Medical College of Virginia/Virginia Commonwealth University, Richmond, Virginia 23298

DHEA is a precursor steroid which gives rise to the sex steroids as well as etiocholanolone which have profound effects on a wide variety of physiologic or pathophysiologic events. DHEA modulates diabetes, obesity, carcinogenesis, tumor growth, neurite outgrowth, virus and bacterial infection, stress, pregnancy, hypertension, collagen and skin integrity, fatigue, depression, memory and immune responses. In the past, we ascribed this action to DHEA's role in "state dependency" and likened its action to "buffer hormones" in an effort to explain the multiple physiologic effects of this single hormone (1). "State dependency" explains the varied action of a hormone based upon the expression of its action only within particular physiologic settings.

Examples of "state-dependent" action or "buffer hormones" with widely varied physiologic effects, other than DHEA, are thyroid or thyrotropin releasing hormones (TRH) and melatonin. Their action affects widely divergent systems, depending on physiologic state, i.e., shock, immune status, circadian or seasonal rhythms. In this regard, DHEA is perhaps our best example of a "buffer steroid" as it appears to work in widely divergent systems against a variety of targets, depending on

Dehydroepiandrosterone (DHEA)

the state of the host. The known metabolic actions of DHEA or its sulfated derivative (DHEA-S) are inhibition of glucose-6-phosphate dehydrogenase (G-6-PD) and the pentose shunt, ornithine decarboxylase, K-channel blockade or via its action on Na^+K^+ ATPase, insulin, peroxisomal induction, and lipid metabolism or regulation of specific or non-specific hormone receptors? In summary, this textual review asks if the diverse actions of DHEA can have a unifying feature or if DHEA works through multiple, but in some cases interrelated mechanisms? Physiologists in the past have viewed DHEA as a primary source of more important steroids because DHEA's major decline following adrenalectomy does not produce gross symptoms of clinical withdrawal suggesting its functional role is paraendocrine rather than acting as a targeted excitatory hormone in the classical sense.

Baulieu showed that DHEA is sulfated to DHEA-S and that this is the major form in clinical circulation. The clinical conversion of DHEA to DHEA-S and back again is rapidly mediated by sulfokinases and sulfatases so that DHEA and DHEA-S are thought to be almost interchangeable in most tissues of the body although the ability of DHEA to inhibit G-6-PD and enter cells in contrast to DHEA-S suggest that critical differences exist between DHEA and its sulfated metabolites which still need evaluation.

The broad function of DHEA was reviewed by Sonka (2), in his pioneering monograph where he suggested that the major age related decline in DHEA is responsible for degenerative disease which he called "the hyperproductive syndrome." He suggested that the function of DHEA was to inhibit enzyme systems necessary for building new cells, i.e., nucleic acids, lipids and steroids. Sonka felt that with the age related decline in DHEA levels proliferative or degenerative events develop related to the loss of DHEA as a key regulatory hormone controlling key enzymes. Under certain pathophysiologic circumstances, i.e., carcinogenesis, obesity, diabetogenesis and stress, DHEA functions as an "anti-hormone." The late Norman Applezweig, once suggested that DHEA "cannot serve to 'excite' in the true classical sense of hormone action, but that it acts to 'de-excite' metabolic processes which overproduce when DHEA is in short supply."

Whatever its mechanism of physiologic action, DHEA is perhaps our most significant endocrine biomarker that declines with age. For this reason, we must continue to ask if the age related clinical decrease is pertinent to the pathology or functional changes seen in aging and if the anti-carcinogenesis, or anti-diabetogenic action of DHEA in animal models has relevance to any of the above? We must ask why DHEA is found in the brain and how this might have relevance to memory, mood, neuroendocrine changes, neuroimmunomodulation, and neurite formation? Does DHEA's apparent affect on improving clinical fatigue, mobilizing fat or lowering cholesterol levels and preventing arteriosclerosis in model systems have clinical relevance and do any of these effects have anything in common with one another?

Because of our growing awareness of its wide ranging physiologic effects, it is essential that we determine if DHEA, in a wide variety of targeted organs, functions by direct interaction with as yet fully defined receptors (3) or if DHEA is metabolized to other functional intermediates which have not been thought to have any distinctive targeted role. Based on our recent experience (4), it is our position that we will find that DHEA is very likely metabolized by two completely different pathways. One which we are more familiar with, i.e., via the hepatic circulation to DHEA and DHEA-S, and a second pathway in the cutaneous tissues where DHEA is converted to androstenediol (AED) and androstenetriol where it is racemized, redistributed and sulfated to appear in the circulation as androstenediol sulfate. We feel that the second pathway is the one by which DHEA exerts its immunostimulating effect.

Most importantly, DHEA functions as the progenitor for a wide variety of functional intermediates produced in the synthesis of the sex steroids, each one of which will be found to have modulating or direct effects on targeted tissue or organ sites. This is supported by our work (4) (Loria, et al., this text) wherein DHEA acts as an up regulating steroid enhancing immune resistance to viral and bacterial action which suggests that DHEA is metabolized to a more effective immunostimulatory hormone. This is based on the initial action of DHEA, in vitro, where it is cytocidal or inhibitory to lymphocytes in a mixed lymphocyte reaction (MLR) (Loria, work in progress). Similarly, in Coleman's studies (5) (this text) it is not just DHEA which is anti-diabetogenic, but estrogen and etiocholanolone, two DHEA metabolites which have similar anti-diabetogenic action.

The precursor role of DHEA is not only to produce useful metabolites, but to modulate interactions between steroids. In terms of its immunomodulating action, DHEA or its AED metabolites act as immunostimulators, but also may "buffer" or decrease corticosteroid or stress related thymic involution. DHEA blocks corticosteroid or stress related involution of the thymus in young mice (Riley, et al. this text) and in this regard, DHEA shows how a single hormone can have diverse effects dependent on the physiologic or metabolic state of the host independent of the commonly ascribed very specific targeted role for that hormone.

Based on Loria et al. (4) and Coleman's et al's (5) work, we have to reappraise DHEA and see it as the source of metabolites that were largely ignored because they were felt to be incidental in the synthesis of dihydrotestosterone, testosterone, estrogen and etiocholanolone. This review of the biologic action of DHEA seeks to stimulate us to examine the widely diverse physiologic expression of DHEA and DHEA-S as a function of its metabolites in similar fashion to studies which have led to our current understanding of peptide hormone action where differing peptide fractions have varied functions that differ from the parent hormone.

In this volume, we present specific data regarding the biologic action of DHEA as well as concepts and speculation regarding its mechanism of action. From observations of the variety of physiologic expression seen for DHEA, we must look to see if its divergent effects relate to specific receptor action, or in addition, can act by providing a buffering, modulating or state dependency effect.

In regard to the above mechanisms, it is our considered opinion that many of the actions of DHEA will eventually be explained by the expression or inhibition of cytokines by DHEA or its metabolites. The model that suggests that modulation of cytokines may explain much of DHEA effects is that of DHEA steroid related inhibition or stimulation of mammary tumor growth that is now frequently related to changes in cytokine production. Cytokine modulation by DHEA, directly or indirectly, could explain how it exerts its wide ranging proliferative or anti-proliferative effects on skin, hair, sebaceous glands liver, spleen, pancreatic islet cells, ovaries and testes. In support of this cytokine concept as a mechanism of DHEA action, the reader is referred to Colemans' review wherein the antidiabetogenic action of DHEA in C57BL1 mouse models is clearly associated with a DHEA induced pancreatic islet cell hyperplasia. Alternatively, the actions of DHEA while diverse, can intersect at key metabolic places, i.e., having an insulin permissive effect and/or a simultaneous blocking action on hypercorticism that would encourage gluconeogenesis and/or block thymic involution.

In summary, because of the multiple roles of DHEA, and its far ranging, biologic effects, we maintain that all its effects can be explained if one looks at it as a precursor that provides us with a host of steroid progeny with which to maintain the broad balance of host response related to species and individual survival. This volume attempts to shed more light on the action of DHEA and to call attention to the role of this hormone on new physiologic targets for the expression of steroid function. We feel DHEA and its metabolites will have useful relevance to clinical function and the treatment or prevention of disease, and we hope that this text can help in stimulating new and productive basic and clinical research.

References

1. Regelson, W., R. Loria, M.K. Kalimi, 1988. Ann. N.Y. Acad. Sci., 521, 260-273.

2. Sonka, J. 1976. ACTA Univ. Carol 71, 1-137, 146-71.

3. Kalimi, M., W. Regelson. 1988. Biochem.Biophys.Res. Commun. 156, 22-29.

4. Loria, R.M., T.H. Inge, S.H. Cook, et al. 1988. J. Med. Virol. 26, 301-314.

5. Coleman, D.L., A. Leiter, R.W. Schwizeb. 1982. Diabetes, 31, 830-833.

THE BIOLOGICAL SIGNIFICANCE OF DEHYDROEPIANDROSTERONE

A.G. Schwartz, D.K. Fairman, L.L. Pashko
Fels Institute for Cancer Research and Molecular Biology, Temple University Medical School, Philadelphia, PA 19140

What is perhaps so striking about dehydroepiandrosterone (DHEA) and dehydroepiandrosterone-sulfate (DHEAS) is that, although it has been known for many years that the human adrenal cortex secretes these steroids in abundant quantities, no definitive biological function has been ascribed to them in the male and non-pregnant female [1]. Although these steriods can be metabolized into active estrogens and androgens, there is little evidence that they serve a primary role as sex hormone precursors in the human.

Glucose-6-Phosphate Dehydrogenase Inhibition

There is now a great deal of experimental evidence that the administration of DHEA (but not DHEAS) to laboratory animals produces an array of biological effects, which, in their diversity, appear remarkable [2,3]. Although it is unlikely that all of these biological effects will be attributed to a single mechanism of action, there is increasing evidence that one effect of the steroid, that is the inhibition of glucose-6-phosphate dehydrogenase (G6PDH), is central to the cancer preventive activity of DHEA (2,3,4).

DHEA is a potent uncompetitive inhibitor of mammalian G6PDH [5]. G6PDH is the rate-controlling enzyme in the pentose-phosphate cycle, a pathway that is a major source of ribose-5-phosphate and extra-

Dehydroepiandrosterone (DHEA)

mitochondrial NADPH. Reduced cellular levels of ribose-5-phosphate and NADPH are apparently responsible for the inhibition of cellular proliferation (NADPH and ribose-5-phosphate are both required for ribonucleotide and deoxyribonucleotide synthesis) as well as for the inhibition in the rate of carcinogen activation (NADPH is required for mixed-function oxidase activity). NADPH is also a necessary co-factor for a membrane-bound oxidase which generates superoxide anion (O_2^-)(6).

Thus, reduced pentose-phosphate cycle activity following DHEA treatment may produce both anti-tumor initiating activity (a reduction in the rate of carcinogen activation) and anti-tumor promoting effects (inhibition of cellular proliferation and O_2^- formation).

It is an interesting possibility that G6PDH inhibition by DHEA could also contribute to the anti-atherogenic effect of the steroid [7,8,9], since both hyperproliferation of smooth muscle cells [10] as well as O_2^- - mediated oxidation of LDL [11] are both believed to be important processes in the development of an atheroma.

Biological Role of DHEA

Although there is now little doubt that the administration of large oral doses of DHEA to laboratory animals produces specific biological effects, some of which are apparently the result of the potent and specific inhibition of G6PDH by the steroid, there is little consensus that these actions of DHEA are associated with the normal biological function of the steroid. The inhibitor constant (Ki) for DHEA against several different purified mammalian G6PDH preparations is approximately 18 µM [4], whereas the concentration of DHEA in human plasma is about 0.01 µM to 0.02 µM [12], a level that would produce negligible inhibition of G6PDH. The concentration of DHEAS in human plasma, approximately 5 µM to 7 µM in the second decade [13], is well within the range that produces G6PDH inhibition for the free steriod, but not for DHEAS, which is essentially inactive as an inhibitor of G6PDH (14).

Oertel reported the isolation of an unstable lipoidal conjugate between DHEA and sulfatidic acid (DHEA sulfatide) which was an even more potent inhibitor of G6PDH than DHEA [15]. According to Oertel and Benes [14], DHEA-sulfatide is the predominant form of DHEA circulating in human plasma, at concentrations that could physiologically regulate G6PDH activity. However, this finding of Oertel has not been confirmed by others, and whether or not DHEA, or a conjugate of the steroid, regulates G6PDH activity levels must remain an open question.

Food Restriction

One cannot help but be impressed by the similarity of effects of long-term DHEA treatment of laboratory animals in delaying the development of diseases closely linked with aging (cancer, atherosclerosis, auto-immune diseases, etc.) and the beneficial effects of food restriction. Reducing the food intake of laboratory mice and rats produces a remarkable retardation in the rate of appearance of age-related pathologic changes, including spontaneous and experimentally induced tumors [16], autoimmune diseases [17], etc.

Food restricting mice and rats impairs reproductive capacity in favor of maintenance of essential function [18,19]. However, if the food deprived animals are fully fed later in life, they demonstrate a reproductive capacity superior to those fed ad libitum throughout life [18,19], that is the food restricted animals experience a delay in the normal decline in reproductive capacity. Thus, the delay in the rate of aging produced by food restricting laboratory animals may represent an adaptive response to periodic variations in food availability in short-lived species, and may help prevent a decline in population density during such periods.

We have found that the epidermis of mice responds to food restriction in a manner that is remarkably similar to the topical application of DHEA. When [^{3}H]7,12-dimethylbenz(a)anthracene is applied to the backs of either food restricted or DHEA treated mice, there is an inhibition in the amount of carcinogen bound to epidermal DNA over a 12-hour period, and similarly, tetradecanoyl-phorbol-13-acetate stimulation of epidermal [^{3}H]thymidine incorporation is suppressed (20). Both of these biological effects, when produced by topical DHEA treatment, are very likely the result of decreased G6PDH activity, and we did indeed also find that food restricting mice for two weeks depressed epidermal G6PDH activity by 58% (20).

Many years ago Boutwell et al. reported that reducing the food intake of mice increases the activity of the adrenal cortex (21), and one might speculate that elevated levels of DHEA, or a conjugate of the steroid, could possibly mediate some of the age-retarding effects of food restriction.

DHEA Analogs

Irrespective of whether or not the biological effects that DHEA produces in laboratory animals are related to the normal physiological action of the steroid, if some of these beneficial effects can be achieved in the human, important new therapeutic applications would result. Although long-term DHEA treatment of mice and rats produces no apparent toxicity, side-effects have been noted which could limit the use of DHEA as a drug for humans. DHEA can be metabolized into both estrogens and androgens and produces both seminal-vesicle enlarging and a uterotrophic effect in mice and rats (22). In addition, DHEA treatment produces hepatomegaly and stimulates liver catalase activity and peroxisome proliferation (22,23). These side-effects are greatly reduced in the synthetic steroid, 16α-fluoro-5-androsten-17-one, whereas tumor preventive efficacy, as determined in the two-stage skin papilloma model in mice, is enhanced (24). Synthetic analogs of DHEA, such as 16α-fluoro-5-androsten-17one, may find clinical application as drugs for humans.

References

1. Lieberman, S. 1986. J. Endocrinol. 111, 519.

2. Schwartz, A.G., Whitcomb, J.M., Nyce, J.W., Lewbbart, M.L., Pashko, L.L. 1988. Ad. Cancer Res. 51, 391.

3. Gordon, G.B., Shantz, L.M., and Talalay, P. 1987. Ad. Enzyme Regul. 26, 355.

4. Shantz, L.M., Talalay, P., Gordon, G.B. 1989. Proc. Natl. Acad. Sci. USA 86, 3852.

5. Raineri, R. and Levy, H.R. 1970. Biochemistry 9, 2233.

6. Babior, B.M. 1982. Can. J. Physiol. Pharmacol. 60, 1353.

7. Gordon, G.B., Bush, D.E., Weisman, H.F. 1988. J. Clin. Invest. 82, 712.

8. Arad, Y., Badimon, J.J., Badimon, L., Hembree, W.C., Ginsberg, H.N. 1989. Arteriosclerosis 9, 159.

9. Eich, D.M., Johnson, D.E., Dworkin, G.H., Nestler, J.E., Thomson, J.A., Hess, M.L., Wechsler, A.S. 1989. Arteriosclerosis 9, 731a.

10. Ross, R. 1986. New Engl. J. Med. 314, 488.

11. Parthasarathy, S., Wieland, E., Steinberg, D. 1989. Proc. Natl. Acad. Sci. USA 86, 1046.

12. Zumoff, B., Rosenfeld, R.S., Strain, G.W., Levin, J., Fukushima, P.K. 1980. J. Clin. Endocrinol. Metab. 51, 330.

13. Orentreich, N., Brind, J.L., Rizer, R.L., Vogelman, J.H. 1984. J. Clin. Endocrinol. Metab. 59, 551.

14. Oertel, G.W. and Benes, P. 1972. J. Steroid Biochem. 3, 493.

15. Oertel, G.W. 1966. Hoppe-Seylers Z. Physiol. Chem. 343, 276.

16. Tannenbaum, A. and Silverstone, H. 1953. Adv. Cancer Res. 1, 451.

17. Fernades, G., Good, R.A. 1984. Proc. Natl. Acad. Sci. USA 81, 6144.

18. Carr, C.J., King, J.T., Visscher, M.B. 1949. Proc. Fedn. Am. Socs. Exp. Biol. 8, 22.

19. Merry, B.J., Holehan, A.M. 1979. J. Reprod. Fert. 57, 253.

20. Schwartz, A.G. and Pashko, L.L. 1986. Anticancer Res. 6, 1279.

21. Boutwell, R.K., Brush, M.K., Rusch, H.P. 1948. Am. J. Physiol. 154, 517.

22. Schwartz, A.G., Lewbart, M.L., Pashko, L.L. 1988. Cancer Res. 48, 4817.

23. Leighton, B.L., Tagliafero, A.R., Newsholme, E.A. 1987. J. Nutr. 117, 1287.

24. Schwartz, A.G., Fairman, D.K., Polansky, M., Lewbart, M.L., Pashko, L.L. 1989. Carcinogenesis 10, 1809.

DEHYDROEPIANDROSTERONE (DHEA) AND ITS SULFATE (DHEAS) AS NEURAL FACILITATORS: EFFECTS ON BRAIN TISSUE IN CULTURE AND ON MEMORY IN YOUNG AND OLD MICE. A CYCLIC GMP HYPOTHESIS OF ACTION OF DHEA AND DHEAS IN NERVOUS SYSTEM AND OTHER TISSUES

Eugene Roberts

Department of Neurobiochemistry, Beckman Research Institute of the City of Hope, Duarte, CA 91010

Introduction

When it is desired to improve nervous system function, such as enhancing suboptimal learning or accelerating repair of damage as a result of disease, injury, or aging, it is necessary to facilitate adaptive coupling among relevant functional processes by relieving rate-limiting constrictions in mutually shaping interactions among intracellular, intercellular, and extracellular components of the system.

Dehydroepiandrosterone (DHEA) and its sulfate (DHEAS), the major circulating steroids in humans, play important roles in many aspects of cellular controls (1-4). Although the mechanisms of effects observed in particular instances have not yet been identified, the overall results to date may be considered to be examples of the facilitation of processes by which immune, neural, and metabolic systems, separately and together, can cycle freely through their operational modes in solving problems of survival and reproduction and in achieving rebalancing when malfunctioning occurs.

Below are related results of experiments that tested the supposition that raising extracellular levels of DHEA and/or DHEAS might have some beneficial effects on aspects of nervous system function. <u>It is suggested that some of the effects observed may be consequent to facilitation of the operation of the cyclic GMP (cGMP) cycle, possibly by activating the</u>

Dehydroepiandrosterone (DHEA)

guanylyl cyclase, which may be the rate-limiting element in the operation of the cycle.

Low Concentrations of DHEA and DHEAS Reduce Neuronal Death and Enhance Astrocytic Differentiation in Brain Cell Culture (5,6)

Aliquots of suspensions of dissociated brains of 14-day-old mouse embryos were maintained in culture for 5 days, at which time they were found to contain closely similar numbers of cells per dish. Fresh medium alone or medium containing DHEA or DHEAS (10^{-4}-10^{-8} M) then was added, and after another 4 days the cultures were fixed and fluorescently immunostained for neurofilament protein (NF) and glial fibrillary acidic protein (GFAP). The preparations were examined by fluorescence, phase-contrast, and scanning electron microscopy. Positive staining for NF or GFAP allowed cells to be identified either as differentiated neurons or astrocytes, respectively (see ref. 5 and 6 for experimental details and literature citations).

In control cultures, rather sparsely distributed small NF-positive (NF^+) cells with short processes were observed, either isolated from each other or in clusters consisting of a few cells (Fig. 1A). The staining was, at most, of moderate intensity. In media containing DHEA (Fig. 1B) or DHEAS (Fig. 1C), remarkable increases were found in numbers of NF^+ neurons in comparison with control cultures (Fig. 1A). Under higher power (not shown), processes of the NF^+ cells appeared to be greatly extended, thickened, and intertwined. Numerous connections were noted between isolated neurons and greatly enlarged neuronal clusters. The brilliance of the fluorescence shown by the neurons and their processes was greatly enhanced in DHEA or DHEAS-containing media by comparison with the rather dull fluorescent signal observed in the controls, indicating the presence of increased content of NF protein per unit area of cell surface. Many of the NF^+ neurons in the experimental cultures were larger than those seen in the controls. Among the concentrations tested, maximal enhancements were seen at 10^{-7} M DHEA and 10^{-8} M DHEAS, respectively, the neuronal clusters being somewhat larger with DHEAS. In the case of DHEA, higher concentrations (10^{-6} M and 10^{-5} M) were less effective, while a

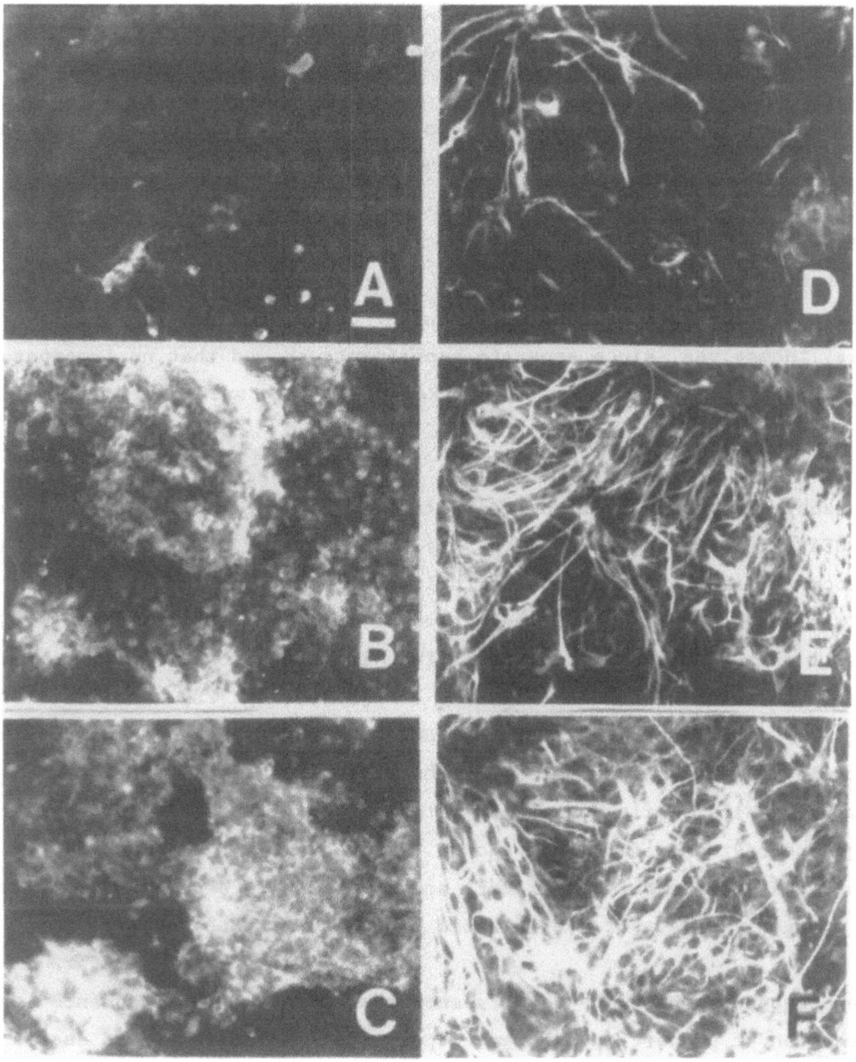

Fig. 1. Effects of DHEA and DHEAS on embryonic mouse brain cultures, 9 days <u>in vitro</u> (see text). Immunostained for NF: A, control; B, 10^{-7} M DHEA; C, 10^{-8} M DHEAS. Immunostained for GFAP: D, control; E, 10^{-7} M DHEA; F, 10^{-8} M DHEAS. Bar = 34.7 μm. From ref. 6.

concentration of 10^{-4} M failed to produce any effect at all. Some enhancing effect of DHEAS was still observed at 10^{-4}M.

Highly significant increases over control values were found in numbers of NF^+ cells (neurons) per 380 mm^2 coverslip ($P < 0.01$; 24 determinations in 3 separate experiments): control, 107±10 X 10^3; DHEA (10^{-7} M), 983±53 X 10^3; DHEAS (10^{-8} M), 997±49 X 10^3. The latter increases could be attributable to enhanced neuronal survival and/or to an acceleration of the differentiation into neurons of precursor cells, but not to mitosis of NF^+ cells, themselves, since separate experiments showed that no incorporation of [^{3}H]thymidine, whatsoever, had taken place into such cells under standard conditions of study.

Remarkable increases over the controls (Fig. 1D) were noted in the presence of DHEA (Fig. 1E) and DHEAS (Fig. 1F) in numbers of GFAP-positive ($GFAP^+$) astrocytes, in the extensions of their processes, and in the brilliance of their fluorescent staining, indicative of increased total amounts of GFAP as well as per unit cell surface. Since incorporation of [^{3}H]thymidine was less in cultures with DHEA (Fig. 2A) and DHEAS (Fig. 2B) than in controls (Fig. 2C), the results suggest that DHEA and DHEAS enhanced the survival of initially plated astrocytes and/or the differentiation of $GFAP^+$ astrocytes from stem cells, while decreasing mitosis. No effects were observed in oligodendrocytes (results not shown). Subsequently published results showed that concentrations of DHEA and DHEAS as low as 10^{-14} M inhibited the mitogenic effect of myelin basic protein on astrocytes (7).

Taken together, the above results show that DHEA and DHEAS enhanced the expression of properties related to postmitotic differentiated states both in neurons and glia. In this regard, it is of great interest that estradiol-17β and testosterone, both metabolites of DHEA, elicited growth and arborization of neurites in cultures of new-born mouse brain hypothalamic/preoptic area and that a neuritic response to estradiol occurred in [^{3}H]estradiol-accumulating cells (8-10).

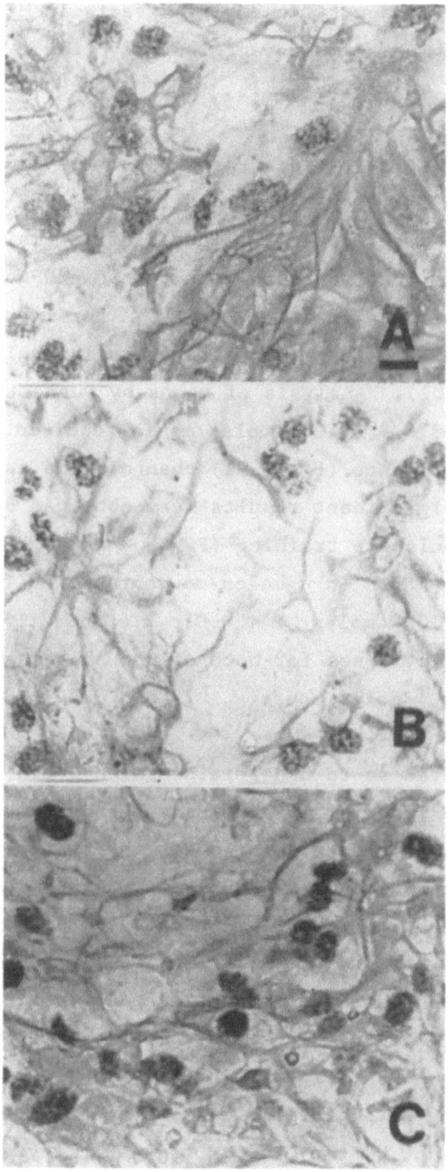

Fig. 2. Combined GFAP immunoperoxidase staining and ^{3}H-thymidine autoradiography of embryonic mouse brain cell cultures, 9 days in vitro. A, 10^{-7} M DHEA; B, 10^{-7} M DHEAS; C, control. Bar = 28.2 μm. From ref. 6.

DHEAS Showed Convincing Memory Enhancing Effects in Mice Whether Administered Intracerebroventricularly (icv), Subcutaneously (sc), or Orally (11)

Employing footshock active avoidance and passive avoidance paradigms in groups of 15-20 mice for each experimental point, DHEAS in saline given icv to undertrained male mice (35 gm) 2 minutes after training improved retention of the task when the mice were tested one week later. In T-maze footshock active avoidance training, an ANOVA for mean trials to first avoidance response showed there to be an overall significant drug effect ($F_{6,98} = 3.18$, $P < 0.01$). Analysis of means by Dunnett's t-test showed that groups receiving 2.18-6.9 X 10^{-10} moles/mouse required significantly fewer trials to achieve avoidance ($P < 0.01$) than did the saline controls. Similarly, overall significant results were obtained on memory retention in step-down passive avoidance training ($F_{8,171} = 4.72$, $P < 0.001$) (Fig. 3A). Groups receiving 1.4-5.5 X 10^{-10} moles/mouse required fewer trials than the controls (2.8-5.5 X 10^{-10} moles, $P < 0.01$; 1.4 and 2.1 X 10^{-10} moles, $P < 0.05$). The optimal icv doses for both of the above tasks were approximately 4 x 10^{-10} moles/mouse.

When DHEAS was injected sc to undertrained mice 2 min. after training in the active avoidance paradigm, significant improvement in retention was found, with the optimal effect being observed at 1-2 X 10^{-6} moles/mouse (Fig. 3B). Thus, approximately 10,000 times as much DHEAS was required to produce the same effect on retention when injected sc as when given icv.

Significant improvement of retention of the active avoidance paradigm was achieved when DHEAS was given in taste-camouflaged drinking water for one week before and one week after training (Fig. 3C). It was established that DHEAS in the drinking water for two weeks did not improve acquisition (11). The maximal enhancement by the oral route was achieved by the daily dose of 2 x 10^{-6} moles/mouse, approximately the same amount as required to give the optimal effect when given in a single sc injection immediately after training.

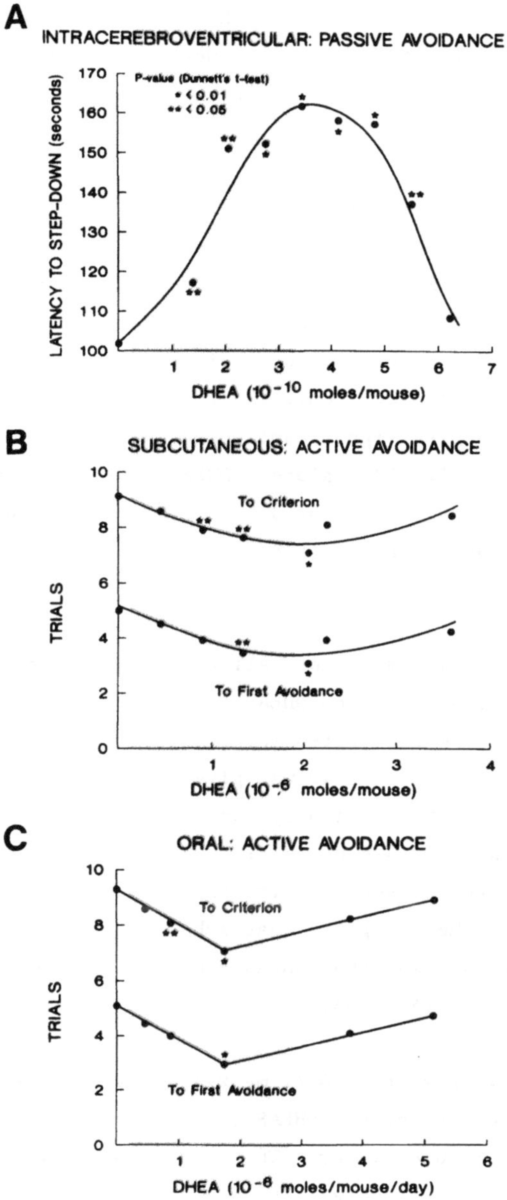

Fig. 3. Effect of icv DHEAS on retention of step-down passive avoidance training (A) and of sc (B) and oral (C) DHEAS on retention of T-maze footshock avoidance training (see ref. 11 and text for details).

DHEA Occluded the Amnestic Effects of a Membrane Perturbant (dimethylsulfoxide; DMSO). DHEAS Blocked Amnesia Induced by an Inhibitor of Protein Synthesis (anisomycin; ANI) and a Blocker of Central Muscarinic Cholinergic Receptors (scopolamine; SCO) (11)

DHEA is water insoluble but freely soluble in DMSO. Pure DMSO (2 μl) administered icv caused amnesia. DHEA dissolved in the DMSO prevented the amnestic effect. Pure DMSO, saline, or 1 of 13 doses of DHEA dissolved in pure DMSO (9×10^{-13} - 5.6×10^{-10} moles/mouse) were injected icv within 2 minutes after training. The results in Fig. 4 show the dose-response curve obtained in well-trained mice. Only the two lowest doses of DHEA and the highest one failed to give significant enhancement of retention over that found in mice treated with DMSO alone. DHEA (icv) in DMSO was active in impressively low amounts, a dose of 3.5×10^{-12} moles/mouse giving significantly improved retention ($P < 0.01$) relative to the control group receiving DMSO alone (see insert in Fig. 4)..

ANI (20 mg/kg) in 0.35 ml saline or saline alone (0.35 ml) was injected sc 15 minutes prior to training, and DHEAS (4.2×10^{-10} moles) in 2 μl saline or saline alone (2 μl) was injected icv immediately after training. A second injection of ANI or vehicle was given 1.75 hours after training. Two such injections of ANI previously were found to inhibit protein synthesis in the brain for 4-6 hours (12). The saline control group showed good retention (80% recall score). Treatment with ANI resulted in amnesia (13% recall score). The ANI group that subsequently was given DHEAS had a recall score of 80%, showing that DHEAS completely counteracted the amnestic effects of ANI (Fig. 5A).

In another set of experiments SCO (1 mg/kg) or saline was injected sc (0.35 ml) immediately after training. DHEAS (4.2×10^{-10} moles) or saline was administered icv 45 minutes later. The saline-saline group remembered well, giving an 80% recall score. The SCO-saline group showed profound amnesia (7% recall score). The group receiving SCO followed by DHEAS had a recall score of 80%, indicating that DHEAS completely prevented SCO-induced amnesia (Fig. 5B).

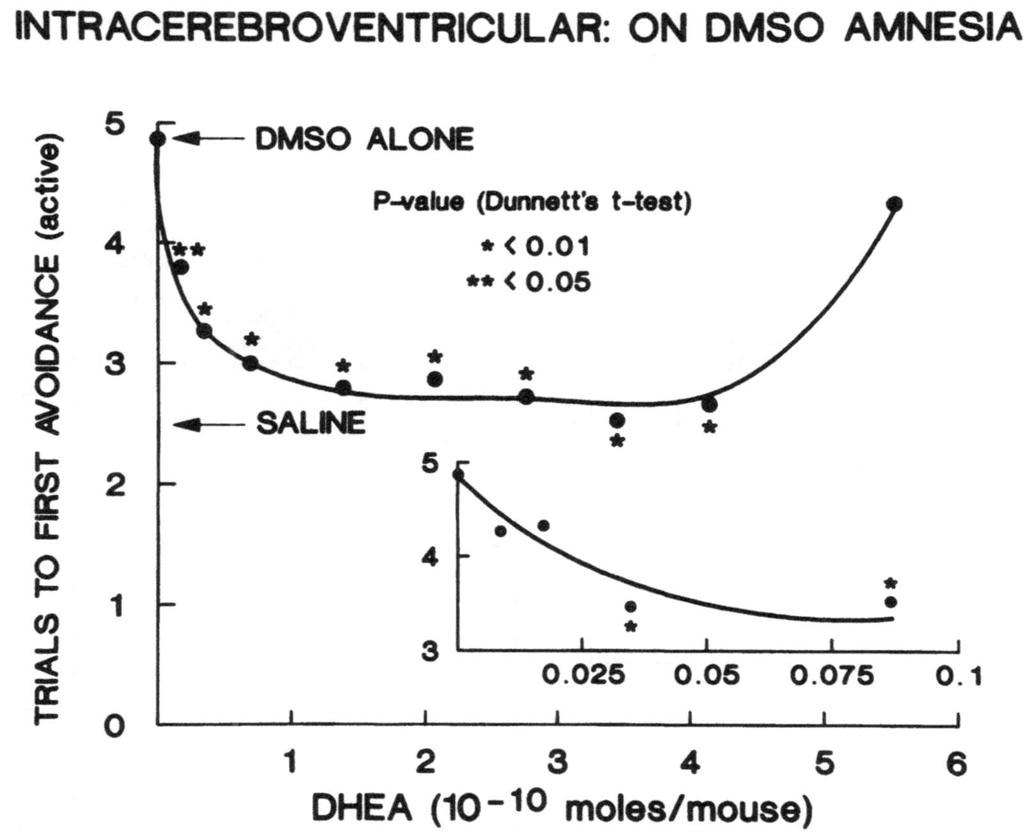

Fig. 4. DHEA prevents amnesia caused by DMSO in T-maze footshock active avoidance training. Results with lowest doses are shown in insert (from ref. 11).

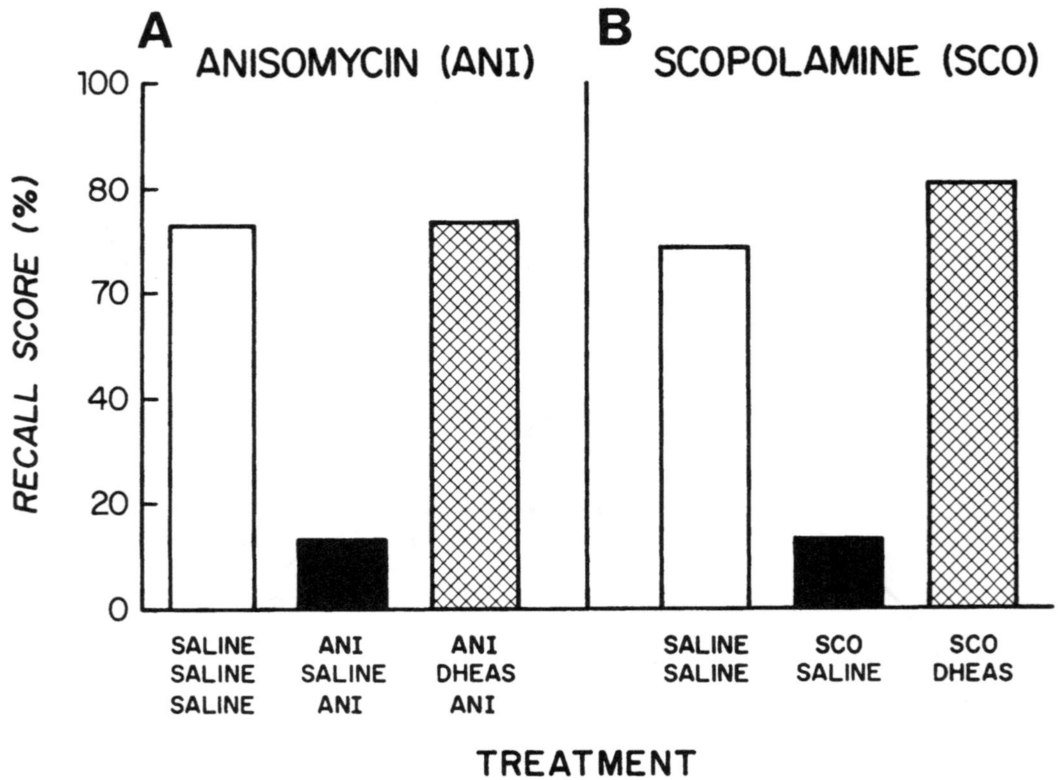

Fig. 5. DHEAS prevents amnesia caused by anisomycin (A) or scopolamine (B) in T-maze footshock active avoidance training (see text and ref. 11).

DHEAS Improved Impaired Memory Processes in Aging Mice (13)

Employing the footshock active avoidance paradigm, we found that middle-aged (18 month old) and old (24 month old) mice showed poorer retention of the task than young mice (2 months old) in a paradigm in which the young mice showed excellent retention (Fig. 6 and 7). However, when a single subcutaneous injection of DHEAS (20 mg/kg) was given within 2 minutes after training and the old and middle-aged mice were tested one week and two months later, respectively, the retention of the older groups was improved remarkably to the high levels observed in young mice. These results are compatible with the idea that inadequate availability of DHEA and/or DHEAS or of the variety of sex-related steroids derived from them (1-3) could be rate-limiting in achievement of plastic changes required for retention of learning to take place and that the age-related defect can be corrected by peripheral administration of DHEAS.

Low Concentrations of H_2O_2 Reduced Neuronal Death and Enhanced Astrocytic Differentiation in Brain Cell Culture in a Manner Similar to DHEA and DHEAS. Is There a cGMP Connection?

The idea of using H_2O_2 in our brain cell culture system (14) arose from the remarkable finding (15) that in the presence of 0.003% H_2O_2 it was possible to maintain vascular-perfused guinea pig cerebellar preparations for many hours in such condition that they showed "normal field potentials following stimulation of either the surface or the underlying white matter at both cerebellar and brain stem levels. Intracellular studies at both these levels indicate cellular activity in every way similar to the in vivo or the slice preparation from the same regions" (15). Only poor and very rapidly deteriorating responses were found when the latter preparation was perfused with Ringer solution oxygenated to saturation but lacking H_2O_2.

Experiments with 14 day-old mouse embryos showed that H_2O_2 (Fig. 8 and 9) gave results that were remarkably similar to those found with DHEA and DHEAS (Fig. 1 and 2). When cultures were maintained in the usual medium at 37°C in an atmosphere of air--5% CO^2, the NF^+ cells (neurons) were small

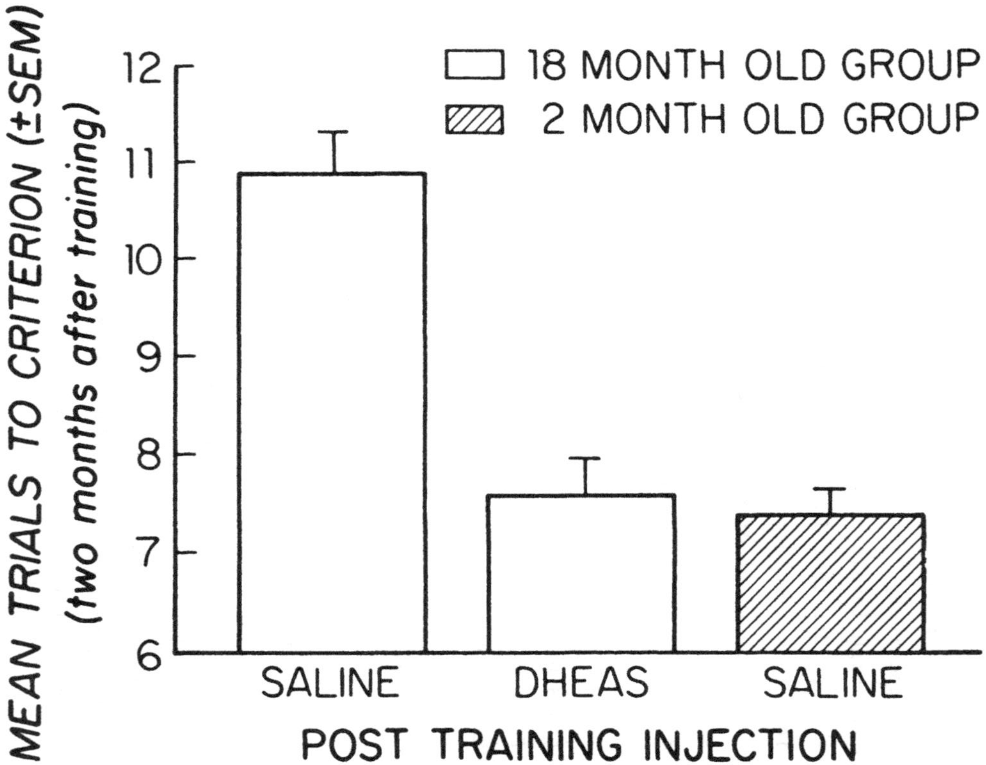

Fig. 6. Effects of subcutaneously injected DHEAS on memory retention in 18 month old mice. Injections of test solutions were made within 2 minutes after initial training and testing was performed two months after training. From ref. 13.

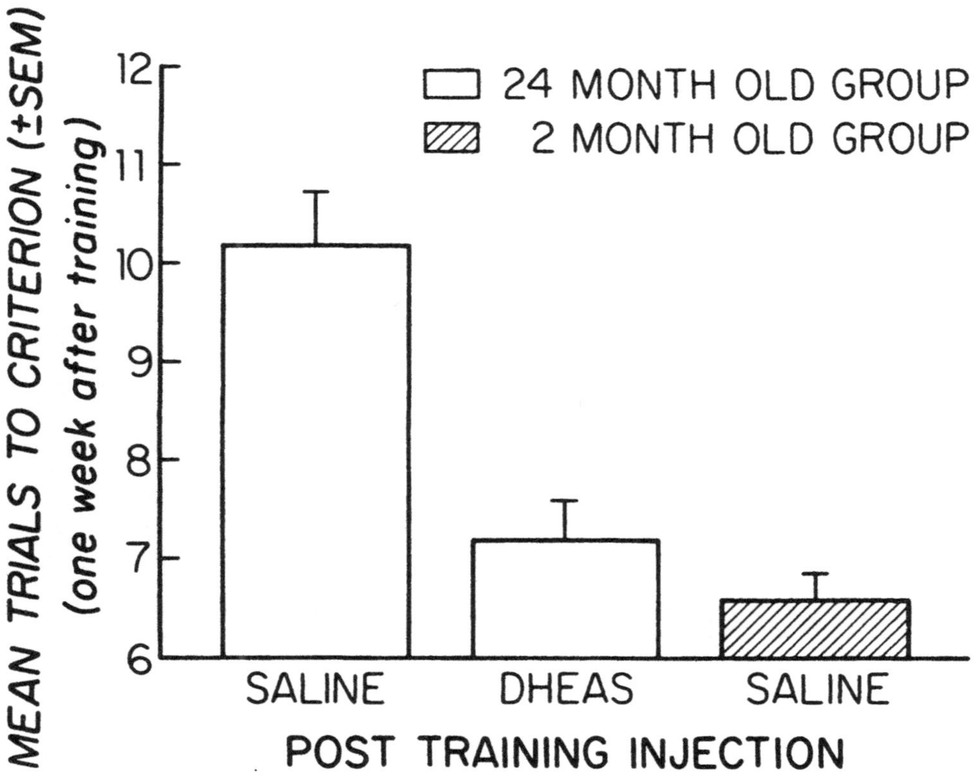

Fig. 7. Effects of subcutaneously injected DHEAS on memory retention in 24 month old mice. Injections of test solutions were made within 2 minutes after initial training and testing was performed one week after training. From ref. 13.

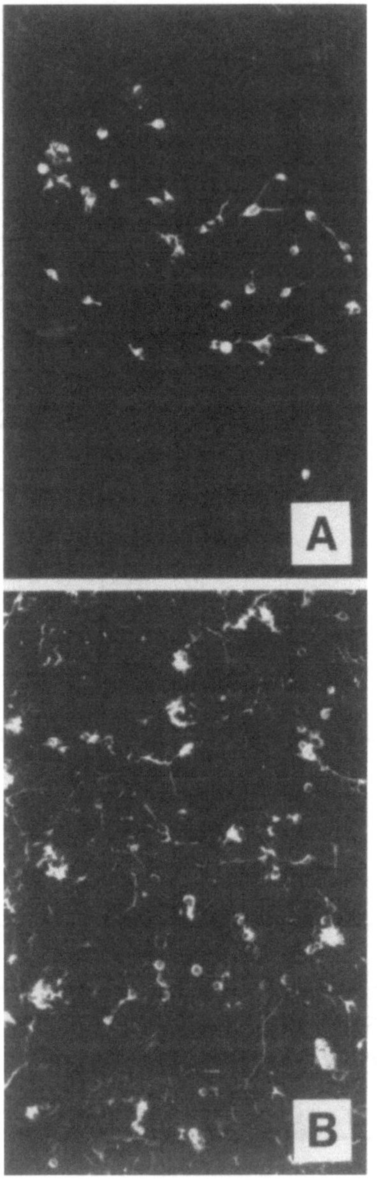

Fig. 8. Low-power micrographs of NF^+ neurons in cultures from brains of 14-day-old mouse embryos, 9 days in vitro. A: control culture. B: 0.003% H_2O_2. Bar represents 34.7 μm. From ref. 14.

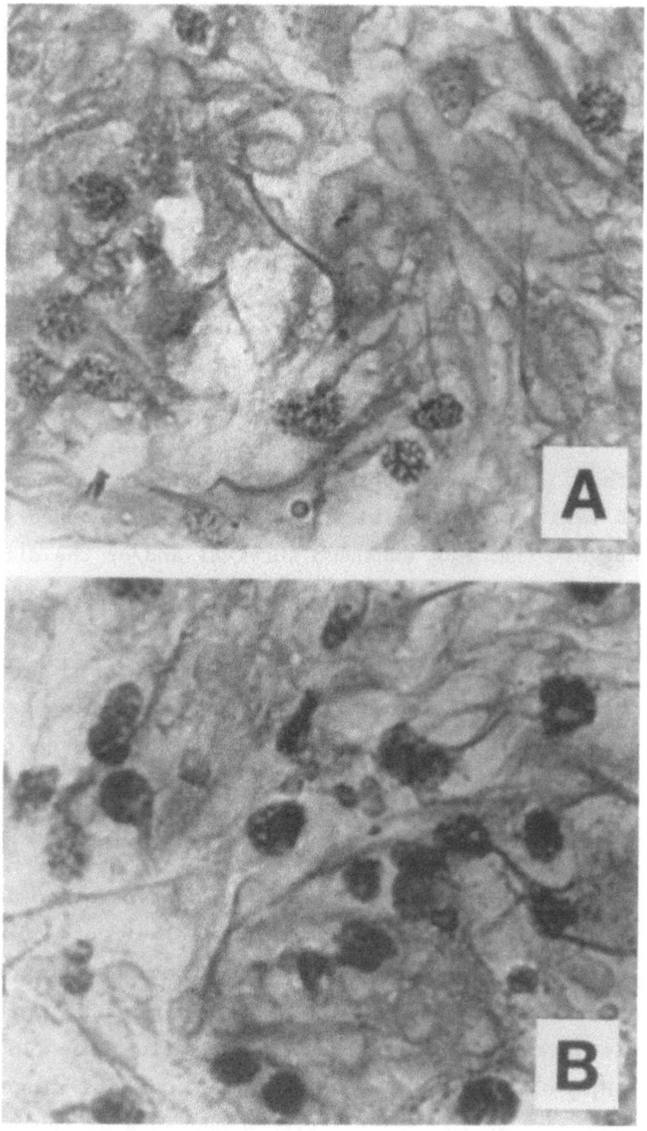

Fig. 9. Immunoperoxidase staining with anti-GFAP serum combined with [^{3}H]thymidine autoradiography of dissociated brain cell cultures prepared from 14-day-old mouse embryos, 9 days in vitro. A, 0.003% H_2O_2; B, control. From ref. 14.

and generally either were isolated from each other or were grouped in small clusters. Only approximately one-tenth of the cells showed even short processes. In medium containing 8.8 X 10^{-4} M H_2O_2 (0.003%) (Fig. 8B) there were approximately three-fold increases over control (Fig. 8A) in numbers of NF^+ neurons per unit of cover glass surface (control, 107±6.5; H_2O_2, 343±9.3). These increases might be attributable to enhanced neuronal survival and/or to an acceleration of differentiation of precursor cells into neurons, but not to mitosis of NF^+ cells, since no incorporation of [^{3}H]thymidine, whatsoever, took place into such cells under conditions of our experiments. Increases over the controls were noted in numbers of $GFAP^+$ astrocytes in the presence of H_2O_2 and in their extensions (not shown), indicative of increased amounts of total GFAP as well as GFAP per unit cell surface. Since incorporation of [^{3}H]thymidine into nuclei of astrocytes was less in cultures with H_2O_2 than in the controls (Fig. 9), the results suggest that H_2O_2 enhanced survival and/or the differentiation of $GFAP^+$ astrocytes from stem cells, while decreasing DNA synthesis and mitosis.

At the time that the above experiments were performed, the effects of low concentrations of H_2O_2 were a complete puzzle to both groups of investigators, since the levels of oxygenation in the respective experimental systems were more than sufficient for all known metabolic purposes. After the above work was completed, light was shed on the H_2O_2 effect by a study (16) in which were observed concentration-dependent, endothelium-independent (direct on smooth muscle) effects of H_2O_2 on the extent of relaxation of precontracted isolated bovine intrapulmonary arterial rings and on increases in tissue levels of cGMP, which is known to play a key role as a mediator of vascular relaxation (see (16) for references). H_2O_2 elicited a relaxation of the arterial rings, the maximal effect being observed at 10^{-4} M H_2O_2, with a decrease or reversal to a contraction at 10^{-3} M H_2O_2. The maximal increase in tissue content of cGMP also was achieved at 10^{-4} M H_2O_2. It is of great interest in this connection that in the previously cited studies, initial concentrations of H_2O_2 in the perfusion fluid or tissue culture medium were 8.8 X 10^{-4} M; but the <u>ambient</u> concentrations surely were somewhat lower because of exposure of the H_2O_2-containing fluids to ubiquitously occurring catalase activity in the tissues. The <u>effective</u> concentrations for all three systems probably were similar. Experiments on the enhancement of the H_2O_2 arterial relaxing effect by catalase were consistent with the suggestion that the activation of soluble guanylyl cyclase occurs by a species of catalase (compound I) formed during metabolism of H_2O_2 (16).

The marked similarity in our tissue culture experiments of the effects of H_2O_2 (14) to those obtained with DHEA and DHEAS (5,6) led me to query whether or not the latter substances also had exerted their effects by activation of soluble guanylyl cyclase. A search of the literature revealed that it had been reported that DHEA and a number of its precursors and metabolites in nanomolar concentrations activated soluble guanylyl cyclases in various tissues (17). This was a valuable supportive hint, although the measurements of guanylyl cyclase activity in crude tissue extracts can be affected by many substances and by arbitrarily chosen assay conditions in complex fashions (18). Perhaps the question as to whether or not guanylyl cyclase activity is influenced by steroids in target tissues could best be answered in study of relatively intact responding systems, such as brain slices, perfused brain regions, etc.

Discussion

Introduction

The results presented above show that DHEA and DHEAS exert effects on nervous tissue at structural and functional levels. Is it possible that many of these can be traced to effects on a single rate-limiting process? I would like to answer provisionally in the affirmative and to suggest that the process involved is the function of the cGMP metabolic system. Critical experiments to test the hypothesis are yet to be performed.

The dynamic state (19-21)

The relative constancies of structural, compositional, and functional features of cells, tissues, and organs in mature animals under various environmental conditions are indicative of the existence of remarkable servomechanisms (biochemical, physiological, endocrinological, and immunological) which coordinate a variety of complex pathways and which continuously adjust the rates of flow of activities and substances between the organism and its environment, between extracellular and intracellular compartments of tissues, and between membranes, cytosol, and organelles and among cytosolic aggregates (molecular ensembles) in individual cells. The mechanisms for the regulation of rates of turnover and of amounts of individual substances, organic and inorganic and simple or complex, may interact with one another at one or more points. There are examples of coordinate changes in metabolically related constituents as well as of changes in some cellular constituents without marked changes in others with which the ones that change have known metabolic relations (amino acids; ref. 20). Although our knowledge of possible metabolic transformations is extensive, our comprehension of the switching mechanisms employed and of the kinetics of most of these processes in living organisms still is negligible.

When ligands attach to specific receptors on membranes of cells or when the cells are otherwise effectively stimulated, as by purely physical means, nested cascades of reactions are released in a coordinated way so that in a brief period the systems of which they are a part react rapidly and appropriately in a manner compatible with their individual behavioral repertoires. In many instances only minor changes may occur in steady-state concentrations of a particular constituent in a cell or tissue under conditions that have great effects on its turnover rates. More than transient early increases or decreases in steady state concentrations of cellular constituents or of their intracellular redistributions often may be indicative of pathological processes (20).

It is particularly difficult to characterize aspects of normal and abnormal function in the nervous system in molecular-kinetic terms because there are complex mixtures of cells, many of which are postmitotic neurons while others, viz., glial, endothelial, ependymal cells, still are capable of mitosis and differentiation. Under normal conditions the different cell types communicate with each other physically through direct contact as well as chemically by liberation and uptake of substances ranging in size from protons to macromolecular organelles. These may act by initiating or inhibiting biochemical events directly or may release cascades of reactions by attaching to specific ligand-binding sites on membranes, on cytosolic receptors, or on genomic promoter sites in some types of cells and not in others.

DHEA and related steroids facilitate adaptive coupling between environmental pressures and biological responses by activating guanylyl cyclase, thereby increasing cGMP flux and elevating cGMP in second messenger compartments. A hypothesis and some relevant observations.

At all levels of observation, from genomic expression to freeway driving, progressive disinhibition is coupled to increased variability generation in healthy organisms (22-27). A released system, like a wound spring, has the tactical advantage over a driven one in that it does not have to overcome the inertia of start-up when it switches from an inactive or minimally active to a fully active state. Metaphorically, metabolically generated energy is used to wind the biological springs. Ever-present tonically active inhibitory influences, together with phasically active ones, maintain barriers to physicochemical perturbations, so that the interactions within the system in Yin-Yang fashion produce asymmetric graded local changes. Transient signals are transduced by a variety of devices at hand to release processes that govern amounts and turnovers of substances, their locations, and their relations to each other.

Some physiological and biochemical considerations. The physiological events that take place after effective perturbation by physical or chemical means may be extremely rapid, occurring on a submillisecond or millisecond time scale. An ingenious ^{18}O-labeling approach was applied to the study of the dynamics of cGMP metabolism in retina (28-30). In the dark-adapted state in photoreceptors in intact isolated retina, cGMP metabolic flux was minimal at a time that total cGMP was near maximal (30) (Fig. 10). Progressive increases in extent of photobleaching at low light intensities resulted in rapid increases in cGMP turnover to a maximum and, at most, only in small increases in cGMP content. Under these conditions, substrate-enzyme relations must be altered for all of the components; but the rate-limiting step in the whole process was deduced to be guanylyl cyclase, which may be activated rapidly by a light-sensitive process (30), the activation possibly resulting from interaction of the cyclase with free radicals formed during the isomerization of retinaldehyde from the 11-*cis* to the all-*trans* configuration, the only action of light in vision in the vertebrate eye. There is tight coupling among components of the cGMP system, cGMP hydrolysis and synthesis both becoming accelerated, the steady-state concentration of cGMP being only slightly changed. At higher light intensities there are decrements of cGMP flux and of cGMP levels (Fig. 10). This attenuation at high light intensities is attributable to down-regulation of guanylyl cyclase, the probable rate-limiting entity. Since all of the other steps in the visual process, which include further chemical changes and their physiological and psychological consequences, occur in the dark, the flux through the cGMP pathway probably is a key regulator of cell dynamics in the retina.

Employing the ^{18}O technique, greatly enhanced Ca^{2+}-dependent turnover of cAMP lasting for only 1-3 minutes was observed in the absence of increase in cAMP content when parotid gland secretion was stimulated by cholinergic or β-adrenergic agents (31).

The above results are compatible with the idea that "hydrolysis of energy-rich cyclic nucleotides by certain species of phosphodiesterase...represent an energy-linked catalysis that subserves signal-transduction processes" (30). I would like to suggest that in such cellular compartments, as defined above, the cyclic nucleotide flux subserves signal-transduction machinery by rapidly and efficiently "rewinding its springs," so to speak, so that once again it could be released instantaneously in response to appropriate stimuli.

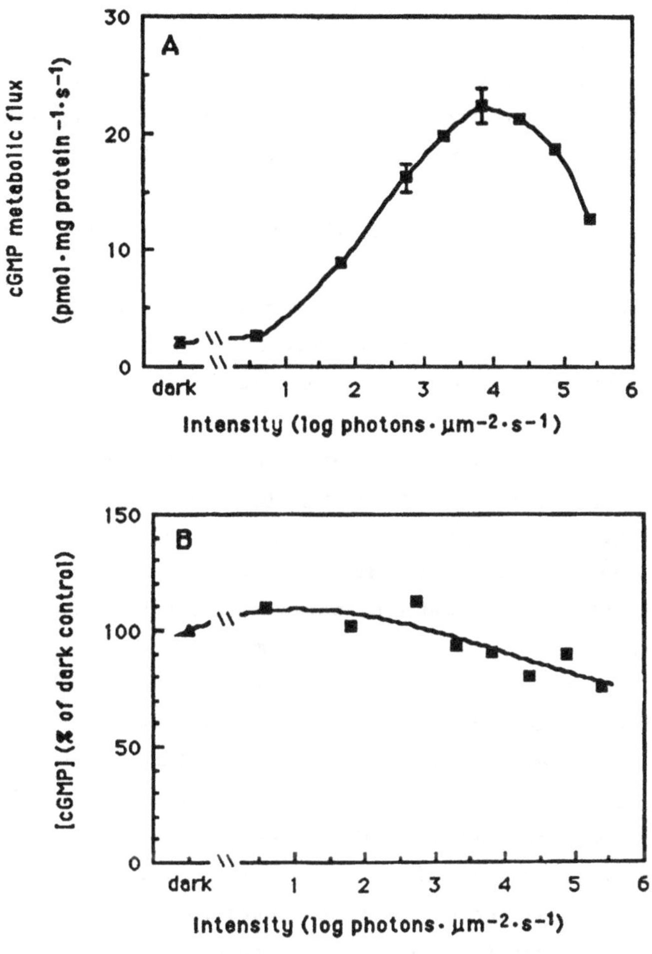

Fig. 10. Relationship of the intensity of 500 nm light stimulation to cGMP metabolic flux and cGMP concentration in toad retinas. The experimental details are given in ref. 30. A, the cGMP hydrolytic fluxes calculated from the rates of labeling of guanine nucleotide α-phosphoryls with ^{18}O. The dark control rate represents the mean of four retinas exposed to infrared lighting only. B, the concentrations of cGMP in retinas illuminated for 20 seconds. Figure kindly furnished by Dr. Nelson D. Goldberg.

Excitatory effects on membranes increase their permeability to cations, thus decreasing the potential across the membranes (depolarization). Inward sodium current is responsible for most of the observed depolarization, although calcium flows in as well. Rapid increases in free intracellular Ca^{2+} may occur from release from calcium binding entities and slower increases by secondary influx of extracellular Ca^{2+} (32). The opening of potassium channels is activated by the changes in membrane potential. Outward potassium currents then serve to repolarize the cell and, in some instances, to produce a hyperpolarization before the calcium balance is restored via the action of Ca^{2+}-Mg^{2+} ATPase, mitochondrial uptake and reassociation with calcium binding proteins and the potassium channels are closed. The action of Na^{+}-K^{+} ATPase restores the monocation balance.

The time-sequence coordination of the latter events is such that there are, in effect, pulsatile increases in free Ca^{2+} which reflect accurately the amounts of depolarization to which the membranes have been subjected. The free Ca^{2+} either directly or via its interaction with special calcium binding proteins, of which calmodulin, parvalbumin, troponin C, S-100, and calbinden are examples, releases cascades of many intracellular and intramembranous processes which continue until free Ca^{2+} is reduced to the resting level by combinations of pumping mechanisms, intracellular binding to calcium binding proteins, and sequestration by mitochondria. During this period, various enzymes related to cyclic nucleotide metabolism are activated. The levels of cAMP and cGMP become elevated, existing in various compartments, and these nucleotides serve their well-known roles as second messengers. Local increases in concentrations of cGMP and cAMP ordinarily lead to changes in activities of protein kinases, which by actions on specific substrate proteins establish new cell balances. Phosphorylation controls the activity of many enzymes and the conformational states of non-enzymatic proteins (33,34). The control of the degrees of protein phosphorylation is the "principle mechanism by which all cell functions in eukaryotic cells are controlled by extracellular signals. Processes as diverse as metabolism, membrane transport and permeability, secretion, contractility, the transcription and translation of genes, cell division and fertilization, neurotransmission and even memory, are all regulated by this versatile post-translational modification" (33). There are important interplays of phosphorylation-dephosphorylation events with other post-translational protein modifications, viz., proteolytic cleavage, acylation, carbamylation, hydroxylation, methylation, glycosylation, and sulfation.

The various components of the cAMP and cGMP systems are distributed ubiquitously among cellular components, viz., plasma membranes, cytoplasm, and cytoskeleton, and can be regulated by transmitters, ions, free radicals, steroids, peptides, proteins, etc. The cyclic nucleotide systems interact importantly with aspects of virtually all events that are initiated when membrane depolarization occurs and that their relative activities and the ratios of amounts of cAMP to cGMP that derive from them in the various subcellular compartments may determine the balances of activities therein and ultimately the behaviors of cells.

At low stimulus intensities and frequencies, free Ca^{2+} levels and, therefore, cell activities return rapidly to pre-stimulus levels. At higher levels of stimulation, genomic transcriptional activation occurs. There may be rapid transient increases in mRNAs of putative transcription factor genes, followed by secondary waves of gene expression (26,27). With the latter, permanent changes could occur in structures and operations of intracellular machinery, and the behaviors of cells could be changed permanently. This might be occurring when stimuli are sufficiently great so as to produce long-term potentiation or learning in neural systems. When intensity of stimulation exceeds a certain level, the series of interlocked processes leading to coordinate changes at lower intensities might be disrupted because of inability to maintain fine regulation of free calcium levels. This, in turn, would lead to disorganization of the orderly management of released reaction cascades. Persistent elevations of free calcium may result in malfunction and eventual cell death. "...differences in cell calcium-regulating systems may determine whether a neuron lives or degenerates in the face of an excitatory challenge" (35).

The Cyclic Nucleotides and Cell Maintenance and Memory

Dual effects were observed in our tissue culture and memory experiments, in the study of effects of H_2O_2 on relaxation of contracted arterial strips, and in many studies of memory enhancement by excitatory substances (36). In the effective dose ranges, progressively increasing doses first increased responses to a maximum, beyond which decreasing responses were observed until a dose was reached at which no significant effects were seen over the controls. A compendium of results in the literature showed that this phenomenon was observed for "27 memory-enhancing compounds given by seven routes of administration to six organisms in 15 laboratories" (36). A large body of evidence supports the idea that excitatory substances, among which are the known memory enhancers, have cGMP as an intracellular messenger for their action (37).

Experiments with glutamate, a major excitatory transmitter in mammalian brain, show that cGMP probably is the intracellular messenger for its postsynaptic action (38-41). Administration of L-glutamate and other NMDA agonists icv to mice in the same learning paradigms as used for DHEA and DHEAS showed enhancement of retention in a dose-dependent manner (42). NMDA receptor antagonists caused amnesia. In no instance has it been possible to show that excitatory substances that elevate cGMP in brain could stimulate soluble guanylyl cyclase in cell-free systems, suggesting that a diffusible factor is released that activates the enzyme. This now has been identified as nitric oxide (NO), which is formed from the guanidino group of arginine in a Ca^{2+}-, NADPH-, tetrahydrobiopterin-dependent and receptor-mediated fashion (38-41,43,44). It appears that NO mediates the stimulation of cGMP formation by glutamate, by other NMDA receptor agonists, and by a variety of other excitants in brain and other tissues. Indeed, the formation of NO from L-arginine is a widespread transduction mechanism for the stimulation of the soluble guanylyl cyclase.

H_2O_2 probably enhances cGMP formation by activating the guanylyl cyclase directly by a mechanism similar to that of NO. DHEA, DHEAS, and the steroids formed from them may activate soluble guanylyl cyclases directly (allosterically?) and possibly may intervene at several points in the formation of NO from arginine, as well. It is of the greatest interest that nitric oxide has cytostatic properties, inhibiting DNA synthesis and mitochondrial respiration (45,46). We observed that DHEA, DHEAS, and H_2O_2 all inhibited DNA synthesis in astrocytes (Fig. 2 and Fig. 9). With regard to neural transmission, release of NO might be a way of preventing overgrowth of non-neural cells during nerve stimulation.

Substances that improve retention decline in effectiveness when administered at progressively longer times after training (47). When mice were trained under conditions giving poor retention in saline-injected controls, administration of DHEAS immediately after training (2 minutes) or at 30 or 60 minutes gave significant improvement ($p < 0.01$) of retention over that found when saline alone was injected (Table 1).

Table 1
Effect of DHEAS on Retention of Step-down Passive Avoidance Training (from ref. 11)

	Saline	DHEAS (162 ng icv)					Saline
Time (min)[a]	0-2	0-2[b]	30	60	90	120	120
Mean trials	4.73	3.00	3.20	3.60	4.47	4.93	4.87
±S.E.M.	0.33	0.30	0.37	0.38	0.44	0.33	0.38
Recall score (%)	20	73	60	47	33	20	13
P value[c]	-	<0.01	<0.01	<0.05	n.s.	n.s.	-

[a]Time of saline or DHEAS administration after training.
[b]Immediately after training.
[c]Statistical difference from either saline control group as determined by Dunnett's t-test; n.s. = not significant

When injected at later times, 90 and 120 minutes, DHEAS had no statistically significant effect. What makes memorial processes amenable to enhancement long after the fast neural events during learning have occurred? It does not appear to be perseverative neuronal activity consisting of reverberation of impulses in neuronal circuits that outlasts the stimulating events (48). Acute cerebral ischaemia (bilateral carotid occlusion) induced in goats 20 seconds after subjecting them to a classical conditioning paradigm for 1.5 minutes, although resulting in an isoelectric cortical and subcortical EEG, did not impair learning by comparison with controls.

However, the continuum of biochemical processes within cells initiated by stimulation, coincident with free exchanges of materials which take place among components of "synaptic units" that consist of vascular, extracellular, glial, and pre- and postsynaptic components, is essential for coordinated generation of energy for repolarization of membranes and resynthesis of transmitters and synthesis and utilization of rate-limiting materials required for synaptic remodeling. Pharmacologically active substances with excitatory and inhibitory effects would be expected to increase or decrease, respectively, the probability of synapses attaining optimal refashioning in relation to their preceding activity. Substances, conditions, or procedures which can enhance the rate of energy production, the flow of precursors for biosynthesis of cellular constituents, and the rates of genetic information transcription and translation would be expected to increase the rate and extent of plastic change (memory).

Inhibition of synaptic information transfer (GABA, scopolamine, NMDA antagonists), of genetic information transcription (actinomycin D) or translation (puromycin and anisomycin), and of macromolecular synthesis (amino acid and purine and pyrimidine antimetabolites) and generally unfavorable circumstances for cell growth and development such as aminoacidopathies (hereditary and acquired), nutritional deficiencies, and circulatory insufficiency (blood vessel disease and arteriosclerosis) would decrease the rate and extent of plastic changes required for memory formation. There is much literature in support of the latter.

Why is it that <u>all</u> of the memory enhancing substances overcome amnesia caused by different means? For example, in our studies DHEA and DHEAS in small amounts occluded the amnestic effects of a membrane perturbant (DMSO)and DHEAS blocked amnesia induced by an inhibitor of protein synthesis (anisomycin) and a central muscarinic cholinergic antagonist (scopolamine). I would like to suggest that all of these amnestic treatments together with aging and a variety of nervous system disorders may result in decreased capacities of one or more components of responding systems to manage their Ca^{2+} economies effectively and that these impediments, in ways as yet unknown, may be overcome by increasing cGMP turnover and cGMP levels by a variety of means.

Comment

Cascades of processes are generated by actions of environmental factors that reduce transmembrane potential and/or interact with specific membrane receptors. Cascades of other processes then are released by expression of genetic potential. When environmental pressures are increased above some background level, the numbers of such countercurrent cascades and their extents are increased in such a way that the probability of their meshing to give system-typical adaptive response patterns remains approximately constant. Disease may be said to occur in a particular living system when constrictions or disruptions in communication channels between members of its interlocking components become sufficiently extensive so as to result in continued uncoupling between environmental pressures on the system and its ability to adapt to them. A primary task in dealing with disease is to identify the rate-limiting processes and to facilitate their relief. The remarkable self-organizing properties of biological systems often enable them to begin to function adaptively at all levels, from intracellular to international, once effective communications are reinstituted.

Interpretation of work in my laboratory cited in this paper and relevant data in the literature have suggested that the results observed with DHEA and DHEAS may be related to enhancing effects of these substances on cGMP metabolism, which I believe to be a key rate-limiting process in many aspects of cell life.

Experiments at several levels are being planned to test this hypothesis critically. Meanwhile, realizing that the ideas expressed herein are radically different from the prevailing views of mechanisms of action of

DHEA and DHEAS, I await with eager anticipation the vigorous discussion that will ensue.

Acknowledgement

Supported in part by grants from the G. Harold and Lily Y. Mathers Foundation, the Ziskin Family Trust and from National Metal Steel Foundation. I am particularly grateful to Prof. Nelson D. Goldberg of the University of Minnesota for discussions of cGMP metabolism. However, he is not responsible for any of the statements made in the text.

References

1. Sonka, J. 1976. Acta Univ. Carol [Med.] (Praha) 71, 1-171.

2. Parker, L. N. 1989. Adrenal Androgens in Clinical Medicine. Academic Press · New York, 615 pp.

3. Roberts, E. 1986. In: Treatment Development Strategies for Alzheimer's Disease (T. Crook, R. T. Bartus, S. Ferris and S. Gershon, eds.). Mark Powley and Associates · Madison, CT. pp. 173-219.

4. Regelson, W., R. Loria, and M. Kalimi. 1988. Ann. N. Y. Acad. Sci. 521, 260-273.

5. Roberts, E., L. Bologa, J. F. Flood, G. E. Smith. 1987. Brain Res. 406, 357-362.

6. Bologa, L., J. Sharma, E. Roberts. 1987. J. Neurosci. Res. 17, 225-234.

7. Muntwyler, R. and L. Bologa. 1989. Schweize Arch. Neurol. Psychiatr. 140, 29-33.

8. Toran-Allerand, C. D. 1976. Brain Res. 106, 407-412.

9. Toran-Allerand, C. D. 1980. Brain Res. 189, 413-427.

10. Toran-Allerand, C. D., J. L. Gerlach, and B. S. McEwen. 1980. Brain Res. 184, 517-522.

11. Flood, J. F., G. E. Smith, and E. Roberts. 1988. Brain Res. 447, 269-278.

12. Flood, J. F., D. Vidal, E. L. Bennett, A. E. Orme, S. Vasquez, M. E. Jarvik. 1978, Pharmacol. Biochem. Behav. 8, 81-87.

13. Flood, J. F. and E. Roberts. 1988. Brain Res. 448, 178-181.

14. Bologa, L., J. Sharma, D. Dahl, and E. Roberts. 1987. Brain Res. 411, 282-290.

15. Llinas, R., Y. Yarom, and M. Sugimori. 1981. Fed. Proc. 40, 2240-2245.

16. Burke, T. M. and M. S. Wolin. 1987. Am. J. Physiol. 252, H721-H732.

17. Vesely, D. L. 1979. Proc. Natl. Acad. Sci. USA 76, 3491-3494.

18. Tremblay, J., R. Gerzer, and P. Hamet. 1988. In: Advances in Second Messenger and Phosphoprotein Research, Vol. 22 (P. Greengard and G. A. Robison, eds.). Raven Press · New York. pp. 319-383.

19. Ratner, S. 1979. Ann. N. Y. Acad. Sci. 325, 188-209.

20. Roberts, E. and D. G. Simonsen. 1962. In: Amino Acid Pools (J. T. Holden, ed.). Elsevier · Amsterdam. pp. 284-349.

21. Schoenheimer, R. 1942. The Dynamic State of Body Constituents. Harvard University Press · Cambridge, MA. 78 pp.

22. Roberts, E. and S. Matthysse. 1970. Annu. Rev. Biochem. 39, 777-820.

23. Roberts, E. 1976. In: GABA in Nervous System Function (E. Roberts, T. N. Chase, and D. B. Tower, eds.). Raven Press · New York. pp. 515-539.

24. Hikosaka, O. and R. H. Wurtz. 1983. J. Neurophysiol. 49, 1285-1301.

25. Roberts, E. 1986. Exp. Neurol. 93, 279-290.

26. Morgan, J. I., D. R. Cohen, J. L. Hempstead, and T. Curran. 1987. Science 237, 192-197.

27. Saffen, D. W., A. J. Cole, P. F. Worley, B. A. Christy, K. Ryder, and J. M. Baraban. 1988. Proc. Natl. Acad. Sci. USA. 85, 7795-7799.

28. Goldberg, N. D., A. Ames, III, J. E. Gander, and T. F. Walseth. 1983. J. Biol. Chem. 258, 9213-9219.

29. Ames, III, A., T. F. Walseth, R. A. Heyman, M. Barad, R. M. Graeff, and N. D. Goldberg. 1986. J. Biol. Chem. 261, 13034-13042.

30. Dawis, S. M., R. M. Graeff, R. A. Heyman, T. F. Walseth, and N. D. Goldberg. 1988. J. Biol. Chem. 263, 8771-8785.

31. Deeg, M. A., R. M. Graeff, T. F. Walseth, and N. D. Goldberg. 1988. Proc. Natl. Acad. Sci. USA 85, 7867-7871.

32. Bennett, N., M. Ildefonse, S. Crouzy, Y. Chapron, and A. Clerc. 1989. Proc. Natl. Acad. Sci. USA 86, 3634-3638.

33. Cohen, P. 1988. Proc. R. Soc. Lond. [Biol.] 234, 115-144.

34. Sprang, S. R., K. R. Acharya, E. J. Goldsmith, D. I. Stuart, K. Varvill, R. J. Fletterick, N. B. Madsen, and L. N. Johnson. 1988. Nature 336, 215-221.

35. Mattson, M. P., P. B. Guthrie, and S. B. Kater. 1989. FASEB J. 3, 2519-2526.

36. Cherkin, A. and J. F. Flood. 1988. In: Cellular Mechanisms of Conditioning and Behavioral Plasticity (C. D. Woody, D. L. Alkon, and J. L. McGaugh, eds.). Plenum Press · New York and London, pp. 343-354.

37. Drummond, G. I. 1983. Adv. Cyclic Nucleotide Res. 15, 373-494.

38. Moncada, S., R. M. J. Palmer, and E. A. Higgs. 1989. Biochem. Pharmacol. 38, 1709-1715.

39. Knowles, R. G., M. Palacios, R. M. J. Palmer, and S. Moncada. 1989. Proc. Natl. Acad. Sci. USA 86, 5159-5162.

40. Garthwaite, J., E. Southam, and M. Anderton. 1989. J. Neurochem. 53, 1952-1954.

41. Bredt, D. S. and S. H. Snyder. 1989. Proc. Natl. Acad. Sci. USA 86, 9030-9033.

42. Flood, J. F., M. L. Baker, and J. L. Davis. In Press. Brain Res.

43. Tayeh, M. A. and M. A. Marletta. 1989. J. Biol. Chem. 264, 19654-19658.

44. Marletta, M. A., P. S. Yoon, R. Iyengar, C. D. Leaf, and J. S. Wishnok. 1988. Biochemistry 27, 8706-8711.

45. Hibbs, Jr., J. B., R. R. Taintor, Z. Vavrin, and E. M. Rachlin. 1988. Biochem. Biophys. Res. Commun. 157, 87-94.

46. Stuehr, D. J. and C. F. Nathan. 1989. J. Exp. Med. 169, 1543-1555.

47. Flood, J. F., G. E. Smith, and A. Cherkin. 1985. J. Pharmacol. 16, suppl. III, 39-49.

48. Baldwin, B. A. and S. S. Soltysik. 1966. Brain Res. 2, 71-84.

Serum Steroid Levels in Two Old Men with Alzheimer's Disease (AD) Before, During, and After Oral Administration of Dehydroepiandrosterone (DHEA). Pregnenolone Synthesis May Become Rate-Limiting in Aging.

Eugene Roberts

Department of Neurobiochemistry, Beckman Research Institute of the City of Hope, Duarte, CA 91010

L. Jaime Fitten

Department of Psychiatry (Geropsychiatry Unit), VA Medical Center, Sepulveda, CA 91343

Introduction

General

This work is part of an ongoing effort to identify manipulable rate-limiting processes in age-related deterioration of nervous system function, in general, with special emphasis on senile dementia of the Alzheimer's type (AD) (1, 2). Deterioration is observed in many brain regions in AD, including the cholinergic system of the brainstem and basal forebrain core (3), which is of decisive importance in maintaining the waking state, consciousness, and the ability to learn, and the hippocampus (4), which is known to be important in memory processes. Incoordinations in a number of bodily systems with age, including the nervous system, are correlated with decrements in sex hormone levels, generally, and most specifically with uniquely monotonic decreases after puberty in both males and females in serum levels of dehydroepiandrosterone (DHEA) and dehydroepiandrosterone sulfate (DHEAS), largely adrenally derived substances which serve as precursors for both androgens and estrogens in tissues of the body (5, 6). In the aging male, losses in muscle mass and other androgen-related characteristics often precede frank signs of

Dehydroepiandrosterone (DHEA)

mental deterioration. Administration of DHEA and DHEAS may exert ameliorative effects in such different conditions as diabetes, obesity, autoimmune disease, cancer, and connective tissue disorder, possibly by releasing diverse aspects of metabolic machinery necessary for effective intracellular and extracellular communication to take place (1, 6-8).

DHEA and DHEAS in low concentrations had remarkable effects in enhancing survival of neurons in explants of fetal mouse brain in culture in the outgrowth of neuritic processes from them, and in the establishment of connections between them (9, 10). These substances also enhanced memory in young and old mice (9, 11, 12). This suggested the possibility that DHEA and DHEAS also might protect nervous tissue in vivo against damaging influences and might facilitate regeneration when injury occurs. Because there are many metabolic crossroads at which DHEA and DHEAS and/or their metabolites may play important roles, one possibility being facilitation of cGMP metabolism (13), administration of these substances might exert remarkable recyberneticizing effects on the nervous system and other bodily systems and on their relations to each other. Recently, it was reported that serum levels of DHEAS were on average 48% lower in AD patients than in age-matched controls, which, in turn, were approximately 50% lower than those of younger controls (14).

In the course of our studies of AD, it became possible to determine serum levels of precursors and metabolites of DHEA in two men (72 and 88 years) before and at several times after preprandial oral ingestion of a single dose of DHEA (400 mg) and at weekly intervals during successive biweekly periods of daily DHEA ingestion of 1.2 gm/day (400 mg TID), 2.4 gm/day (800 mg TID), and placebo, respectively. This work was performed with a view to delineation for further study processes that may become rate-limiting in the maintenance of normal adult levels of these substances in serum. Although our studies were performed on AD patients, probably the results are typical of those that would usually be observed in aging, non-demented individuals and are not specific for AD.

Some aspects of steroid metabolism

In Fig. 1 is shown a partial outline of the steroid metabolic scheme as it may occur in the human organism as a whole (6). If one were to homogenize the entire organism and extract the homogenate appropriately, one could expect to be able to find the substances shown in Fig. 1 and to demonstrate the activities of enzymes that catalyze the indicated interconversions. However, it is highly unlikely that any single tissue contains all of the substances and/or enzymes; and among those tissues in which they do exist, marked differences in levels would be found. Probably all cells in the body require for regulation of their functions some or all of these steroids, whose activities range from modulation of membrane excitability to regulation of genomic transcription. The blood furnishes the common reservoir from which these substances can be withdrawn by tissues to meet their needs and to which they can contribute excesses that are produced. Serum levels of these substances are a resultant of synthesis, secretion and transport in different cell types, which processes are controlled by varieties of feedback and feed-forward signals. These relations may be complexly affected by greatly increasing serum levels of DHEA by exogenous administration. Elegant double-labeling, constant infusion procedures are employed in the steroid field to answer specific metabolic questions. Our data are only on serum steroid levels and, therefore, are subject to limited interpretation.

The biosynthesis of steroid hormones begins with cholesterol, from which the sex steroids, glucocorticoids, and mineralocorticoids all eventually derive (Fig. 1). Pregnenolone, a key cholesterol metabolite, is the major precursor for the steroid hormones. Its formation is rate-limiting and is regulated by pituitary hormones, such as luteinizing hormone (LH) and follicle-stimulating hormone (FSH) in ovaries and testes and ACTH and possibly a non-ACTH pituitary peptide in adrenals.

Pregnenolone can go directly to progesterone and thence to aldosterone (route A, Fig. 1) or to 17α-OH pregnenolone, which is a precursor for cortisol formation (route B, Fig. 1) and for sex-related steroids (route C, Fig. 1). Route A can contribute to route B and route B to route C, as

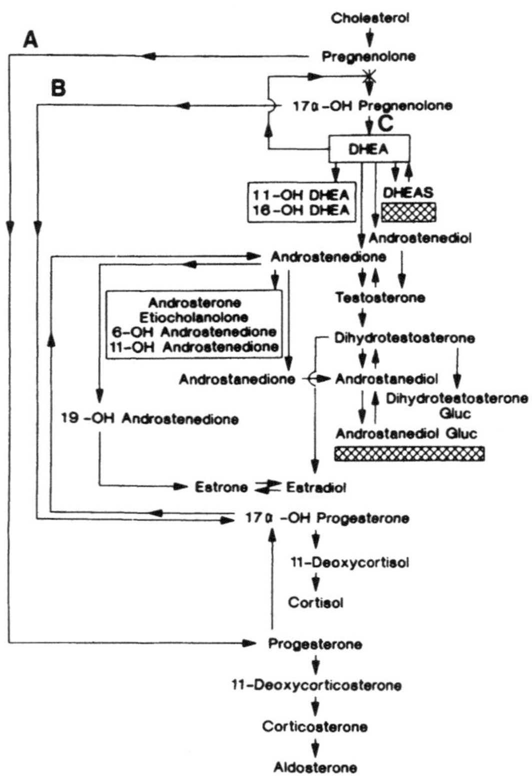

Population Controls (X)	Subject	Baseline (Y)	Y/X	Post-ingestion (ng/DL) Low Dose Wk. 1	Low Dose Wk. 2	High Dose Wk 1	High Dose Wk 2	Post Drug Wk 1	Post Drug Wk 2
ng/DL									
111(46-225)	A	68	0 61	75	71	127	85	67	58
	B	84	0.76	72	75	69	83	82	54
124(53-357)	A	36	0 29	35	22	21	16	27	22
	B	42	0 34	109	34	<5	<5	6	54
455(160-800)	A	112	0.25	703	673	10.320	-	141	79
	B	164	0.36	1.103	1.165	2.524	6.221	144	165
270.000 (100.000-450.000)	A	63.000	0.23	1.230.000	1.230.000	2.590.000	-	211.000	90.000
	B	68.000	0 25	1.270.000	1.240.000	1.820.000	3.715.000	101.000	114.000
115(75-105)	A	69	0 60	401	448	10.534	-	143	84
	B	92	0 80	434	542	1.331	7.000	111	111
627(350-1030)	A	305	0 49	223	199	942	-	156	156
	B	243	0 57	737	210	166	150	135	207
440(190-900)	A	164	0 37	2.224	2.785	11.600	6.140	277	123
	B	288	0 65	5.370	6.880	18.100	22.000	5.280	3.700
5 9(2-8)	A	6	1 02	11	8	11	-	3	4
	B	5	0.85	9	8	10	15	3	3
103(27-199)	A	25	0 24	19	23	9	-	27	31
	B	93	0 90	137	64	25	27	49	58
100(30-200)	A	30	0 30	19	46	28	49	26	23
	B	51	0 51	88	45	32	39	45	48
11.000 (8.000-19.000)	A	16.000	1 45	16.000	20.000	16.000	20.000	17.000	17 000
	B	15.000	1 36	19.000	12.000	8.600	9.300	12.000	12.000
33(13-97)	A	21	0 64	9	<5	14	8	8	<5
	B	19	0 58	50	31	11	38	15	<10
425(130-820)	A	223	0 52	217	740	302	480	245	239
	B	243	0 57	737	210	166	150	135	207
13(7-30)	A	6	0 46	12	6	6	10	4	4
	B	8	0 62	6	7	6	6	7	6

Fig. 1. Outline of steroid metabolism and serum levels of steroids in two AD patients before, during, and after chronic oral ingestion of DHEA. Low dose, 1.2 gm/day; high dose, 2.4 gm/day.

shown. DHEA, the first product in route C may inhibit the flow through routes B and C by inhibiting conversion of pregnenolone to 17α-OH pregnenolone.

Pregnenolone is formed from cholesterol in mitochondria of those tissues that produce steroid hormones. The rate of steroid synthesis in the zona fasciculata of the adrenal cortex is controlled by the delivery of cholesterol from cytoplasmic inclusion droplets to the inner mitochondrial membrane, where steroidogenesis begins by production of pregnenolone from cholesterol by action of cytochrome P-450scc (side-chain cleavage enzyme), a reaction also found in brain. Recently an adrenal protein, des-(Gly-Ile)-endozepine, has been isolated and characterized, the synthesis of which is induced by ACTH and which stimulates delivery of cholesterol to the inner mitochondrial membrane and possibly assists in the mobilization of inner membrane cholesterol (15; see also for pertinent references).

DHEA and DHEAS fit into the steroid metabolic scheme in the following fashion. In the human, pregnenolone goes via 17α-OH pregnenolone to DHEA, from which is formed the androgen, testosterone, in the testes, ovaries, adrenal cortex, and placenta. Testosterone is a precursor of the equally potent androgen, dihydrotestosterone, in the prostate, skin, hair follicles, and brain. Testosterone also is the precursor for estrogen formation by aromatization. In the female, the latter takes place largely in the ovaries, but estrogens also are formed from androgenic precursors in both males and females in other tissues, including muscle, adipose tissues, liver, and brain. The rate of estrogen formation from circulating androgens is increased in hyperthyroidism, certain forms of liver disease, and obesity; it may be decreased in some other pathological states (6, 7).

DHEAS is the most abundant steroid in blood, its levels being far higher than those of DHEA, from which it can be formed by a sulfotransferase and to which it can be converted by a sulfatase (6, 16-19). Serum levels of DHEA and DHEAS normally largely are determined by synthesis and secretion from the adrenal cortex, but also to some extent from other tissues (6).

Low serum values of DHEA and DHEAS found at birth persist through the sixth year of life and then rise abruptly at the seventh year (20). Increases continue until approximately the 16th year in both sexes, but only in males do they increase thereafter, maximal levels being attained between 20 and 24 years of age (5). From puberty on, the blood levels of men are significantly higher than those of women at all ages from 20-69 years, probably reflecting a testicular contribution to the serum pool (5). Most important for present consideration is the observation that progressive declines in serum concentrations in these steroids take place in both sexes at all subsequent ages, values at 70 years of age being approximately only 20% of those found at the peak. This appears to be unique among the classes of steroid hormones. DHEAS and DHEA levels are close to minimal at ages when the incidence of AD (sporadic form) begins to increase significantly.

The fact that there is a relative lack of significant circadian, monthly, seasonal, or annual rhythmicity in serum levels of DHEA and DHEAS (5, 6) suggests that their release into the blood may serve to buffer the organism against the vicissitudes of large fluctuations of these substances, which may be necessary for many important ongoing tissue functions. In this light, the decreases in their serum levels with age could presage changes in metabolism in many tissues that might have far-reaching consequences. Age-related progressive decreases in serum levels of DHEA and DHEAS may reflect gradual decrements in output of pituitary hormones which control the production of pregnenolone, their precursor. The release of pituitary hormones may, in turn, be controlled by hypothalamic hormones and also by feedback inhibition by serum levels of DHEA and DHEAS and other steroids. A decreased release of LHRH from the hypothalamus, for example, could result in decreased pituitary release of LH which, in turn, could lead to a tendency to decreased formation of DHEA and DHEAS in gonadal tissue, as a result of which there might be decreased rates of formation of androgens and estrogens (21). However, the latter tendency might be counteracted by increases in secretion of chorionic gonadotropin.

Probably all of the above hormones act on tissues by binding to specific receptors, whose activation leads, often through effects on cyclic nucleotide mechanisms, to the release of rate-limiting reactions at consequent steps of relevant cascades. Even if normal levels of hypothalamic and pituitary hormones were to be maintained with aging, the receptor sensitivity or metabolic aspects of the ensuing reaction cascade of the responding tissues might be decreased. For example, a study with aging humans showed there to be no impairment of the ability of the adrenals to respond to acute ACTH stimulation with production and secretion of cortisol, but there was a significant decrease in stimulatability of secretion of DHEA and DHEAS (22). Another study showed that in older individuals (72-102 years old, average 85 years) basal levels of cortisol and aldosterone and the ACTH-stimulatability of their formation and release were essentially the same as in younger individuals (35-62 years old, average 51 years); but the basal levels of DHEA and DHEAS were significantly higher in the younger group than in the older one and no stimulation at all by ACTH of increase in serum levels of these substances was observed in the older group (23). Thus, not only do the serum levels of DHEA and DHEAS fall with age similarly in both sexes, but also the stimulatability of the release of these substances by ACTH from the adrenal cortex, their chief serum source, is markedly reduced in the aging human organism.

Methods

Patient selection

The two patients in this study, aged 88 and 72 years, designated as subjects A and B, respectively, met the criteria of the NINCDS and the ADRDA for probable AD (24) and were admitted as inpatients to the geropsychiatry unit of the Sepulveda VA Medical Center. Informed consent was obtained from the patients or their conservators, psychotropic drugs were withdrawn, and the patients were stabilized behaviorally over a 2-week period, during which time they received baseline neuropsychological and laboratory evaluations. All laboratory findings were within the normal range. On the basis of neuropsychological evaluation, the patients satisfied all of the inclusion criteria for such studies (24).

In some instances it was necessary to include in this report data from other comparable patients because amounts of serum for steroid determinations became limiting for patients A and B.

Plan of study

DHEA was given orally in 400 mg capsules. In the single-dose studies, one capsule was given preprandially in the morning after control blood samples were drawn, and additional blood samples were taken at 1, 2, 6, and 24 hours. During chronic ingestion, control bloods were taken prior to the first dose of DHEA and the patients were given 1.2 gm DHEA daily (400 mg TID) and 2.4 gm DHEA daily (800 mg TID) for successive two week periods followed immediately by two weeks of intake of placebo capsules. Blood samples for steroid determinations and other laboratory measurements were taken preprandially on the morning following each 7 day period.

The DHEA employed had satisfied all FDA criteria for purity and stability. RIA determinations of all of the steroids were made by Endocrine Sciences, Tarzana, California.

Results and Discussion

The hematological, blood chemical, and urinary findings will not be presented since there were no trends that correlated with ingestion of DHEA. No abnormalities attributable to DHEA intake were found in weekly laboratory measurements or in daily clinical observations. The neuropsychological findings will be discussed elsewhere.

Under "population controls" (PC) in Fig. 1 are given the means and the ranges for the substances of interest in sera obtained at the Endocrine Sciences laboratories from 50 healthy employees of both sexes, 20 to 50 years of age. The control (subject baseline (SB)) and experimental values for the two patients were compared with the PC.

The precursors of DHEA: pregnenolone and 17α-OH pregnenolone

In both subjects of the present study, SB levels of pregnenolone were below the PC mean. In 4 other male AD patients, to be reported on in detail elsewhere, the SB values for pregnenolone were 0, 29, 35, and 75 ng/DL, respectively. All of these were below the PC mean and the first 3 were below the lowest value of the range for the PC. With the exception of the estrogens and cortisol, all of the other pregnenolone-derived steroids were lower in subjects A and B than in the PC. Estrogens were normal, and the cortisol levels were elevated.

The above results suggest that the decreased levels of serum pregnenolone in these patients reflect decreased formation of pregnenolone only in some steroid-synthesizing compartments. That this cannot be the case everywhere is indicated by the elevated serum levels of cortisol and the normal levels of estrogen, for both of which pregnenolone is an obligatory precursor. From what is known about control mechanisms involved, it appears likely that there may be decreases in one or more hypothalamic and/or pituitary peptides specifically required for pregnenolone synthesis in androgen-forming cells in the adrenal cortex. For example, a proopiomelanocortin (POMC)-derived peptide other than ACTH may play an important role in pregnenolone formation in DHEA-synthesizing cells (25). A pituitary peptide identical with the first 18 amino acids of the hinge peptide of the N-terminus of POMC, adjacent to ACTH, recently was reported to be active by itself and synergistic with ACTH in stimulating DHEA secretion strongly in dispersed human adrenal cells; but it stimulated cortisol secretion only minimally. If these observations are confirmed, the existence of this peptide may help explain the many instances in which the metabolism of adrenal androgens and cortisol diverge (6), even though they share a common precursor for their formation. It will be of great interest to study plasma and tissue levels of this new peptide and of other POMC-related peptides in stress and illness and during aging and in AD.

Insufficient plasma was available from the single dose experiments to measure levels of pregnenolone and 17α-OH pregnenolone at various times after DHEA intake in patients A and B. Therefore, data will be presented for 3 other AD patients similarly studied (Table 1). In all three patients, control values for pregnenolone and 17α-OH pregnenolone not only were below the means for the PC, but they were lower than the lowest values for the respective ranges. In patient C pregnenolone levels fell below the patient's control levels at 3, 6, and 12 hours after ingestion of DHEA and the levels for 17α-OH pregnenolone were lower at all times. In patients D and E there were marked elevations in pregnenolone and reductions in 17α-OH pregnenolone after DHEA. In the latter two instances the changes observed were consistent with an inhibition in 17α-hydroxylation of pregnenolone. The increased levels of 17α-OH pregnenolone at 24 hours in both of the latter cases suggests that a rebound synthesis and secretion occurred when serum levels of DHEA fell (see Fig. 2).

Samples which were taken approximately 12 hours after the last dose of DHEA in the chronic study (Fig. 1) showed small fluctuations in pregnenolone throughout the period of study. However, after the first week of DHEA ingestion, there were marked reductions in 17α-OH pregnenolone that persisted through the first post-drug week in patient B and for both post-drug weeks in patient A. Taken together with the data from the single dose experiments discussed above, the results suggest that a prolonged inhibition of 17α-hydroxylation occurred as a result of continued DHEA intake. Reductions of 17α-OH progesterone levels (Fig. 1) also are congruent with this suggestion. If, indeed, such plastic changes can occur as a result of chronic intake of high levels of DHEA, it might be the better part of wisdom to administer DHEA only in amounts sufficient to bring serum DHEA and DHEAS to young-adult levels.

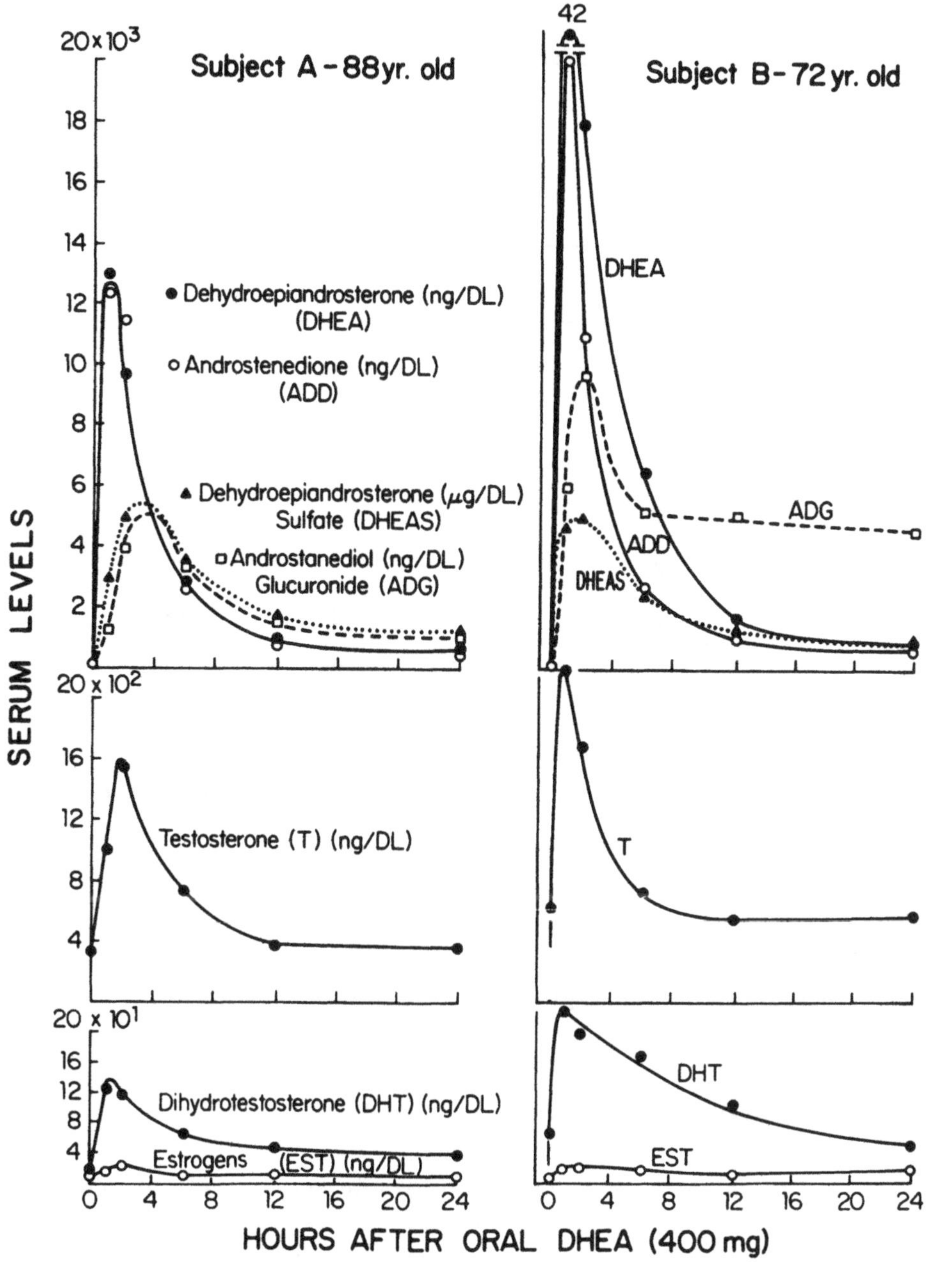

Fig. 2. Time-course of serum levels of DHEA-derived steroids after a single oral dose of DHEA. DHEAS, μg/DL; all others, ng/DL.

Table 1

Pregnenolone and 17α-OH Pregnenolone Levels (ng/DL) Before and After a Single Preprandial Dose of DHEA (400 mg)[a]

Time hr	Patient C (73 yr)		Patient D (75 yr)		Patient E (67 yr)	
	Preg.[b]	17α-OH[c]	Preg.	17α-OH	Preg.	17α-OH
0	25	34	33	25	35	37
1	38	27	220	0[d]	81	20
2	25	31	201	22	125	19
3	21	26	-	-	221	37
6	15	22	218	0	183	0
12	5	19	158	0	73	0
24	24	17	74	53	54	130

[a]Mean and range for population controls (PC): pregnenolone, 111 (46-225); 17α-OH pregnenolone, 124 (53-357).

[b]Pregnenolone

[c]17α-OH pregnenolone

[d]Not detectable by method employed

DHEA and the steroids formed from it

Let us now focus on the kinetics of changes in serum levels of DHEA and some of the key steroids that derive from it after ingestion of a single dose of DHEA (Fig. 2). It should be noted that the values of DHEAS, expressed in micrograms (μg), are approximately 500 to 100 times those of DHEA, expressed in nanograms (ng). There were extremely rapid and large elevations in levels of DHEA and androstenedione, indicating a fast rate of absorption of DHEA and rapid equilibration with androstenedione, probably in the intestinal wall and/or in the liver. The rises in testosterone, dihydrotestosterone, estrogens, DHEAS, and androstenediol glucuronide were somewhat slower. Much of the metabolism of exogenously derived DHEA may be extrasplanchnic, a good part possibly occurring in the skin, the largest organ in the body, by the pathways shown between DHEA and androstanediol glucuronide in Fig. 1 (6, 26, 27). Increases of serum levels of DHEAS and androstanediol glucuronide persisted for longer periods than for other metabolites of DHEA. DHEAS probably serves as a reservoir for DHEA, which can be formed from it by desulfation. Androstanediol glucuronide possibly may be converted to unconjugated androstanediol by action of β-glucuronidase, an enzyme localized in renal proximal tubule epithelial cells and inducible by androgen (28). The androstanediol thus formed could serve as a precursor for formation of dihydrotestosterone (see Fig. 1).

The bloods samples in the chronic study (Fig. 1) were drawn approximately 12 hours after the last dose of DHEA, taken the night before. From the curves in Fig. 2 it is seen that some of the constituents returned almost to pre-drug levels at 12 hours after a single 400 mg dose of DHEA. Testosterone levels in subject B actually dipped below the control values at this time, which also was observed in several other individuals not reported here. Were only the data in Table 1 available, it would have been missed that several-fold elevations in testosterone had occurred early on. In future similar studies or in therapeutic trials it would be preferable to employ time-release capsules or transdermal patches to give more uniform rates of delivery of DHEA so as to maintain relatively constant blood levels throughout the day.

Effects of chronic DHEA ingestion on serum levels of prolactin, luteinizing hormone (LH) and testosterone binding protein

Chronic administration of DHEA produced no particular trend in prolactin levels (Table 2). However, at the high dose there were marked reductions

Table 2

Prolactin, LH, and Testosterone Binding Protein Levels Before, During, and After Chronic Oral Intake of DHEA

	Population Controls (X)	Subject Baseline (Y)		Y/X	Post-ingestion					
					Low Dose		High Dose		Post Drug	
	ng/DL				Wk. 1	Wk. 2	Wk. 1	Wk. 2	Wk. 1	Wk. 2
Prolactin	8.8(3-18) ng/ml	A	18.0	2.05	28.0	20.0	16.0	24.0	19.0	16.0
		B	9.4	1.07	16.0	7.8	7.2	5.6	8.8	7.8
Luteinizing hormone	5.2(3-10) nlV/ml	A	6.2	1.19	5.2	5.2	< 1	1.4	10.0	7.0
		B	7.4	1.42	5.2	3.4	< 1	2.0	9.6	3.8
Testosterone binding protein	1.4(0.1-1.9) µg/DL	A	0.7	0.50	0.5	0.4	0.4	-	0.3	0.4
		B	1.3	0.93	1.1	0.8	0.6	0.4	0.8	0.6

in LH levels. The latter probably is attributable to feedback inhibition of hypothalamic LHRH release by increased estrogen (Fig. 1) and dihydrotestosterone levels. Although insufficient samples were available for the measurement of dihydrotestosterone in bloods of subjects A and B, determinations in 6 other subjects on the same protocol showed, on the average, a doubling of dihydrotestosterone levels. Testosterone binding protein was reduced in both patients, possibly reflecting regularly recurring elevations of testosterone with each dose of DHEA (Fig. 2) and the elevations of dihydrotestosterone (29, 30).

Further comment

Throughout the adult organism the cGMP system may be rate-limiting, the bottle-neck so to speak, in processes that allow differentiated cell systems under particular circumstances to achieve optimal functional states, whatever they may be. In the preceding paper (13), the hypothesis was advanced that DHEA, DHEAS, and steroids that derive from them, in addition to affecting genomic transcription, may relieve a rate-limiting constriction in cGMP metabolism by activating guanyl cyclase. This would allow sufficient increases to take place in cGMP turnover and amounts so that intracellular calcium management can be optimally coupled to those processes that are required for constituent cells in particular tissue systems to relate to each other in a manner compatible with survival, health, and reproduction of the organism of which they are a part. The cGMP and steroids can be constantly interactive insofar as the steroids, either formed indigenously or circulating in the blood, are available to facilitate cGMP metabolism wherever or whenever a need arises. All tissues generally can draw upon a plethoric supply of the steroids in young adulthood.

When the above steroids become inadequately available, as during aging, stress, or disease (6, 31, 32, 33), degenerative processes may be initiated upon occurrence of functional demands that require enhanced metabolic responses and genomic transcription to take place for recovery from activity to occur. Degenerative processes may not become irreversible as long as affected cells still possess receptors for the substances in question. Restoring adequate levels of rate-limiting hormones might allow some measure of recovery to take place so that cellular and organismic death might be delayed and, in some instances, prevented.

There was no natural selection for longevity. Time's arrow always has pointed in the direction of selection of those characteristics that adapted a species to survival in its particular environment until it could at least reproduce its numbers. There may even have been selection of self-destruct mechanisms that come into play at the cessation of the reproductive period which would serve to remove such organisms from competition for limited resources with the breeding population. When chemical and neural signals are broadcast within an organism, and to it from its environment, that its reproductive period is coming to an end, there may be initiated the sequence of events that leads to decreases in levels of DHEA and DHEAS in serum and tissues. The consequently progressive incoordination in metabolism and in communication patterns among the cellular elements of the immune networks may hasten the final demise of the organism. It is from this perspective that preceding discussions about the cyberneticizing roles of DHEA and DHEAS and their progressive decrements in aging (1, 6-8) makes so much sense and recommend these substances for much further study in aging and AD and for serious consideration as potentially prophylactic and therapeutic agents.

DHEA and DHEAS are beyond the metabolic branch point of pregnenolone, at which decisions are made as to how the subsequent metabolic flow should be fractionated between the mineralocorticoid, androgen-estrogen, and glucocorticoid pathways (Fig. 1). In the presence of tightly coupled control mechanisms, each of the above pathways might get all the precursor that it needs to produce sufficient quantities of the relevant substances to facilitate key processes at all levels of existence. This is what one generally would expect to see in the vigor of youth. However, under conditions of stress, real or perceived, in young or old organisms, and in old or sick organisms priorities would have to be established which would first maintain essential functions for survival of an individual organism and only then for its reproductive activities.

Without generation of sufficient metabolic energy and its appropriate coupling to all cellular functions, death would ensue rapidly, the nervous system being especially sensitive to such deprivation. In stress related to serious illness there is a shift in pregnenolone metabolism away from DHEA and DHEAS production to that of glucocorticoids (31, 32-33), which are acutely necessary for survival by virtue of their roles in the control of energetics. Determination made before and after surgical removal of adrenocortical adenomas in patients in whom ACTH secretion presumably was suppressed by feedback inhibition by excess serum cortisol during the presence in them of the tumors suggested that a deficiency of ACTH may result in a longer-lasting loss of the ability to secrete DHEA than of the ability to secrete cortisol (34).

Although there are as yet few relevant comparative data, one might expect the mineralocorticoids to be second to the glucocorticoids in line of priority because of their role in the maintenance of the delicate ionic balances between the cellular and extracellular environments which, after detachment from the placenta, are so essential for mammalian life either on the ionically unfriendly surfaces of the earth or in the depths of the sea. Survival in states of ionic imbalance may be longer than when energy supplies become limiting.

The last in the line of priorities would be the androgen-estrogen pathway, since reproductive activities tend to be much more attenuated, being crucial for the survival of the species, but not at all for any particular individual. In the latter regard it is of interest to list the following SB to PC ratios from the values in Fig. 1 for patients A and B, respectively: cortisol, 1.45 and 1.36; aldosterone, 0.46 and 0.62; DHEA, 0.25 and 0.36; and DHEAS, 0.23 and 0.25.

Although many mechanisms for establishment of such priorities as postulated above are conceivable, the most likely candidates are pituitary peptide hormones; and leading among them is the system of POMC-derived peptides, for reasons cited previously (25). Even without knowledge of the possible action of the hinge peptide, a rationale was developed for the concept of "pro-opiomelanocortin as a homeostatic control system" (35).

Determinations were made in several regions of human brains and in cranial nerves of the contents of pregnenolone, prenenolone sulfate, DHEA, DHEAS, androstenedione, testosterone, dihydrotestosterone, estrone, and estradiol (36). The highest values in brain were found for the 3 steroids mentioned in Table 3. The orders of contents in brain and blood are reversed: brain, pregnenolone > DHEA > DHEAS; blood, DHEAS >> DHEA > pregnenolone.

In rats there appear to be separate regulatory mechanisms for the above 3 substances in brain and blood, their metabolism being largely indigenous to brain. Cytochrome P-450scc, the enzyme involved in pregnenolone formation, is largely localized to the white matter (37). However, it is possible in the human that, because of the large concentration gradient between blood and brain, sufficient DHEAS may enter to be an important source of cerebral DHEAS and DHEA.

Table 3

Pregnenolone, DHEA, and DHEAS in Several Regions of Human Brain (ng/100 gm; from ref. 36).[a] Comparison with Blood levels (ng/100 ml; from Fig. 1).[b]

Brain region	Pregnenolone	DHEA	DHEAS
Frontal cortex	4074	1962 (0.48)[c]	483 (0.12)
Hippocampus	3749	1680 (0.45)	459 (0.12)
Parietal cortex	3620	1930 (0.53)	264 (0.07)
Temporal cortex	3485	1568 (0.45)	305 (0.09)
Amygdala	3201	1578 (0.49)	483 (0.15)
Hypothalamus	2786	1235 (0.44)	364 (0.13)
Blood	111	455 (4.10)	270,000 (2,432)

[a]Means of 4 to 5 brains from individuals of both sexes.

[b]Means from 50 healthy individuals of both sexes, 20 to 50 years of age.

[c]Numbers in parentheses are ratios to pregnenolone

DHEA, DHEAS, and the sex-related steroids that derive form them can enter endothelial cells and exert effects upon them (38 and references therein). A drastic differential reduction in steroid availability could have deleterious effects on endothelial cells. Neurologic and behavioral disorders observed during aging and in AD and various other neurologic disorders, or predispositions to them, may arise as a result of defects that begin in the microvasculature in specific regions of the brain (39, 40). In this regard, administration of DHEA or DHEAS may be helpful in protecting the microvasculature against degenerative changes when low blood levels occur.

Although a number of studies have suggested neuroactive functions for pregnenolone, pregnenolone sulfate and related substances, closure has not yet been achieved between the experimental observations and occurrence of these steroids in brain.

Acknowledgement

Supported in part by grants from the G. Harold and Lily Y. Mathers Foundation, the Ziskin Family Trust, and the National Metal Steel Foundation (to ER) and a Veterans Administration Merit Review grant 560566113 and the Sepulveda Medical center (to LJF).

References

1. Roberts, E. 1986. In: Treatment Development Strategies for Alzheimer's Disease (T. Crook, R. T. Bartus, S. Ferris and S. Gershon, eds.). Mark Powley and Associates · Madison, CT. pp. 173-219.
2. Roberts, E. 1986. Neurobiol. Aging 7, 561-567; author's response to commentaries. Neurobiol. Aging 7, 587-590.
3. Whitehouse, P. J., D. L. Price, A. W. Clark, J. T. Coyle, and M. R. DeLong. 1981. Ann. Neurol. 10, 122-126.
4. Hyman, B. T., G. W. Van Hoesen, A. R. Damasio, and C. L. Barnes. 1984. Science 225, 1168-1170.
5. Orentreich, N., J. L. Brind, R. L. Rizer, and J. H. Vogelman. 1984. J. Clin. Endocrinol. Metab. 59, 551-555.
6. Parker, L. N. 1989. Adrenal Androgens in Clinical Medicine. Academic Press · New York, 615 pp.
7. Sonka, J. 1976. Acta Univ. Carol [Med.] (Praha) 71, 1-171.
8. Regelson, W., R. Loria, and M. Kalimi. 1988. Ann. N. Y. Acad. Sci. 521, 260-273.
9. Roberts, E., L. Bologa, J. F. Flood, and G. E. Smith. 1987. Brain Res. 406, 357-362.

10. Bologa, L., J. Sharma, and E. Roberts. 1987. J. Neurosci. Res. 17, 225-234.
11. Flood, J. F., G. E. Smith, and E. Roberts. 1988. Brain Res. 447, 269-278.
12. Flood, J. F. and E. Roberts. 1988. Brain Res. 448, 178-181.
13. Roberts, E. Preceding paper, This Volume.
14. Sunderland, T., C. R. Merril, M. G. Harrington, B. A. Lawlor, S. E. Molchan, R. Martinez, and D. L. Murphy. 1989. The Lancet 2, 570.
15. Besman, M. J., K. Yanagibashi, T. D. Lee, M. Kawamura, P. F. Hall, and J. E. Shively. 1989. Proc. Natl. Acad. Sci. USA 86, 4897-4901.
16. Lobo, R. A., W. L. Paul, and U. Goebelsmann. 1981. Obstet. Gynecol. 57, 69-73.
17. Haning, Jr., R. V., C. W. Austin, I. H. Carlson, D. L. Kuzma, and W. J. Zweibel. 1985. Obstet. Gynceol. 65, 199-205.
18. Zumoff, B. and H. L. Bradlow. 1980. J. Clin. Endocrinol. Metab. 51, 334-336.
19. Bird, C. E., V. Masters, and A. F. Clark. 1984. Clin. Invest. Med. 7, 119-122.
20. de Peretti, E. and M. G. Forest. 1978. J. Clin. Endocrinol. Metab. 47, 572-577.
21. Kuhl, H., C. Rosniatowski, and H.-D. Taubert. Acta Endocrinol. 87, 476-484.
22. Vermeulen, A., J. P. Deslypere, W. Schelfhout, L. Verdonck, and R. Rubens. 1982. J. Clin. Endocrinol. Metab. 54, 187-191.
23. Parker, L., T. Gral, V. Perrigo, and R. Skowsky. 1981. Metabolism 30, 601-604.
24. Fitten, L. J., K. M. Perryman, P. L. Gross, H. Fine, J. Cummins, and C. Marshall. 1990. Am. J. Psychiatry 147, 239-242.
25. Parker, L., E. Lifrak, J. Shively, T. Lee, B. Kaplan, P. Walker, J. Calaycay, W. Florsheim, and P. Soon-Shiong. 1989. Program and Abstracts 71st Annual Meeting of the Endocrine Society. Abstract 299, p. 97.
26. Moghissi, E., F. Ablan, and R. Horton. 1984. J. Clin. Endocrinol. Metab. 59, 417-421.
27. Morimoto, I., A. Edmiston, D. Hawks, and R. Horton. 1981. J. Clin. Endocrinol. Metab. 52, 772-778.

28. Pfister, K., G. Watson, V. Chapman, and K. Paigen. 1984. J. Biol. Chem. 259, 5816-5820.

29. Vermeulen, A., L. Verdonck, M. Van Der Straeten, and N. Orie. 1969. J. Clin. Endocrinol. Metab. 29, 1470-1480.

30. Tochimoto, S., J. Olivo, A. L. Southern, and G. G. Gordon. 1970. Proc. Soc. Exp. Biol. Med. 134, 700-702.

31. Parker, L. N., E. R. Levin, and E. T. Lifrak. 1985. J. Clin. Endocrinol. Metab. 60, 947-952.

32. Parker, L., J. Eugene, D. Farber, E. Lifrak, M. Lai, and G. Juler. 1985. Horm. Metab. Res. 17, 209-212.

33. Parker, L. N. and W. D. Odell. 1980. Endocr. Rev. 1, 392-410.

34. Yamaji, T., M. Ishibashi, H. Sekihara, A. Itabashi, and T. Yanaihara. 1984. J. Clin. Endocrinol. Metab. 59, 1164-1168.

35. De Wied, D. and E. R. De Kloet. 1987. Ann. N. Y. Acad. Sci. 512, 328-337.

36. Lanthier, A. and V. V. Patwardhan. 1986. J. Steroid Biochem. 25, 445-449.

37. Le Goascogne, C., P. Robel, M. Gouezou, N. Sananes, E.-E. Baulieu, and M. Waterman. 1987. Science 237, 1212-1215.

38. Roberts, E. 1986. In: The Biological Substrates of Alzheimer's Disease (A. B. Scheibel and A. F. Wechsler, eds.). Academic Press, Inc., pp. 241-265.

39. Scheibel, A. B., T. Duong, and U. Tomiyasu. 1986. In: The Biological Substrates of Alzheimer's Disease (A. B. Scheibel and A. F. Wechsler, eds.) Academic Press, Inc., pp. 177-192.

40. Selkoe, D. J. 1989. Neurobiol. of Aging 10, 387-395.

COGNITIVE EFFECTS OF DHEA REPLACEMENT THERAPY

Kenneth A. Bonnet

Department of Psychiatry, New York University School of Medicine,
New York, New York 10016

Richard P. Brown

Department of Psychiatry, Columbia University College of Physicians and Surgeons and New York State Psychiatric Institute, New York, New York 10032

Introduction

The role of dehydroepiandrosterone (DHEA) in psychiatry has been explored many years ago (1,2). DHEA has been identified in the brain, and exhibits local metabolism in the brain (3,4). Tourney and Hatfield (5) have reported that schizophrenic males have lower plasma DHEA levels than controls. The role of DHEA in immune, cancer, endocrine and many other physiological and pathophysiological functions has been extensively studied more recently (6). A role for DHEA function in memory and cognition has only recently become recognized. Roberts (7) had begun to explore the relationship between DHEA and multiple metabolic processes in neuronal functioning and intracellular communication. There are a number of steroids, lymphocytes and potassium channel functions that are directly affected by DHEA and DHEA-S and that are known to decline with aging in high correlation with decreasing DHEA and DHEA-S circulating levels. DHEA and DHEA-S are reported to enhance neuronal and glial survival and differentiation in culture, and to enhance memory retention in adult mice (8). The administration of DHEA-S was reported to improve memory retention in aging mice without significantly affecting acquisition (9). These results

Dehydroepiandrosterone (DHEA)

encouraged consideration of clinical measurement of DHEA and DHEA-S to assess the indications for replacement therapy in a patient reporting twenty years of memory and cognitive inconsistencies. We had previously studied the patient neuropsychologically, and with computer-analyzed EEG, evoked potentials and regional cerebral blood flow.

The patient is a 47 year old nulliparous female with a lifelong history of specific learning disabilities. There were complaints of memory dysfunction, learning disabilities, and recurrent headache for a period of at least ten years prior this study. The patient was diagnosed variously as temporal-mandibular joint syndrome, depressed, manic-depressive, and learning disabled by various treating professionals prior to the study. These "diagnoses" are more a reflection of the constellation of symptoms of the individual patient, and not of the individual phases or episodes of one or the other diagnoses. Indeed, none of these are adequate as a diagnosis, but serve to characterize the symptom cluster that persisted for about twenty years, since early adulthood. Headache was difficult to treat and responded at times to Atenolol effectively. Cognition and memory functioning responded efficaciously for a time to Mysoline with eventual tolerance development, to nitrazepam with improved cognition but with severe attending paranoia development, and to Baclofen with transiently improved cognition. Every treatment trial was met with enthusiasm by the client, and with early claims of prodigious effectiveness of whatever the new treatment may be, with gradual waning of reported efficacy and eventual decline of reported benefits followed by requests for new medication. Not all of these events were "psychodynamic", however, and some of the early benefits were documentable, with subsequent side-effect or tolerance development that merited discontinuation and search for a new treatment. Most prominent among the presenting symptoms was recurrent headache and difficulty with concentration and memory functions.

The family history showed a mother with middle-aged onset of familial head tremor and Cushing's disease. The only sibling, a consanguine brother, is completely healthy and a high functioning businessman with a middle-aged development of mild head tremor reminiscent of that of the mother. The father was a high functioning professional who developed a mild stroke affecting the left third nerve in his late sixties while jogging. The father has since developed moderate multiple infarct dementia with no other symptoms.

Some strategies for symptom relief in the index patient was based upon computer analyzed EEG and evoked potentials records taken in the patient with a medication-free state. A series of anticonvulsant treatments was beneficial, but showed improvement in cognition and less so in memory functioning, and each resulted in a gradual waning of benefit.

A unifying feature of the compounds providing cognitive benefits was studied from a structure-activity relationship, and from a receptor subtype binding profile determined from the literature, and determined to be the potentiation of chloride ion channel-coupled GABA functioning. The quest was undertaken for a GABA-potentiating compound with minimal liability for metabolic tolerance development. A trial undertaken with nitrazepam was highly effective in improving mental functioning uniformly, but resulted in rapid and significant paranoia development as a paradoxical side effect that became almost incapacitating to the patient, and gradual discontinuation was instituted immediately.

A further quest for GABA-potentiating strategies lead to the discussion of the case with Dr. Eugene Roberts, a pioneer in

GABA research. Dr. Roberts alerted us to his studies, and others, of the role of adrenal steroids in memory decline. We undertook measurements of plasma DHEA and DHEA-sulfate levels in this patient and found her levels to be deficient. The laboratory results showed DHEA levels of 99 ng/dl, DHEA-S levels of 73 ng/ml, and cortisol levels of 13.5 μg/dl that were rather consistent in two successive measurements. The challenge with Cortrosyn resulted in increased DHEA production that peaked at 263 ng/dl and DHEA-S that increased to 412 ng/ml at the same sampling time; DHEA-S peaked at 666 ng/dl at a slightly later time than the peak for DHEA. It is apparent, then, that the sulfating enzyme function was intact and that replacement therapy was justified as a trial to determine cognitive improvement. Because of the subjective nature of the improvement that accompanied so many other treatment trials with this patient, a test series of clinical memory assessments was begun with the first trial prior to the beginning of DHEA maintenance. In addition, complex EEG and evoked potentials were studied by computer-based analysis before and at various intervals after the beginning of DHEA replacement therapy treatment. A schedule of symptom ratings was provided to the patient for daily symptom reporting and evaluation, as well.

The memory function measurement was carried out using the Randt Memory Test (10). The Randt battery was designed to test short-term, intermediate and long-term memory of several types in clinical populations. Subtests include general information, noun list acquisition and recall, number lists forward and backward, paired word acquisition and recall, short story acquisition and recall, picture series acquisition and non-verbal recall response, and incidental learning. Recall is tested at two and twenty minutes. The test is available in five separate but equivalent forms that have been validated and that have norms for all parts. The five forms were designed for valid repeat testing of the same individual. The norms are developed for each decade of age. The patient was tested using all five forms, as was designed in this test set, at intervals of DHEA study and maintenance.

The baseline electrophysiological measures were carried out with a standardized procedure employing the 10/20 system rules for placement of 28 monopolar electrodes, midfrontal ground lead and linked ears reference electrodes as reported elsewhere (11). Electrodes were prepared to 2 Kohm impedance by direct measurement of each electrode. EEG and evoked potentials studies were carried out using a NeuroScience EEG and evoked potentials system. The system provides amplification of thirty two channels of EEG and physiological channels at the chair-side, transmission to the main system seven meter distant from the patient, and further amplification. Waveforms are stored on optical disk medium for archival storage and review. On-line Fourier transform, post-hoc editing facilities and statistical averaging and analysis are contained within the system. The evoked potentials are presented in standardized format and procedure, and response is recorded in all 32 channels simultaneously for averaging and standardized analysis and interpretation. EEG was recorded for 3-5 minutes of artifact-free eyes-open resting and for 3-5 minutes of artifact-free eyes-closed resting record at each session. Long-latency auditory evoked potentials were recorded at 95 db tone and 65 db masking white noise in the opposite ear, for stimulation with each ear individually for 32 trials with 300 millisecond record epochs averaged. The P300 auditory evoked potentials procedure used 600ms record epochs, with 32 "rare", 2 kHz tones occurring at 15% of the time randomly interspersed between 194-220 "frequent" 1 kHz tones. In every record session, the long-latency evoked potential was recorded with stimulation of each ear to ensure integrity of unilateral functioning before beginning the P300 procedure binaurally.

DHEA: Was obtained from Berlichem, Wayne, N.J..
Dosage: Low: 12.5mg/kg/day; High: 37mg/kg/day; Orally in divided dosage for periods as indicated. The patient has been on DHEA with varied dosage for two years.

Results
Chemistries, EEG and evoked potentials and neuropsychological testing were conducted just prior to the point of beginning DHEA replacement therapy.

Baseline chemistries

Blood chemistries and hematology were grossly within normal limits. However, the white blood cell count has been elevated over 13,000 throughout the previous ten years, and occasionally falls to high-normal values episodically. No indications of immune deficiency, specific viral or bacterial titers, or of unusual immunoglobulin levels could be found in this patient. CPK, FSH, LH, ferritin, folic acid, GGTP, prolactin and thyroid were all within normal limits, as were all urinalysis values. Cortisol levels were 13.5 μg/dl at baseline. DHEA was 99 ng/dl and DHEA-S was 59-73 ng/ml in several measures prior to DHEA replacement treatment.

Memory testing

General information and mental status was intact in all respects at all times of testing, prior and during DHEA maintenance. Noun list acquisition was direct and errorless, but recall was completely failed that is abnormal for someone of her age. Reacquisition was in nine trials. Digit span forward was 5 digits, and backward was three digits and impaired. Paired words acquisition showed 8 errors and recall 66% errors with poor reacquisition. Short story acquisition was only 9 of 14 elements, and recall was only 6 elements that was

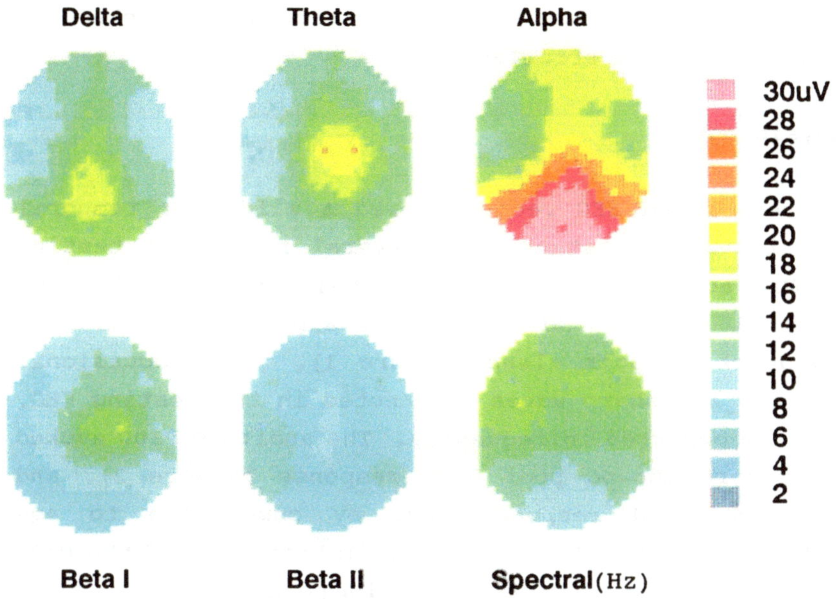

Figure 1. Quantitative summary of patient's baseline EEG. Topographic representations of the data derived from Fourier tranformation of each traditional EEG waveband. Nose at the top of each figure.

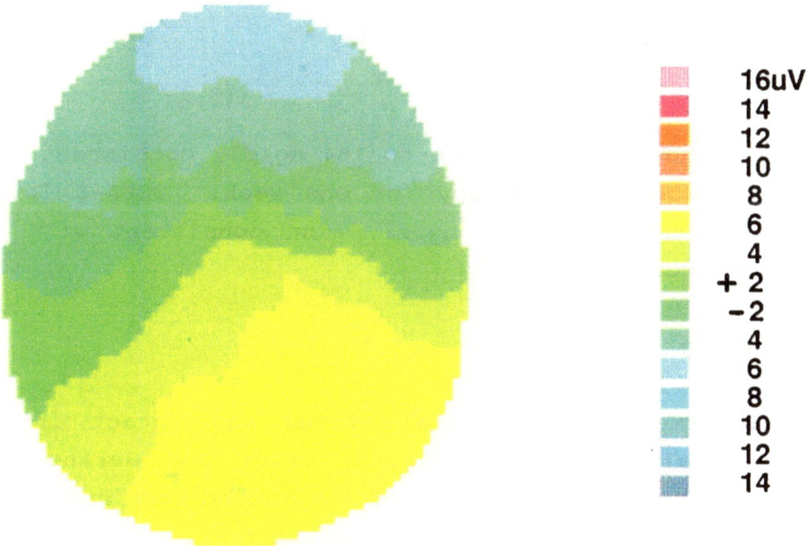

Figure 2. Anatomical distribution of patient's baseline P300 peak response, before DHEA replacement therapy.(Peak latency 420 milliseconds.)

disordered. Picture recognition acquisition was errorless in 7 items and recall was errorless.

Electrophysiology

Resting, eyes-closed EEG record showed a mild left temporal deficit that was persistent over several months of testing prior to DHEA administration. High level alpha was evident, and occasional episodes of high level alpha in the midline showed evidence of poor sleep (Figure 1). Some occasional hypersynchrony and sharp waves were noted in the resting EEG, although these occurred infrequently. The auditory P300 evoked potential record showed poor P300 response (Figure 2), and unusually high P200 response wave by comparison to age appropriate responses. There was poor distribution of the P300 wave, anatomically, and the wave showed a peak latency that was abnormally late in several records, and that was of low amplitude.

Low-Dose DHEA Effects: 12.5 mg/kg/day
Chemistries

DHEA levels showed an increase to 184 ng/dl, consistent with the daily replacement therapy for one week. DHEA-S levels showed an increase to 2966 ng/ml, also consistent with the daily replacement therapy.

Memory testing

General information and mental status was intact in all respects at all times of testing, prior and during DHEA maintenance. The five noun list of the second test module was acquired without error, and recalled without error, an improvement over baseline recall. Digit span was 5 forward and 3 backward, unchanged from baseline. Word pairs were acquired

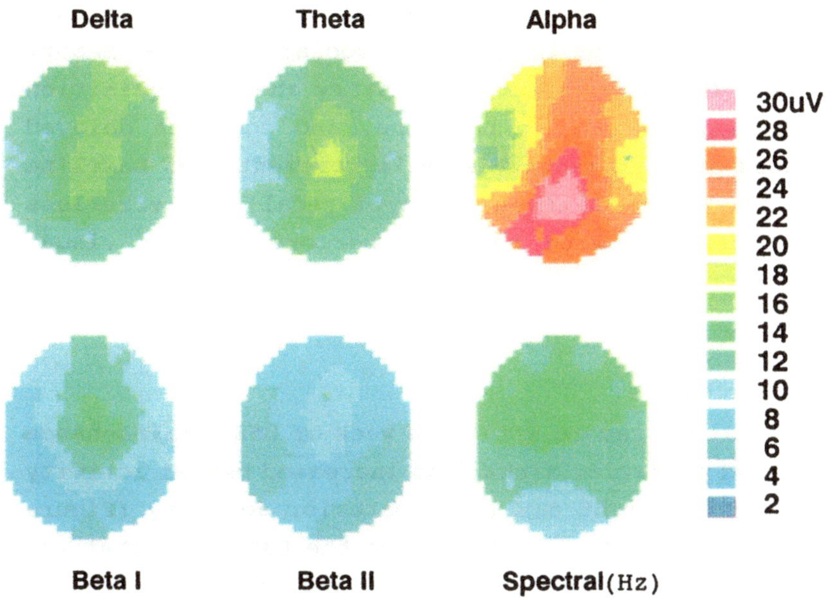

Figure 3. Quantitative summary of patient's low-dose DHEA EEG after one week of maintenance. Topographic representations of the data derived from Fourier tranformation of each traditional EEG waveband.

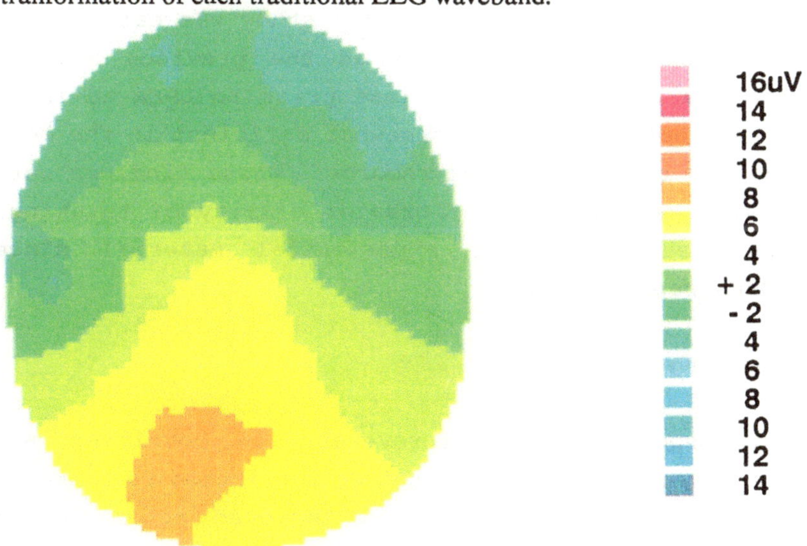

Figure 4. Anatomical distribution of patient's P300 peak response, after low-dose DHEA replacement therapy maintenance for one week. (Peak latency 376 milliseconds.)

in with only 2 errors, and delayed recall was errorless that represented an improvement in "paired associates" verbal memory over baseline. The high-impact short story showed acquisition of 9 elements was similar to baseline, and complete delayed recall of those nine elements was an improvement over baseline recall. Picture recognition and delayed recall was errorless as had been the case with baseline performance for visual retention.

Electrophysiology

Resting, eyes-closed record after one week of DHEA maintenance showed increased posterior alpha, and increased beta-1 activity with an increase in right and left posterior activity (Figure 3). There was increased distribution of the P300 wave that was also returned to a more normal latency, and higher amplitude, compared to several baseline records taken prior to DHEA administration (Figure 4). The P300 peak amplitude showed an increase that was unexpected since this study was carried out with the same time and conditions as the previous several studies, and the baseline record just prior to DHEA administration. Distribution remained somewhat deficient in the left lateral head, as in all previous records. These findings were consistent with self-reports of greater clarity of thinking, and of better retention of experiential material since beginning DHEA replacement.

Symptom reporting

The patient reported some improvement in cognition and "possible improvement in memory" that we attributed to improved assimilation abilities, rather than improved retention or recall. Yet, the patient indicated having better ability to "remember".

Higher-dose DHEA effects: 37 mg/kg/day

Chemistries

DHEA levels after one month at moderate dosage were elevated to a point that reached the level of 370 ng/dl, DHEA-S was 5250 ng/ml and cortisol was 6.6 μg/dl. This clearly showed a normalizing of the levels of DHEA and DHEA-S by replacement therapy. All other chemistries were with normal limits.

Memory testing

General information and mental status was intact in all respects at all times of testing, prior and during DHEA maintenance. Acquisition of a five noun list was errorless and delayed recall was 80% with rapid errorless reacquisition; this remained improved over baseline. Digit span was 5 forward and 4 backward, not substantially changed from baseline. Acquisition of word pairs was improved to only 3 errors, and delayed recall was complete and errorless that also represented a significant improvement. The high-impact short story module showed retention of 8 elements and delayed recall of 4 elements that was not much different from baseline performance. Picture recognition again showed errorless acquisition and errorless delayed recall as in baseline performance.

Electrophysiology

Resting Eyes-Closed Record showed decreased beta activity in comparison to the resting baseline record. There was increased activity in the resting record in the posterior alpha that was more normally distributed than in any previous record (Figure 5). The P300 response was forward and midline in distribution, compared to the earlier records. The P300 response showed more normal latency, compared to very late latency in the pre-DHEA

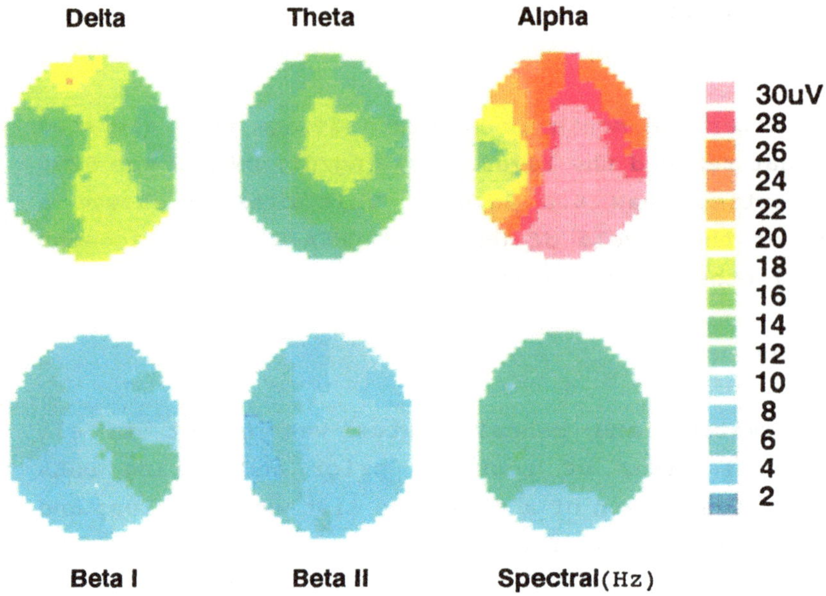

Figure 5. Quantitative summary of patient's medium-dose DHEA EEG after one month of maintenance. Topographic representations of the data derived from Fourier tranformation of each traditional EEG waveband.

Figure 6. Anatomical distribution of patient's P300 peak response, after medium-dose DHEA replacement therapy maintenance for one month. (Peak latency 366 milliseconds.)

records, and showed increased amplitude that was more normalized, as well (Figure 6).

Discussion

The beneficial effects of DHEA replacement trials in this individual with recurrent complaints of memory dysfunction and cognitive inconsistencies indicates the possibility of a new strategy in the diagnosis and indications for DHEA-related treatment of subtypes of cognitive dysfunctions. The beneficial effects in this single-case study must be examined critically since it is not a controlled or double-blind study. It further merits caution since it is only one patient. Nonetheless, the testing of the individual patient for DHEA and DHEA-S deficiency appears to solidify the selection criteria for consideration of this replacement therapy. The testing of the response to replacement therapy in this study is based upon both psychological testing, that depends upon verbal responses, and upon objective EEG and evoked potentials that are considerably less subjective. The beneficial effects are most clear in the EEG and evoked potentials records that avoid psychodynamic involvement in response production by the patient.

The patient described here showed baseline left temporal deficit that is consistent with her deficits in verbal memory tasks while retaining high ability in visual memory. It is evident that benefits were gained from DHEA replacement was in the ability to assimilate material and in the ability to apply cognitive "work" upon that material compared to her baseline condition. We are conducting studies of the pharmacokinetics of DHEA action to more effectively track those variables that show relevance to this endpoint of cognitive enhancement but are too complex to include here.

The replacement therapy of DHEA insufficiency has been applied for cognitive dysfunction in largely elderly populations showing gradual memory failure (Roberts, et al, personal communication). In those individuals, evidence of DHEA-S insufficiency is indicative for replacement therapy. The present single-case study indicates the advantages of testing younger individuals presenting with complaints of persistent memory dysfunction or cognitive dysfunction that is not related to head trauma or tangible insult. It must be emphasized that the patient in this study benefitted in cognitive enhancement, but that some memory deficit remained unaltered in spite of DHEA replacement. This strategy for clinical chemistry measurement and replacement therapy based upon the merit of those chemistries should not be expected to overcome pathological memory dysfunction, but rather appears in this case to have enhanced the ability to attend and to assimilate information that has previous associative experiential memory into which it can be integrated. It is encouraging to note that the individual in this study has begun a self-sufficient business for the first time, and has remained beneficially maintained with modest dosage oral DHEA without the development of tolerance or the waning benefits so characteristic of our other trials with pharmacological agents in this individual.

This study is part of our ongoing work in the area of cognitive dysfunction and cognitive enhancement pharmacology and pharmacokinetics using electrophysiological and neuropsychological measures (11,12).

This work supported in part by the Courtney Block Fund for Brain Chemistry Research (KAB) and the Mallinckrodt Foundation Fund (RPB). Grateful appreciation to Dr. William Regelson and to Dr. Eugene Roberts for encouragement and for the stimulation to explore new ground.

References

1. Strauss, E.B., D.E. Sands, and A.M. Robinson. 1952. Brit. Med. J., 12, 64-66.

2. Strauss,E.B. and W.A.H. Stevenson. 1955. J. of Neurology, Neurosurgery and Psychiatry 18, 137-144.

3. Baulieu, E.E. 1981. Steroid Hormones of the Brain. Pergammon Press. New York.

4. Kishimoto,Y. and M.Hosi. 1972. J. of Neurochem. 19, 2207-2215.

5. Tourney, G. and L.M. Hatfield. 1973. Biol. Psychiatry, 6, 23-36.

6. Regelson, W., R.Loria, and M.Kalimi. 1988. Annals of the N.Y. Acad. of Sciences, 260-273.

7. Roberts, E. 1986. In: Treatment Development Strategies for Alzheimer's Disease, (T. Crook, R.Bartus, S.Ferris and S.Gershon, eds.). Mark Powley Associates, Inc. Madison, Connecticut.

8. Roberts, E., L. Bologa, J.F.Flood and G.E.Smith. 1987. Brain Res. 406, 357-362.

9. Flood, J.F. and Roberts, E. 1988. Brain Res. 10, 178-181.

10. Randt, C., E.R. Brown, and D.P.Osborne Jr. 1980. Clinical Neuropsychology. 2, 184-194.

11. Bonnet, K.A., J.M.Luccioni, C.Walsh and A.J. Friedhoff. 1989. Biological Psychiatry 25, 45.

12. Bonnet, K.A. 1988. In: Disorders of the Developing Nervous System, (J. W.Swann and A. Messer, eds.). Alan R. Liss, Inc. New York. pp.267.

ORAL DEHYDROEPIANDROSTERONE IN MULTIPLE SCLEROSIS. RESULTS OF A PHASE ONE, OPEN STUDY

Eugene Roberts

Department of Neurobiochemistry, Beckman Research Institute of the City of Hope, Duarte, CA 91010

Thomas J. Fauble

117 E. Live Oak Ave., Arcadia, CA 91006

Introduction

Multiple sclerosis (MS) is a disease of the central nervous system of currently unknown etiology (1-6). It is characterized pathologically by demyelination of white matter associated with a mononuclear-cell infiltrate that consists predominantly of T-cells and macrophages. There is plaque formation, and an infiltration of mononuclear cells occurs predominantly in a periventricular distribution. MS is thought by some to be triggered by effects of an infection, viral (7) or microbial (spirochaetal) (8-12), which may result in aberrant immune reactions involved in thymic lymphocytes, macrophages, and excessive production of immunoglobulin in the central nervous system (CNS). There may be T-cell mediated autoimmune processes associated with the abnormalities in immunoregulation as well as humoral immune abnormalities. As a result of the plaque formation, nerve conduction is impaired, resulting in a number of the disabilities seen in MS.

The natural course of rapidly remitting-relapsing MS is complex. The course of chronic MS is relatively predictable. There is continuing worsening, and spontaneous remission is rare. Stabilization or improvement with any form of therapy would be a significant outcome. Numerous therapeutic strategies have been employed in the past (13), among them have been immunosuppression by chemicals (14-17) and radiation (18),

Dehydroepiandrosterone (DHEA)

plasmapheresis (19), and the administration of a K^+ channel blocker (20) or polymers that may affect macrophagic functions (21,22). To date none of the above has proven sufficiently efficacious so as to become standard therapy for MS. To our knowledge, high dose, long term antibiotic therapy suggested by the spirochaetal hypothesis (9-11) has not been tested.

Several recent studies have suggested that dehydroepiandrosterone (DHEA) and its sulfate (DHEAS), the major circulating steroids in humans, play important roles in many aspects of cellular controls, in immunoregulation, in maintenance of metabolic and structural integrity of nervous tissue, and in the achievement of the plastic changes required for learning and retention to occur (23-30). These substances also may have antiviral effects (31). Although the mechanisms of the effects observed still are not known, the results to date may be considered to be examples of rebalancing by the above substances of complex malfunctioning immune, neural, and metabolic systems. We considered the above to be cogent evidence supporting the supposition that administration of DHEA might be useful in the treatment of patients with MS. Below are presented the results of an open, pilot study of the effects of oral DHEA in patients with MS.

Methods

Patient selection

The subject population consisted of 21 patients, 9 females and 12 males, all of whom already were well known to their physician (T.J.F.) and had been thoroughly diagnosed and evaluated by acceptable criteria (3-5). To minimize the well-known difficulties of evaluation of efficacy of therapies in MS, the patients studied all were in a chronic stage of MS, i.e., free of any marked exacerbations for at least 6 months prior to initiation of the study. Six of the female patients and 5 of the male patients were taking immunosuppressive therapy (100 mg/day Imuran (azathioprine)). There were no major medical illnesses other than MS, and the life circumstances of the individual patients were judged to be compatible with potentially good compliance in ingestion of the experimental medication. All of the patients gave informed consent after the nature of the study had been fully explained. The work was on an outpatient basis and carried out within the limits of an office practice at no cost to the patient for the drug, visits or procedures related to the study. It was performed under FDA approval.

Medication

No untoward effects were reported in 5 young men (average age 25 yr.) taking 1.6 gm DHEA/day for 28 days (32). In a recently completed open inpatient study with 8 patients with Alzheimer's disease averaging 73 years of age, no toxicity, whatsoever, was observed during 2 weeks of DHEA at 1.2 gm/day followed by 2 weeks at 2.4 gm/day (Fitten and Roberts, in preparation). Because of the known potential sensitivity of MS patients to environmental changes and medications and our desire to achieve only high normal or, at most, slightly supernormal serum levels, DHEA was given orally in the present study in the much lower daily doses of 90 mg, 30 mg TID, and 180 mg, 60 mg TID. The DHEA employed had satisfied all FDA criteria for purity and stability.

Plan of study

In all instances patients were examined, evaluated, and blood and urine samples taken before they were given a supply of DHEA to last until the next visit. This was repeated at 2 weeks after the beginning of the low dose (90 mg/day) and at 3 monthly intervals thereafter on this dose. The higher dose (180/mg/day) was begun at the end of the latter 14 week period and continued for 3 months, during which time there were 2 or 3 visits to the physician for examination and for taking of samples. An exit visit relevant to the study took place between 1 to 2 months after cessation of DHEA intake. In a few instances there were visits to the physician one month prior to the one at which the first doses of DHEA were begun, at which time blood and urine samples also were taken. Between 8 and 10 blood and urine samples were obtained from each patient during the course of the study.

Measurements

In addition to neurological evaluation and estimation of ambulation and mental and communication status (3-5) by the physician at every visit, observations made by the patients and their care persons were recorded. After completion of the study a questionnaire was asked to be filled out by each patient, seeking to elicit in retrospective fashion a subjective impression as to whether or not improvement had been experienced.

The following laboratory tests were performed on the samples obtained at all of the intervals cited above: complete blood count, including differential and platelet count; a panel of 20 blood chemistries including enzyme measurements; urinalysis, including routine microscopic and macroscopic; and flow cytometric determination of T cell subsets (USC Clinical Laboratories under the direction of John W. Parker, M.D.) (33). RIA determinations of DHEA and DHEAS levels were made on control blood samples and on those obtained at the end of the 14 week period on the 90 mg/day dose of DHEA (Endocrine Sciences, Tarzana, California). Cortisol levels were determined only in the control samples.

Results

In Fig. 1 and 2 are shown the values for DHEA and DHEAS before and after 14 weeks on a daily oral dose of 90 mg of DHEA. It should be noted that the values of DHEAS (expressed in micrograms (μg)) are approximately 500 to 1000 times those for DHEA (expressed in nanograms (ng)). Population means and ranges were determined at the Endocrine Sciences laboratories on blood samples from approximately 50 healthy employees of both sexes, all of whom were between 20 and 50 years old. The results in Fig. 1 are arranged in order of progressively increasing control values of DHEA and the patients are identified. The same order of patients was used in the presentation of the data for DHEAS (Fig. 2). Appended to Fig. 1 are several items that will facilitate interpretation of the findings.

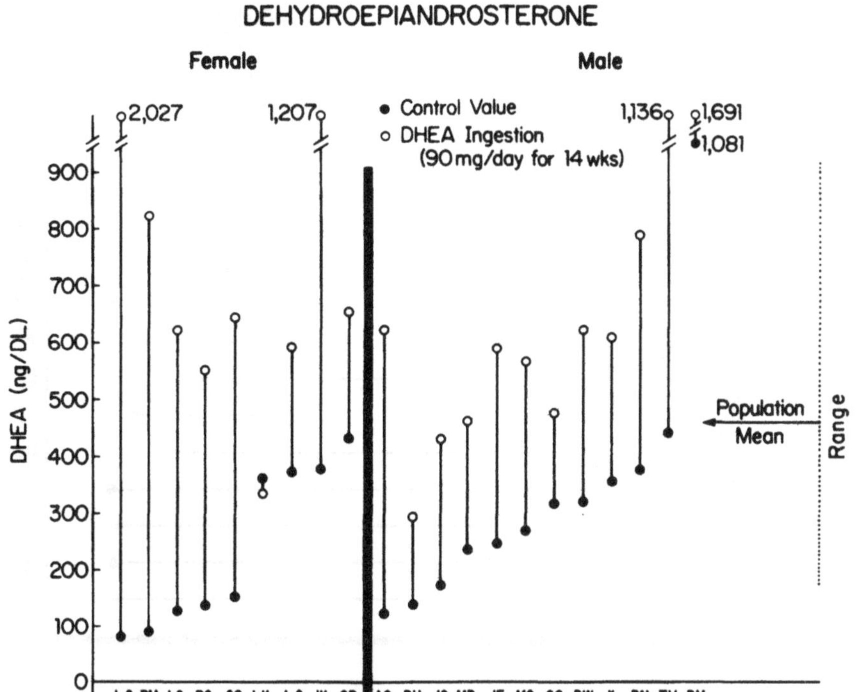

Patient		JeC	PM	LO	PG	OS	LK	JoC	JK	GP	AS	RH	JO	MD	JF	MG	GS	DW	JL	DN	TV	RM
Age (yrs)		35	41	39	60	46	38	47	55	47	56	60	62	65	37	53	44	53	38	40	44	36
Imuran				+	+	+	+	+		+	+	+								+	+	+
Cortisol (μg/DL)		5	7	9	13	10	14	12	13	27	12	11	13	14	13	13	13	9	9	10	10	12
T4 helper	high	0	2	0	0	0	0	0	1	0	2	0	0	0	0	0	0	0	0	1	0	0
	normal	8	5	9	9	7	8	8	8	10	7	6	10	8	8	9	8	9	0	8	9	9
	low	1	2	0	0	2	2	0	0	0	0	2	0	0	0	0	0	0	9	0	0	0
T8 suppressor	high	0	0	0	0	0	0	0	0	0	0	0	0	0	0	0	0	2	0	0	0	0
	normal	7	7	0	1	7	5	2	3	1	2	8	8	6	8	9	7	7	0	9	2	5
	low	2	2	9	8	2	5	6	6	9	7	0	2	2	0	0	1	0	9	0	7	4
T4/T8	high	0	0	5	8	0	1	1	7	9	7	0	3	1	1	0	5	0	0	0	5	3
	normal	8	8	4	1	9	9	7	2	1	2	8	7	7	7	9	3	9	8	9	4	6
	low	1	1	0	0	0	0	0	0	0	0	0	0	0	0	0	0	0	1	0	0	0
Subjective improvement (see text)			+			+		+			+	+	+	+	+			+	+			
Kurtzke scale improvement		NONE									NONE											

Figure 1. Serum DHEA levels in MS patients before and after 14 week oral intake of DHEA (90 mg/day). Relevant items of data are appended (see text).

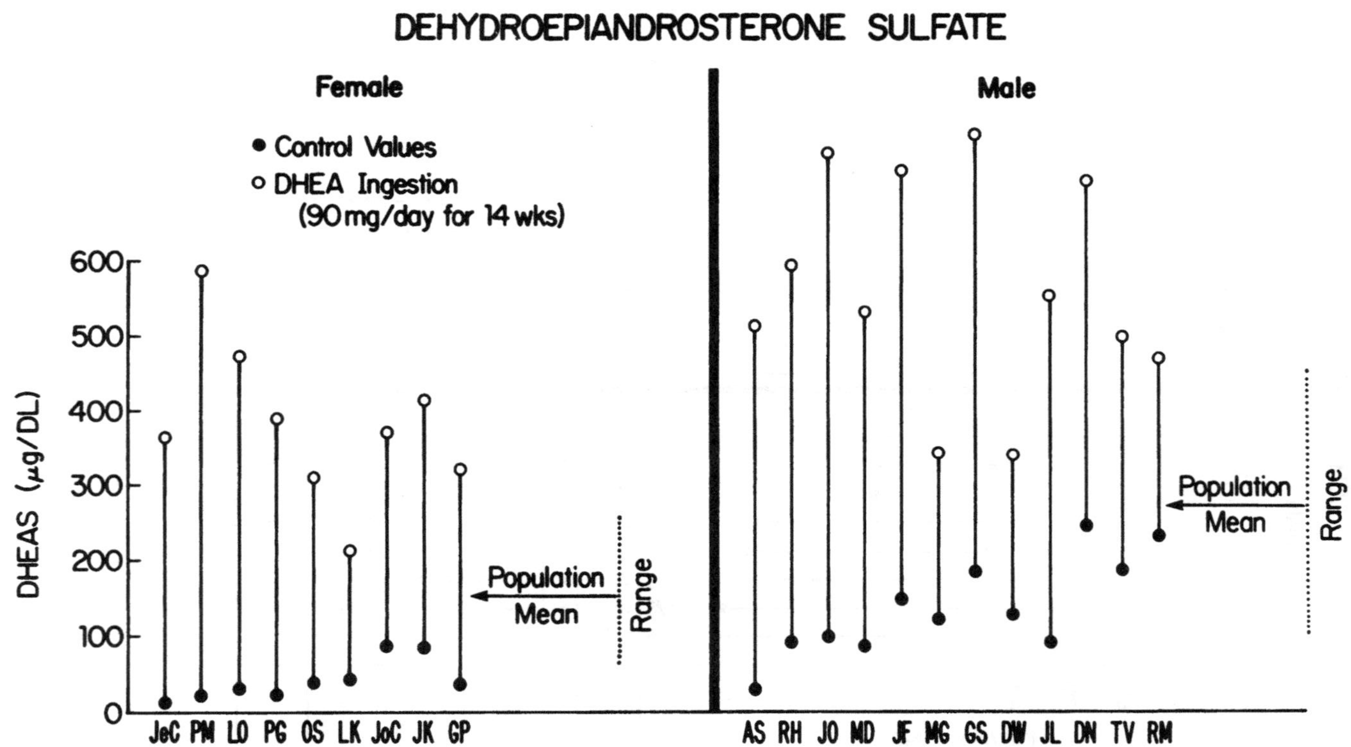

Figure 2. Serum DHEAS levels in MS patients before and after 14 week oral intake of DHEA (90 mg/day).

Pre-experimental values of DHEA in the MS patients all were below the population mean, with the exception of male patient RM, who showed a remarkably high value (Fig. 1). Five of the values for females and 2 for males fell below the lowest value recorded for the control population. At 14 weeks after the beginning of daily ingestion of DHEA, all but one of the female values and 2 of the male values were above the population mean, with two values for both sexes exceeding the highest ones found in the normal population. The latter shows that a dose of 90 mg/day was sufficient to elevate blood DHEA to normal or supernormal levels in the individuals in this study.

Although levels of DHEA generally tend to decrease with age (24), there was no strong age-related trend in the patient population, nor was there a relationship to Imuran therapy. In contrast to DHEA levels, with the exception of two low values (patients JeC and PM) and of one elevated level of 27 μg/DL (patient GP), pretreatment cortisol values were within the range for the control population (mean, 11 μg/DL; range, 8-19 μg/DL).

DHEA and DHEAS are readily interconvertible (24). Normally both substances largely are furnished by the adrenal glands. However, when DHEA is given orally, sulfation to DHEAS may take place in the intestine, liver, skin, and possibly other tissues during the absorption of DHEA and the levels of DHEAS probably are determined to a considerable extent by rates of sulfation and desulfation in various tissues. Not only did the oral intake of DHEA result in the elevation of serum levels of DHEAS but also in an increased variability (Fig. 2). All of the control values for both sexes were below the population means (values for males generally were higher than for females), 7 of 9 for females and 5 of 12 for males being below the lowest values recorded in the control population. At the end of the 14 week period of DHEA ingestion (90 mg/day), all of the DHEAS values for both sexes were above the respective population means, 7 of 9 for females and 10 of 12 for males being above the highest values recorded in the control population.

Normal mean absolute values and ranges for T4 (helper) and T8 (suppressor) cells and their ratio are 921/μl (518-1605); 460/μl (367-1072); and 2.0

(0.9-2.9), respectively (33). During the course of the study, individual patients showed no trends in the above that correlated with DHEA level or any of the other measured variables. However, values above (high) or below (low) the above stated ranges (normal) occurred at times during the study. All values (before, during, and after DHEA ingestion) were combined and the numbers falling into the above 3 categories recorded. Two or more abnormally low levels of T8 cells were found among the samples of every female patient and one or more in 7 of the 12 male patients. The T4/T8 ratios were high at least once in 6 of 9 female patients and in 7 of 12 males. There appeared to be no correlation of the occurrence of low T8 numbers or high T4/T8 ratios with DHEA levels or with clinical status. None of the hematological, blood chemical, or urinary findings showed any trends that correlated with DHEA ingestion. Whatever few abnormalities were observed were seen in the patients at some time before the study and usually were attributable to adventitious happenings, such as urinary tract infections, etc.

No significant numerical improvements were judged to have taken place for any of the patients during the entire course of the study by the rather stringent criteria of the Kurtzke scale. However, a thorough review of patients' and care persons' reports and results of physician-patient interviews showed there to be discernible improvement in some aspects of daily living in 3 of the 9 female patients and 7 of the 12 male patients. There follow some of the items thought to be indicative of improvement in the individual patients.

Female patients

PM: Began to show increased energy level at about 2 weeks after beginning DHEA. At 14 weeks there was greater limb strength, power, and dexterity; and at 18 weeks there was further gain in power, particularly in the lower limbs. An increase in sexuality was noted at two weeks after cessation of DHEA intake. Increased fatiguability and numbness of feet and fingers and spasticity set in four weeks later. Possibly a mild exacerbation of the disease was occurring.

JoC: Increased stamina was noted during DHEA ingestion. The patient was able to make transfers more easily, requiring less aid from her husband. She was able to stand with full support, which was not possible previously. These improvements over her previous condition disappeared after cessation of DHEA. Her appearance became more masculine, facial hair increased, and her hair thinned somewhat.
OS: Endurance improved and there was increase in grip power during DHEA ingestion. The patient was able to live a fuller life, for example, by shopping for herself. There generally was an increase in energy level and extension of the effective day. The above improvements were largely reversed and a gain in weight occurred within one week after cessation of DHEA intake. There appeared to be a thinning of hair.

Male patients

AS: No changes were noted at 2 weeks after beginning DHEA. AT 6 weeks the patient reported more energy and considerable increase in ability to pedal his stationary bicycle. He began experiencing penile erections during the day, but was not able to perform sexually. He last remembered having an erection 5 years previously. Increased energy level also was noted at 10 weeks and for the remainder of the study, enabling the patient to put in a longer work day. No additional effect was observed when daily intake of DHEA was doubled. The above improvements were maintained for a month after DHEA was stopped. At 2 months the patient reported decreased energy, endurance, mental alertness, and tolerance to heat.

RH: Increased endurance was first reported at 6 weeks after beginning DHEA. At 10 weeks the patient reported a feeling of well-being and increased frequency of sexual intercourse. A continuing increase in energy enabled the patient to extend his effective day by approximately 2.5 hours. There were decreased nocturia, fewer urinary accidents due to incontinence during the day, and decreased spasticity in the left lower extremity on the higher dose of DHEA. Improvement was sustained for at least 2 months after DHEA intake was stopped.

JO: The patient was urinating more freely and more completely at 6 weeks after beginning DHEA than before. He was able to rise from a chair and to walk pushing it along ahead of him more readily than before and generally was less lethargic. He could descend from the examination table more readily and walk further than before. His wife noted that he made transfers more readily. There was decrease in numbness that had been experienced previously. He slept better at night. The patient reported a decrease in sexual potency at 2 months after stopping DHEA.

MD: During DHEA treatment this severely afflicted patient showed increased finger flexibility and movement, enabling him to manipulate an important electrical switch more readily. He was able to sit up in a chair for 2-5 hours per day longer than before.

JF: This patient showed more energy generally and experienced less fatigue at his computer-related job, enabling him to work more efficiently while on DHEA. There was a progressive decrease in energy level after cessation of DHEA.

DW: During DHEA treatment the patient was mentally more alert. He was able to walk more readily with greater endurance and to stay up longer to experience a more effective day. His sexual performance increased over that of the pretreatment period. The patient regressed physically after cessation of DHEA, becoming less energetic. His spirits were lower and he was not able to stand as well or to move about as readily as when he was receiving DHEA.

JL: There was an increase of motor energy level and extension of the effective day by several hours during DHEA ingestion. An increase of strength, particularly in the upper body, enabled transfers to be made without help, representing an important improvement. All signs of improvement disappeared shortly after cessation of DHEA.

Discussion

Levels of serum DHEA and DHEAS generally were lower than normal in our group of 21 MS patients of both sexes. These were found to be increased to normal or supernormal levels after 14 weeks of oral ingestion of 90 mg/day. No untoward effects attributable to the medication were observed at any time during the 6 month period of administration, with the exception of a possible slight masculinizing effect in 2 of the female patients. Although on DHEA 3 of 9 female patients and 7 of 12 male patients showed some signs of improvement in activities of daily living, which in most instances disappeared gradually with cessation of intake, no major improvements in existing neurological disabilities were noted. Repeated estimation of T cell subsets showed some evidence of immune dysregulation in most cases, but there appeared to be no relationship between numbers of T4 and T8 cells or T4/T8 ratios with DHEA intake or with serum levels of DHEA and DHEAS.

In the current study, 10 of 21 patients reported improvements in some aspects of their disability that were not reflected in neurological assessments during office visits or in any of the laboratory measurements made on blood and urine. We believe that this warrants the mounting of more sophisticated controlled studies in which will be included the measurement of more variables (a broad spectrum of serum steroid determinations, CSF proteins and cells, brain imaging, evoked potentials, functional lymphocyte measurements, search for spirochaetal antigens, etc.).

Acknowledgement

Supported in part by grants from the G. Harold and Lily Y. Mathers Foundation and Mr. Julian Loewe.

References

1. Kurtzke, J. F. 1988. Neurology 38, 309-314.

2. Tourtellote, W. W. and Pick, P. W. 1989. Mayo Clin. Proc. 64, 592-596.

3. Swanson, J. W. 1989. Mayo Clin. Proc. 64, 577-586.

4. McDonald W. I. and Halliday, A. M. 1977. Br. Med. Bull. 33, 4-9.

5. Kurtzke, J. F. 1983. Neurology 33, 1444-1452.

6. Arnason, B. G. W. 1985. In: The Autoimmune Diseases (N. R. Rose and I. R. Mackay, eds.). Academic Press, Inc. · New York. pp. 399-427.

7. Wynn, D. R., Rodriguez, M., O'Fallon, W.M. et al. 1989. Mayo Clin. Proc. 64, 808-817.

8. Steiner, G. 1954. J. Neuropath. 13, 221-229.

9. Gay, D., Dick, G., and Upton, G. 1986. Lancet i, 815-819.

10. Gay, D. and Dick, G. 1986. Lancet i, 75-77.

11. Dick, G. and Gay, G. 1988. J. Infect. Dis. 16, 25-35.

12. Marshall, V. 1988. Med. Hypoth. 25, 89-92.

13. Waksman, B. H. 1987. Nature 328, 664-665.

14. Weiner, H. L. and Hafler, D. A. 1988. Ann. Neurol. 23, 211-222.

15. Rosen, J. A. 1988. Transplant. Proc. XX, 318-319.

16. Hauser, S. L., Dawson, D. M., Lebrich, J. R., et al. 1983. N. Eng. J. Med. 308, 173-180.

17. Kappos, L., Patzold, U., Dommasch, D., et al. 1988. Ann. Neurol. 23, 56-63.

18. Devereux, C., Troiano, R., Zito, G., et al. 1988. Neurology 38 suppl. 2, 32-37.

19. Khatri, B. O., McQuillen, M. P., Harrington, G. J. et al. 1985. Neurology 35, 312-319.

20. Stefoski, D., Davis, F. A., Faut, M. et al. 1987. Ann. Neurol. 21, 71-77.

21. Campbell, B., Vogel, P. J., Fisher, E. et al. 1973. Arch. Neurol. 29, 10-15.

22. Bornstein, M. B., Miller, A., Slagle, S., et al. 1987. N. Engl. J. Med. 317, 408-414.

23. Sonka, J. 1976. Acta Univ. Carol [Med. Monogr.] (Praha) 71, 1-171.

24. Parker, L. N. 1989. Adrenal Androgens in Clinical Medicine. Academic Press, Inc. · New York, 615 pp.

25. Roberts, E. 1986. In: Treatment Development Strategies for Alzheimer's Disease (T. Crook, R. T. Bartus, S. Ferris and S. Gershon, eds.). Mark Powley and Associates · Madison, CT. pp 173-219.

26. Regelson, W., Loria, R., and Kalimi, M. 1988. Ann. N. Y. Acad. Sci. 521, 260-273.

27. Roberts, E., Bologa, L., Flood, J. F. et al. 1987. Brain Res. 406, 357-362.

28. Bologa, L., Sharma, J., and Roberts, E. 1987. J. Neurosci. Res. 17, 225-234.

29. Flood, J. F., Smith, G. E. and Roberts, E. 1988. Brain Res. 447, 269-278.

30. Flood, J. F. and Roberts, E. 1988. Brain Res. 448, 178-181.

31. Loria, R. M., Inge, T. H., Cook, S. S. et al. 1988. J. Med. Virol. 26, 301-314.

32. Nestler, J. E., Barlascini, C, O., Clore, J. N. et al. 1988. J. Clin. Endrocrinol. Metab. 66, 57-61.

33. Parker, J. W. 1985. Lab. Med. 16, 21-31.

DEHYDROEPIANDOSTERONE IN MULTIPLE SCLEROSIS: POSITIVE EFFECTS ON THE FATIGUE SYNDROME IN A NON-RANDOMIZED STUDY

Vincent P. Calabrese, M.D.

Departments of Neurology and Biochemistry and Molecular Biophysics, Medical College of Virginia, Richmond, Virginia 23298

Edward R. Isaacs, M.D.

Richmond Neurological Consultants, Inc., Richmond, VA 23229

William Regelson, M.D.

Departments of Medicine and Microbiology, Medical College of Virginia, Richmond, Virginia 23298

Introduction

Multiple Sclerosis (MS) is a common neurologic disorder affecting primarily young adults. It is not usually fatal but often becomes a chronic debilitating disease leaving patients with variable amounts of disability and an uncertain future (1). The disabilities consist of weakness, sensory loss, incoordination, autonomic dysfunctions and a peculiar fatigue syndrome which is unrelated to exertion or lack of rest (2,3).

The pathophysiology consists of patchy loss of central nervous system myelin with little regard to established fiber tracks, accompanied by a reactive gliosis. The neuron perikaryon and axon are generally spared until late in the disease process. This loss of myelin leads to conduction failure which is thought to be secondary to potassium leakage from the demyelinated axon.

The etiology of MS is unclear but the suggestion of immune abnormalities and viral infections have raised the question of the use of drugs such as dehydroepiandrosterone (DHEA) since this steroid has the property of an immunomodulator including the ability to prevent significant disease in mice following infection with Coxsackie B virus (4) and the ability to prevent a lupus like syndrome in NZB mice (5). These findings may be of importance since there is no doubt that there is an altered immune response in MS, both humoral and cellular, and there is evidence that particular immunoregulatory genes may be present in MS patients. It is also unclear where the immune abnormalities fit into the

Dehydroepiandrosterone (DHEA)

The role in potassium leakage as a major factor in producing neuron conduction loss is another factor in considering DHEA in therapy. There is circumstantial evidence that DHEA has an effect on potassium channels in cells (6) which would make it important as a symptomatic therapy. This concept is pertinent to observations that DHEA in vivo enhances T cell maze learning in mice and stimulates in vitro neurite formation in neuroblasts (7). DHEA is also found in the brain of rodents (8) at concentrations equal to that found in the adrenals, although studies of its presence in the spinal cord or peripheral nerves have not been made.

At present medical therapy of MS consists of immunosuppressive drugs such as corticosteroids and ACTH. More recently drugs such as cyclophosphamide (9), azathioprine (10), Cyclosporin A (11) and bone marrow irradiation (12) have been found to modify the course of the disease to varying degrees. These drugs have only a moderate effect of the fatigue syndrome however. The latter symptom can be treated with drugs such as symmetral or amphetamine drugs such as pemoline, though only a small number of patients respond well to these drugs (13,14). Preliminary studies using potassium channel blockers such as 4-amino pyridine have been shown to be effective (15) but so far their toxicity and short duration preclude their use.

Our previous experience in treating advanced cancer patients (17) for periods as long as 2 1/2 years showed DHEA to be well tolerated at 40 mg/kg/day (17). In view of these studies we instituted a small open study of oral DHEA in well established MS patients. We initially intended to study primarily the effects on motor and sensory disability but soon found it had an effect on the fatigue syndrome.

Patient Population

Seventeen patients with long standing MS and who had become poorly responsive to steroid therapy were given a trial of oral DHEA. All the patients had a chronic neurologic disability and either a chronic or exacerbating remitting course. After obtaining informed consent, the patients were started on DHEA up to 40 mg/kg/day in four divided doses. Patients on chronic steroid were allowed to remain on their maintenance dose. (These were steroid dependent patients who would develop increased disability if the dose of steroid was decreased.) The patients had neurologic and laboratory evaluations at weekly intervals for the first month and monthly thereafter. The patients were maintained on DHEA for a minimum of three months when possible. The demographics of the patients are shown in Table 1.

TABLE 1
Patients entered into the oral Dehydroepiandosterone Study

Number of Pts	Sex	Age at Entry into Study	Duration of Disease
17	13F/4M	45.5 $\pm$ 2.4 yr	3.1 $\pm$ 1.6 yrs

Age and duration given as years $\pm$ SEM

Results

Ten patients completed a three month course of therapy. Three patients took the medication for only two months and then stopped; one because of increased spasticity, another because of persistent nausea and the third because of a very complex course with severe seizures and apnea spells (both of which preceded the use of DHEA). Two patients discontinued the drug after two weeks; one because of a general feeling of malaise and the other because she became frightened of using an experimental drug. Finally, one patient dropped out after one week because of a flu like illness where nausea made it impossible to maintain the drug schedule. After the illness she opted not to re-enter the study. Three female patients had the medication temporarily discontinued because of deepening of their voices. Following a washout period, in which their voices returned to normal, all three asked to have the medication restarted. The DHEA was restarted but at a lower daily dose and was given every other day instead of daily. These patients felt an improvement in their well being with reinstitution of the DHEA.

Table 2 shows that the Expanded Kurtzke Score improved in only one patient whereas in the others it remained stable. Nine of fourteen patients who complained of fatigue said the DHEA significantly improved their stamina and sense of well being. This manifested itself as an ability to carry on activities of daily living without feeling exhausted. In two female patients there was an improvement in libido. Some of the patients said they were able to get out and do things whereas before they felt confined to home because of severe exhaustion with minimal effort.

TABLE 2
Results of Treatment with Oral Dehydroepiandosterone

Patient	Kurtzke Scale Before	After	Fatigue Relief	Increased Spasm	Side Effects
1	6.0	6.0	+	-	Transient voice change
2	7.5	7.5	+	+	
3	7.0	7.0	+	+	
4	6.5	6.5	+		
5*	7.0	7.0	+	+	
6*	7.5	7.5	-	+	Nausea
7	7.5	7.5	+	+	
8*	8.5	8.5	-	-	Increased seizures
9*	6.0	6.0	+	+	Nausea
10*	6.5	6.5	Increased	-	Increased weakness
11	8.0	8.0	-	-	
12*	5.5	5.5	-	-	
13	5.0	5.0	+	-	Transient voice change, hirsutism
14	6.0	5.5	NA	-	
15	5.5	5.5	NA	-	
16	8.0	8.0	NA	-	
17	3.0	3.0	+	-	

* Did not finish the full three month course
+ Yes
- No change
NA Not applicable

Discussion

One major factor in considering DHEA therapy in MS is its ability to prevent viral infections and autoimmune diseases from developing in experimental animals. Obviously, by using patients with well established Multiple Sclerosis this could not be evaluated, nor was the short course of therapy long enough to evaluate the effect of DHEA on the progression of the disease. It was hoped, however, that the ability of DHEA to modulate the immune system and its possible effect on blocking potassium channels would have an immediate effect on the clinical symptoms of MS. From this small open study it does not appear that there was any significant immediate effect on either the motor or sensory symptoms of the disease unless increased spasm can be considered a sign of therapeutic effect despite its discomfort to the patient.

The ability to alleviate the fatigue syndrome in 64% of the patients is felt to be significant. This effect was not stressed when the drug trial was discussed with the patient. Therefore, even though a placebo effect should be considered, the fact that the other major symptoms were not affected makes it more significant.

It is possible that the effect was simply due to testosterone production, especially in view of the fact that hirsutism and voice changes were noted in some patients. DHEA does enhance androgen production with increases in both testosterone and dihydrotestosterone levels (17). Unfortunately serum testosterone levels were measured in only two patients while they were taking the DHEA. Both were females. Both patients had serum testosterone levels above the normal for females, but well within the low normal range for adult males. Treatment of MS with ACTH has been a standard for many years. While it is presumed that the effect of ACTH is due to corticosteroid secretion there has always been the question whether other steroids are important. The results of the present study lend some support to the role of other steroids.

It is to be noted that we started this study independent of the DHEA Multiple Sclerosis treatment program organized by Roberts and Fauble in California (17). It was not until the end of our study that we were aware of the similar experience with DHEA reported in the California study. The implications of this small study is that DHEA may be an effective therapy for the fatigue syndrome in MS. A larger, double blind study using more formal methods of measuring fatigue would certainly be indicated with a crossover comparison to testosterone administration and a monitoring of androgen levels to ascertain if the DHEA effect relates to the psychic energizing action of androgens. The long term effects on the progression of MS using DHEA is still an open question and worthy of further study.

Acknowledgments

We wish to thank Searle, Skokie Ill. and the Elan Corporation, Gainesville Fla. for courteously supplying us with DHEA. An additional commercial source of DHEA was from Berlichem.

References

1. Matthews WB, Acheson ED, Batchelor JR and RO Weller. 1985. McAlpine's Multiple Sclerosis. Churchill Livingston, Edinburgh London Melbourne and New York. 49-72

2. ibid. pp 96-145

3. Krupp LB, Alvarez LA, LaRocca NG, Scheinberg LC. FAtigue in Multiple Sclerosis. Arch. Neurol. 45(1988) 435-440

4. Loria R, Inge T, Cook S, Szakal A and W Regelson. 1988. J. Med. Virol. 26, 301-314

5. Lucas JA, Ansar Ahmed S, Linette Casey M and PC MacDonald. 1985. J. Clin. Invest. 75, 2091- 2093

6. Roberts E. Guides through the labyrinth of Alzheimer's Disease: dehydroepiandosterone, potassium channels and the C_4 component of complement. In Crook T, Bartus RT, Ferris S and Gershon S (Eds) Treatment Development Strategies for Alzheimer's Disease Mark Powley and Associates, Madison CT 1986 pp 173-219

7. Roberts E, Bologna L, Flood TF et al. 1987. Brain Research 406, 357-362

8. Corpechot c, Robel P, Alexon N et al. 1983. Brain Research 270, 119-125

9. Carter JL, Hafter DA, Dawson DM, Orav J and HL Weiner. 1988. Neurol. 38(suppl 2), 9-13

10. Ellison GV, Myers LW, Meckey MR, Graves MC, Tourtellottee WW and MR Nuwer. 1988. Neurol. 1988. Neurol. 38(suppl 2), 20-23

11. Dommasch D. 1988. Neurol. 38(suppl 2), 28-29

12. Devereaux C, Troiano R, Zito G, Deveraux RB, Kopecky KJ, Friedman R, Dowling PC, Hafstein MP, Rokowsky-Kochan C and SD Cook. 1988. Neurol. 38(suppl 2), 32-37

13. Murray TJ. 1985. Can. J. Neurol. Sci. 12, 251-254

14. Rosenberg G and O Appenzeller. 1988. Arch. of Neurol. 45, 1104-1106

15. Stefoski D, Davis FA, Faut M and CL Schauf. 1987. Annals of Neurol. 21, 71-77

16. Regelson W, Loria R and M Kalimi. 1988. Ann N.Y. Acad Sci. 518, 260-273

17. Roberts E. 1989. Personal communication

REDUCED PLASMA DEHYDROEPIANDROSTERONE CONCENTRATIONS IN HIV INFECTION AND ALZHEIMER'S DISEASE

C.R. Merril

Laboratory of Biochemical Genetics, National Institute of Mental Health
Bethesda, Maryland 20892

M.G. Harrington

Division of Biology, California Institute of Technology
Pasadena, California 91125

T. Sunderland

Laboratory of Clinical Science, National Institute of Mental Health
Bethesda, Maryland 20892

Introduction

Alzheimer's disease and autoimmune deficiency syndrome (AIDS) are fatal disorders which have a variable course in terms of length of survival after the onset of initial symptoms. The variables affecting the clinical course of these diseases have not yet been fully differentiated. In this regard, we were prompted to study the plasma levels of the endogenous steroid dehydroepiandrosterone (DHEA). The possibility of a relationship between the length of survival for patients with these diseases and the plasma levels of the steroid DHEA was raised by the results of prior studies(1,2,3). First, a correlation has been demonstrated between decreased levels of plasma dehydroepiandrosterone and all causes of death(1). Second, correlations have been reported in a number of diseases and plasma dehydroepiandrosterone levels. Low plasma levels have been found in women with breast cancer(2,3), while a diet supplemented with dehydroepiandrosterone protects rodents against: breast tumors(4) diabetes(5), obesity(5), auto-immune nephritis(6), and chemically-induced tumors of the colon(7) and lung(8). Furthermore, human cells in culture are also protected against chemical mutagenesis(9) and viral transformation(10) by the addition of DHEA to the growth media.

DHEA in HIV Infection

The length of survival after the onset of initial AIDS symptoms is consistent with the normal plasma distribution of this steroid, in that the survival of older adults and young children is less than that of young adults. The level of plasma DHEA is low in young children, reaching a maximal level in early adulthood, and then it decreases with age. The level of this steroid in 70 year olds is about 20% of that found in young adults(11). In addition, as illustrated in Table 1, we have measured the levels of plasma DHEA sulfate, using a radioimmunoassay in an age matched blind study (12).

Table 1.

Condition of Individuals	Number Studied	Assay Results in µmol/L	P values
HIV negative	8	8.0 ± 2.4	
HIV positive apparently Well	9	5.8 ± 1.7	P<0.05
HIV positive with AIDS	10	3.9 ± 2.3	P<0.001
Intra-assay coefficient of variation = 4.1% Inter-assay coefficient of variation = 12.2%			

DHEA in Alzheimer's Disease

In a study with 10 Alzheimer's disease patients and 10 age and sex matched healthy controls the levels of plasma DHEA sulfate were on the average 48%(P<0.05[one-way]) lower in the Alzheimer patients(13). Intra-assay variability for duplicate samples was 7.5%. In this study the patients and controls were in good medical health and had been off psychotropic medications for at least 3 weeks prior to this study. In addition this study excluded any patients with non-Alzheimer multi-infact dementia. There was not corrletaion between the levels of plasma DHEA sulfate and serum cortisol, the severity of the apparent dementia, or baseline cognition.

Conclusions

At present there is no evidence for an etiologic effect in the negative associations that have been demonstrated between the plasma DHEA level and AIDS and Alzheimer's disease. However, the reports of protective effects of DHEA sulfate in: animals and human cell(4-10), partly degenerated or at-risk brain cells(14,15) and the anti-amnestic effects(16,17) suggest that the role of DHEA in HIV infections and Alzheimer's disease be further investigated. Clinical trials on the effects of the administration of DHEA sulfate to Alzheimer's disease patients are currently being performed(13).

References

1. Barrett-Connor, E., K-T. Khaw and S.S.C. Yen. 1986. New Eng.J.Med. 315, 1519.
2. Bulbrook, R.D., J.L. Hayward, C.C.Spicer. 1971. Lancet 2, 395.
3. Brownsey, B., E.H.D. Cameron, K. Griffiths, E.N. Gleave, A.P.M. Forrest, H Campbell. 1972. Europ.J. Cancer ; 8, 131.
4. Schwartz, A.G.. 1979. Cancer Res. 39, 1129.
5. Coleman, D.L., R.W. Schwizer, E.H. Leiter. 1984. Diabetes 33, 26.
6. Lucas, J.A., S.A. Ahmed, M.L. Casey, P.C. MacDonald. 1985. J.Clin.Invest. 75, 2091.
7. Nyce, J.W., P.N. Magee, G.C. Hard, A.G. Schwartz. 1984. Carcinogenesis 5, 57.

8. Schwartz, A.G., R.H. Tannen. 1981. Carcinogenesis 2, 1335.

9. Pashko, L.L., A.G. Schwartz. 1983. J. Gerontol. 38. 8.

10. Henderson, E., A.G. Schwartz, L. Pashko, M. Abou-Gharbia, D. Swern. 1981. Carcinogenesis 2, 683.

11. Orentreich, N., J.L. Brind, R.L. Rizer, J.H. Vogelman 1984. J.Clin.Endocrinol.Metab. 59, 551.

12. Merril, C.R., M.G. Harrington, T. Sunderland. 1989. J.Am.Med.Assoc. 261, 1149.

13. Sunderland, T., C.R. Merril, M.G. Harrington, B.A. Lawlor, S.E. Molchan, M. Martinez, D.L. Murphy. 1989. Lancet 570.

14. Roberts, E., L. Bologa, J.F. Flood, G.E. Smith. 1987. Brain Res. 406, 357.

15. Bologa, L., J. Sharma, E. Roberts. 1987. Jour.Neurosci.Res. 17, 225.

16. Flood, J.F., E. Roberts. 1988. Brain Res. 448, 178.

17. Flood, J.F., G.E. Smith, E. Roberts. 1988. Brain Res. 447, 267.

IMMUNE RESPONSE FACILITATION AND RESISTANCE TO VIRUS AND BACTERIAL INFECTIONS WITH DEHYDROEPIANDROSTERONE (DHEA).

Roger M. Loria.

Departments of Microbiology and Immunology and Pathology, Schools of Basic Health Science and Medicine, Virginia Commonwealth University,
Richmond, VA 23298-0678, USA.

William Regelson.

Departments of Microbiology and Immunology and Medicine, Schools of Basic Health Science and Medicine, Virginia Commonwealth University,
Richmond, VA 23298-0678, USA.

David A. Padgett.

Department of Microbiology and Immunology, School of Basic Health Science and Medicine, Virginia Commonwealth University,
Richmond, VA 23298-0678, USA.

INTRODUCTION

Dehydroepiandrosterone, (DHEA), is quantitatively one of the major adrenal cortical steroid hormones in humans and other mammals,(1,2) and we reported that DHEA protects the host against acute lethal virus infection. In addition to protecting against an RNA virus (coxsackievirus B4) and a DNA virus (herpes virus type 2), we present results that indicate protection is similarly conferred against a bacterial infection (*Streptococcus faecalis*).

DHEA, 5-Androsten-3ß-ol-17-one or dehydroiso-androsterone is sulfated by an adrenal sulfokinase to DHEA sulfate (DHEAS) and thereby excreted in the urine in humans, and likewise but to a lesser extent in rodents (3). The level of DHEA begins to decline in the second decade of life reaching 5% of the original level in the elderly (4).

Dehydroepiandrosterone (DHEA)

Although DHEA appears to serve as an intermediary in gonadal steroid synthesis, the primary physiological function of this hormone is unclear.

Our interest in DHEA stemmed form our previous studies (5-7) which have shown that the C57BL\KsJ db+\db+ diabetic mutant mouse has an impaired immune response and is markedly more susceptible to a human diabetogenic coxsackievirus B4 (CVB4) infection. It is now recognized that diabetes mellitus in humans may be a virus mediated autoimmune reaction which may result in the destruction of the islets of Langerhans (8,9). Since dietary DHEA was reported to have anti-diabetic effects (10-13) in the diabetic mutant mouse, we examined whether this was mediated in part by an effect on the immune response or by a reduction in the pathogenicity of the infectious virus.

In these studies, two acute virus infection models, with different replicative and pathogenic mechanisms were examined to determine the effects of DHEA on virus-mediated pathophysiology; in addition we now have also examined the effects of DHEA on a lethal bacterial infection. The results show that peroral (p.o.) and subcutaneous (s.c.) administration of DHEA results in immune facilitation and reduces the virulence of an RNA, a DNA virus and a bacterial pathogen that are lethal by widely different mechanisms (14,15). In regard to the mechanism for DHEA's protective action, we also report the effects of dietary DHEA on plasma neutral lipid levels and on resistance to infection. The results depict the overall importance of the base diet on DHEA's effect on plasma neutral lipid levels, and that changes in the base diet, specifically the addition of 1% cholesterol and 0.5% cholic acid, can significantly alter the action of DHEA. Additionally, it is clearly illustrated that in protecting the host from lethal infection, dietary DHEA is considerably less effective than a single subcutaneous dose of the compound. These observations demonstrate that mobilization of the host cutaneous immune response by DHEA effectively achieves systemic protection against lethal infections.

MATERIALS AND METHODS

Viruses, Tissue Culture and Bacterium

Two different human virus isolates were used to challenge C57BL/6J inbred mice, one was the CVB4 Edwards strain and the second was herpes simplex type 2 (HSV2). Details on the passage history of CVB4 and tissue culture procedures have been published previously (5-7). The HSV2 strain MS was obtained from the American Type Culture Collection (ATCC VR-540). This virus was grown and plaqued on Vero cells. For quantitation of HSV2 plaques, virus was serially diluted and plated on confluent monolayers of Vero cells. Titer was determined by staining with a 1% crystal violet solution for 20 minutes followed by plaque counts. The plasmid containing strain of Streptococcus faecalis, x1515.OG1RF, which is highly virulent to mice was grown on blood agar. The number of colony forming units was determined by spectroscopy as compared to standards that were verified by subculture and streaking of plates (16). For i.p. injection of the bacterium, the organism was suspended at the proper dose in a volume of 0.5 mls of saline.

Animals

Male mice have been shown to be more susceptible than female mice to enterovirus infection (17-18) and the reverse is true for HSV2 susceptibility (19-21). Therefore male inbred C57BL/6J mice 6 to 8 weeks old (Jackson Laboratories, Bar Harbor, ME) were infected with CVB4 while female inbred mice of the same age and strain were used for HSV2 infections. Genetically immunodeficient hairless female HRS/J hr/hr inbred mice (Jackson Laboratories, Bar Harbor, ME) at 6-8 weeks of age were used to test the effect of a non-functional immune system (22,23) on the response to DHEA. The average weight of mice was 25 grams. For bacterial challenge experiments, mice of the outbred ICR mouse strain (Harland Sprague Dawley Inc. Indianapolis.In.) were used.

Diet

All animals were maintained on Purina Laboratory Rodent Chow #5001 (Ralston Purina, St. Louis, MO). In experiments where the effects of dietary DHEA were the focus, animals were maintained on a semipurified diet high in animal fat; the diet contained 20% casein, 52.5% sucrose, 18% animal fat (lard), 5% cellulose, 4% salts, 0.2% choline chloride, 0.1% inositol and 0.1% vitamin mix. This semipurified diet has been used extensively (24-27).

Route, Vehicle and Dose

Several routes of DHEA administration were examined. These included feeding as 0.4% of the diet (p.o.), subcutaneous injection (s.c.) or intraperitoneal injection (i.p.). For injection, DHEA (Searle, Chicago IL.) was suspended in 0.2 ml dimethyl sulfoxide (DMSO). The optimal dose of DHEA mediating antiviral activity was determined by injecting animals with DHEA doses of 2g, 1g, 500 mg and 250 mg/kg. In CVB4 experiments, animals were infected with virus one hour after DHEA injection, and each group was challenged i.p. with a dose of CVB4 ranging from 10^2 pfu to 10^5 pfu/animal. In HSV2 infection experiments, DHEA was administered s.c. at a dose of 1g/kg four hours prior to viral infection. Mice were challenged by intracranial (i.c.) injection with a dose of HSV2 ranging from 10^5 to 10^8 pfu/animal. Control animals were injected with virus, and 0.2 ml of DMSO. Protection from lethal CVB4 and HSV2 infection was observed when DHEA was injected s.c. at a dose of 1g/kg. Also, feeding DHEA at 0.4% concentration protected from lethal virus infection. No significant protection from lethal virus infection was evident with any other s.c. dose of DHEA, or with any dose of a closely related compound, etiocholanolone, 3 α-hydroxy-5ß-androstan-17-one (Sigma Chemical Co. St Louis Mo.). In all subsequent experiments a dose of 1g/kg DHEA (s.c. 25 mg/mouse) was used. For experiments designed to determine the effects of DHEA on Streptococcus faecalis induced mortality, the bacterium was injected i.p. at a dose of 1.00 x 10^{10} CFU/animal in 0.5 mls of saline two hours after injection with the steroid.

Enumeration of Spleen Antibody Forming Cells (AFC)

As previously described (8) ten days after CB4 infection, test animals were subjected to an i.p. injection of 5 x 10^8 sheep red blood cells (SRBC), while noninfected control animals were immunized only with SRBC. Four days after SRBC immunization, animals were euthanized with an overdose of methoxyflurane, (Metofane, Pitman-Moore, Inc. Washington Crossing, NJ.) and their spleens were removed. The procedure of Moller et al (28) was used for enumeration of spleen cells secreting IgM antibody.

Peripheral White Blood Cell Counts

Peripheral white blood cells were counted following Diff-Quik (American Scientific Products, McGaw Park, IL.) staining of blood smears. No differentiation of lymphocytes or monocytes by special stains or cell marker was performed.

Histopathological Examinations

For histopathological studies animals were sacrificed via an overdose of methoxyflurane. Tissues were removed and fixed at room temperature in 10% phosphate buffered formaldehyde. Specimens were embedded in paraffin, sectioned and then stained with hematoxylin and eosin.

Statistical Analysis

The General Linear Models Procedure (SAS) was used to determined the significance of the particular changes in a given cell type. A significant difference between the various groups was determined at a $P<0.05$ level using Tukey's studentized range test for each variable. A confirmation of these results was obtained from the non-parametric Wilcoxon Rank Sums test.

RESULTS

The effects of dietary DHEA on plasma neutral lipid levels.

Experiments designed to ascertain the effects of feeding DHEA to inbred C57BL/6J mice using a semipurified diet were undertaken. The rationale of this approach was that many studies had used both a variety of different animal models and a wide range of dietary feeding regimes for DHEA. The complexity of this alone made a comparison of the results difficult at best. Four different variations of a semipurified diet that previously was utilized for the development of a hypercholesteremic regime (26), were used to allow a more useful comparison. The base semipurified diet served as the control diet, while the second group was fed the control diet to which 0.4% DHEA was added. The third group was maintained on the same control-semipurified diet to which 1% cholesterol and 0.5% cholic were added (hyperlipemic diet), while the fourth group was fed this same hyperlipemic diet with 0.4% DHEA added. All groups underwent an initial adaptation to their respective new diets; this adaptation was associated with a loss in Body weight (BW). During the first 2 weeks the control group lost 4.5 g. (16.12%) of their starting BW. In group 2, fed 0.4% DHEA, the adaptation was longer with the decline in BW lasting 3 weeks and amounting to 6.9 grams or 25.74% of BW. The hyperlipemic diet fed group lost 5.5 g. or 19.78% of their BW, while group 4 which was fed the Hyperlipemic diet + DHEA, the decline lasted only 2 weeks but the loss in body weight was similar in magnitude, 7.3 g. (26.6%). After 12 weeks, the experiment was terminated, and the final BW of the animals on the different diets were measured. Even though there was no significant difference in food intake between the various groups, there were significant changes in body weight. These results are listed in table 1.

Table 1

Effects of dietary DHEA on body weight

	Group 1	Group 2	Group 3	Group 4
	Control	DHEA	Cholest	DHEA+Chol
Body weight	32.26^{A}	21.10^{B}	30.86^{A}	23.60^{C}
± SD	4.70	1.50	3.75	1.30
% of control	100.0	65.40	95.66	73.14

Results are presented as grams and are based on 11 animals per group. The superscripts ^{A,B,C} indicate values statistically different at $P < 0.0001$ in groups with different superscripts.

These findings indicate that the addition of 0.4 % DHEA to this set dietary regime will result in a significant reduction in body weight. The addition of 0.4% DHEA (group-2) was associated with approximately 35% lower BW than the control. The feeding of the hyperlipemic diet, which constituted the addition of 1% cholesterol and 0.5% cholic acid to the control diet (group 3), resulted only in a minor insignificant change in BW. The addition of 0.4% DHEA to the hyperlipemic diet (group 4) also resulted in a significant reduction in body weight albeit smaller than that associated with feeding DHEA alone. This clearly demonstrates the effect of this hormone on body weight as similarly reported by other investigators (29,30). Feeding 0.4% DHEA significantly reduced the level of plasma free cholesterol by about 25% as compared with the control, $P < 0.05$, Table 2. Similarly it also led to a reduction in triglyceride levels, however the levels of cholesterol esters and plasma total neutral lipids were not statistically different. In the group fed the hyperlipemic diet supplemented with DHEA, the lowering effect of DHEA on plasma free cholesterol was abolished and conversely there was a significant elevation in plasma cholesterol ester levels as compared to either animals fed the control diet, the hyperlipemic diet, or control+DHEA. Within the same group, plasma total cholesterol and total neutral lipid levels were similarly elevated ($p<0.05$). The elevation of plasma free fatty acid levels was significant when the compared with animals fed control + DHEA, but not the group fed the Hyperlipemic diet.

Table 2

Effects of dietary DHEA on plasma neutral lipid levels

TOTAL Diet	Chol	CHES	TOTAL CHOL	TG	FFA	NEUTRAL LIPID
Control	80.40^{A}	149.08^{A}	229.48^{A}	32.25^{A}	45.45^{AC}	307.18^{A}
± SD	7.27	4.80	11.81	10.70	19.70	28.13
Hyperlipemic	92.50^{A}	156.31^{A}	248.86^{A}	7.13^{B}	49.75^{AC}	305.74^{A}
± SD	18.57	28.45	37.75	5.55	7.75	47.40
Control+DHEA	61.92^{B}	138.96^{A}	200.88^{A}	17.43^{B}	43.27^{A}	261.58^{A}
± SD	6.25	37.80	42.23	8.65	9.47	43.09
Hylip+DHEA	98.39^{A}	229.82^{B}	333.47^{B}	34.70^{AB}	69.24^{BC}	437.42^{B}
± SD	16.58	28.55	44.68	-----	17.04	48.08

Control = semipurified diet. Hyperlipemic = semipurified diet + 1% cholesterol and 0.5% cholic acid (CC). DHEA = Control semipurified diet + 0.4% DHEA. Hyperlipemic diet + 0.4% DHEA. The superscripts ^{A,B,C} indicate that values in groups with different superscripts are statistically different at $P<0.05$. All lipid determinations were done in triplicate on 4 animals in each group. Values are in mg/dl.

Protection against lethal virus infection

Inbred C57BL/6J mice fed the semi-purified control diet + 0.04% DHEA for 12 weeks prior to viral challenge developed a significant resistance to CVB4 infection, $p \leq 0.05$. (Figure 1). At a dose of 1 x 10^3 pfu/animal 79% of mice receiving the Control diet (without DHEA) died, whereas only 47% of those receiving 0.04% DHEA in their diet died.

Figure 1

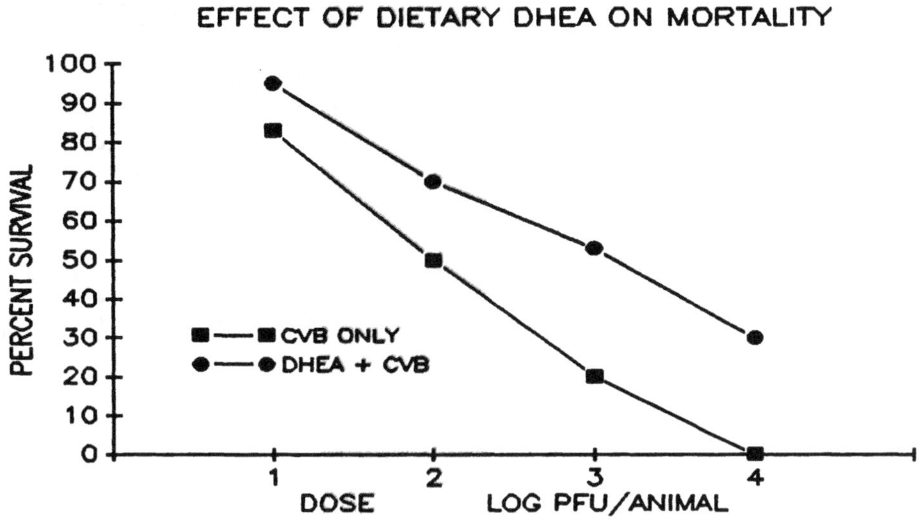

The effect that 1g/kg DHEA injected subcutaneously (S.C.) has on viral mediated mortality due to CVB4 and HSV-2 infection are presented in Figures 2(a) and 2(b), respectively. The percent cumulative mortality of animals following CVB4 infection was close to 42% in the control group that did not receive DHEA. In sharp contrast to this, only 12% of animals in the DHEA treated group died. This is a significant reduction from the control ($P<0.03$), Figure 2(a). Similar observations were made with reference to animals infected with HSV2. Again 42% of the control population not receiving DHEA died, whereas only 8% of those treated with DHEA died, ($P<0.03$) Figure 2(b). These statistically significant observations explicitly illustrate the protective effects of DHEA against viral disease. Furthermore, in reference to increasing host resistance, the results demonstrate that DHEA injected SC was markedly more effective than when fed in the diet.

Figure 2(a)

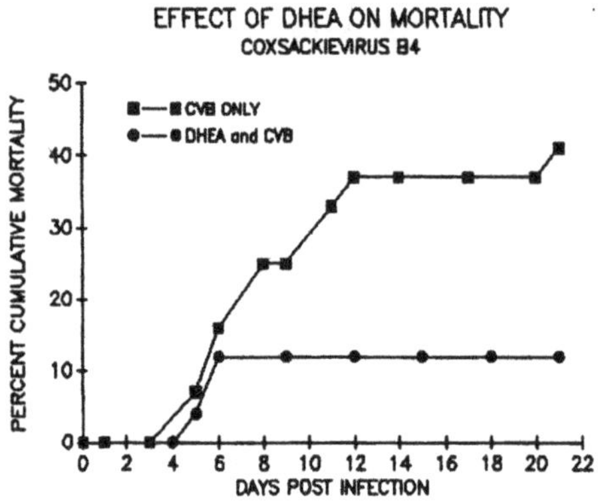

Figure2(b)

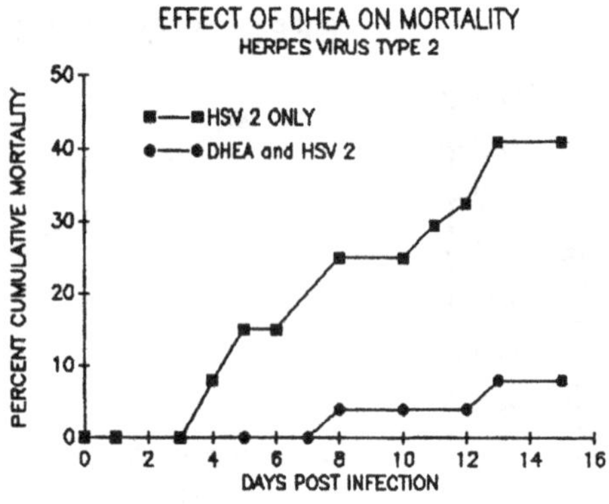

The effects of DHEA on *Streptococcus faecalis* induced mortality are presented in Figure 3. It is apparent that subcutaneous DHEA treatment resulted in a 50 % reduction in cumulative mortality as compared to *S. faecalis* infected and untreated animals. Based on the results in Figure 3 it is clear that in this case the primary protective effects of DHEA are early and are manifested within the first 24 hours following infection.

Figure 3

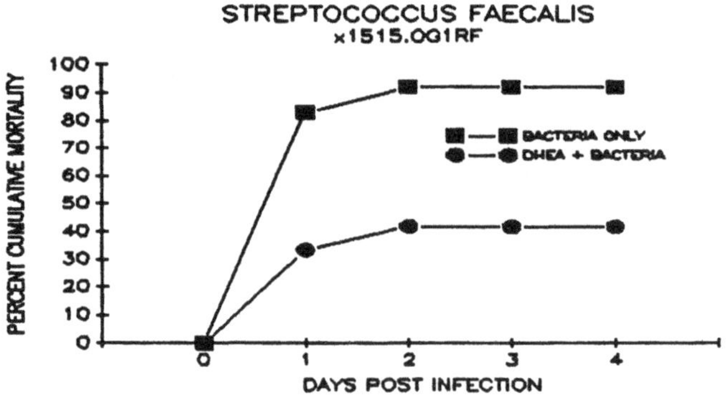

The effects of s.c. injected DHEA on virus-dose dependent mortality following infection with CVB4 and HSV2 are presented in Figure 4(a) and 4(b). In these experiments, 8 animals were used at each virus dose with an average of 82 animals being used for each experiment.

Figure 4(a)

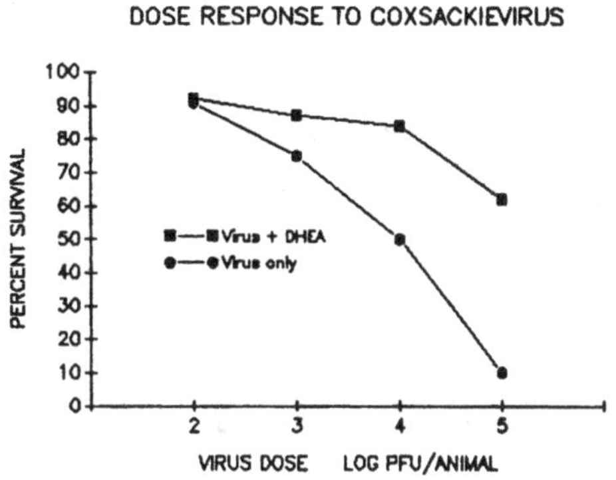

Figure 4(b)

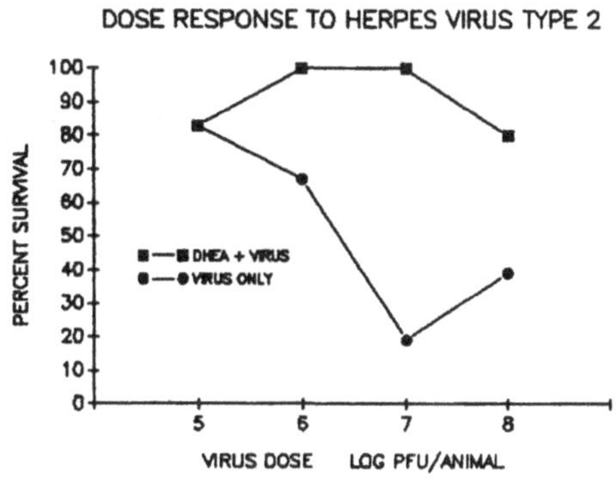

The findings demonstrate that the percent survival of animals treated with DHEA and infected with an LD_{90} dose of CVB4 (10^5 pfu/animal) increases from the expected 10% to 60 % or more. Additionally, at a dose of 10^7 pfu/animals of HSV2 where survival of control mice is 18%, the percent survival of HSV2 infected and DHEA treated mice increases to 100% . This protective effect of DHEA against i.p. CVB4 and i.c. HSV2 infections was statistically significant, $p \leq 0.03$. Tests on the effects of subcutaneous and intraperitoneal administration of sulfated DHEA (DHEAS) as well as the related steroid, etiocholanolone, were also done. At doses similar to the effective level of DHEA, there was no evidence of protection against virus mediated lethality with either compound.

Number of spleen antibody forming cells

The effect of DHEA on the number of spleen antibody forming cells (AFC) in virus infected and uninfected animals was determined by sheep red cell immunization as described previously (7). The number of IgM AFC per 10^6 spleen cells in uninfected and CVB4-infected mice with and without DHEA treatment are presented in Figure 5. As can be seen, the number of IgM AFC per 10^6 spleen cells was 35% higher in uninfected DHEA-treated mice as compared to the uninfected control mice. This increase, however, was not statistically significant. A comparison of infected groups reveals that in CVB4-infected DHEA-treated mice the number of spleen IgM AFC was 80% higher than the number of IgM AFC in CVB4-infected control mice, $p \leq 0.025$. The number of spleen IgG AFC is also presented in Figure 5; in DHEA-treated, CVB4-infected mice a 70% increase in the number of IgG AFC were observed as compared to virus-infected mice not treated with DHEA. This increase, however, was not statistically significant.

Figure 5

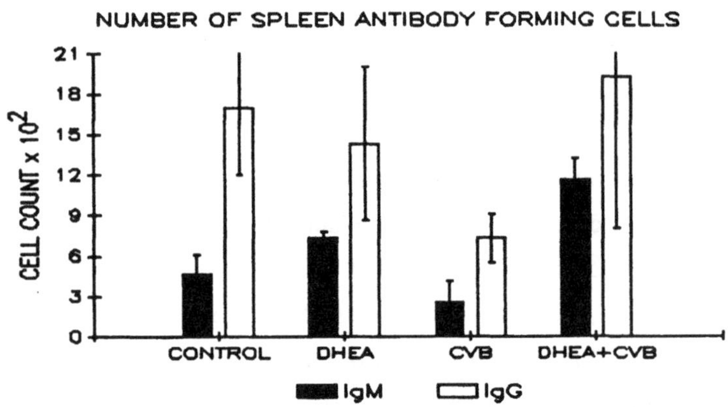

Histopathological Examination

Histopathological studies of hematoxylin and eosin stained spleen sections revealed that the spleen peri-arteriolar sheath of lymphocytes (PALS), which is composed largely of T lymphocytes that are primarily Thy 1.2 cells, were well developed in DHEA-treated, CVB4-infected animals. However, infection with CVB4 was associated with a reduction in the number and size of spleen germinal centers. In contrast, in non-infected, DHEA treated animals, a marked increase in the number and size of splenic germinal centers was evident suggesting B lymphocyte hyperplasia. Additionally, there was a marked increase in the hematopoietic activity in the spleen red pulp. Furthermore, in untreated CVB4-infected animals, the splenic white pulp was characterized by a prominent "starry sky pattern" indicative of phagocytosis of a large number of dead lymphocytes (31,32). In DHEA-treated/CVB4-infected animals, the "starry sky" pathological picture was reduced. These histological observations are suggestive of DHEA-mediated changes in the splenic lymphocyte and the hematopoietic cell populations.

Peripheral Leukocyte Counts

We also evaluated the effect of DHEA on peripheral leukocyte concentrations in the following four groups: (1) control, (2) 1 g/kg DHEA s.c., (3) CVB4-infected, and (4) 1 g/kg DHEA s.c./ CVB4-infected. All data were analyzed for statistical significance and the results are presented in Figure 6. There was no significant effect of DHEA alone on the total leukocyte count as compared to untreated control animals. However, the total leukocyte count three days after infection was significantly lower in the DHEA-treated/CVB4-infected animals as compared to either uninfected controls or DHEA treated controls, $p \leq 0.05$. At subsequent samplings, total leukocyte counts were not significantly different between any of the experimental groups. With regard to monocyte counts, as compared to the noninfected control, three days after infection with CVB4, there was a significant decrease of 50% in group 3 (virus only) and 84.5% in group 4, (virus + DHEA) ($p<0.05$). In contrast, seven days after infection, the monocyte counts in group 4 animals receiving DHEA were 214% higher than similarly infected animals not treated with DHEA (group 3), $p \leq 0.05$. There was no significant difference in the monocyte counts between the CVB4-infected animals not receiving DHEA (group 3) and uninfected controls (group 1), however DHEA injection alone without CVB4 infection (group 2) resulted in a 62% elevation of monocyte counts over the control animals. The results are presented in Figure 6. A biphasic response in peripherally sampled polymorphonuclear leukocyte (PMN) numbers was evident in CVB4-infected animals. Three days after infection the PMN counts reached 3.94×10^3 cells /mm^3 which was 515% higher than the PMN count in the control group of 0.64

x 10^3 cell/ mm^3, $p \leq 0.05$. A second elevation in PMN counts was seen at 14 days in CVB4-infected or DHEA-treated/CVB4-infected animals only. This elevation was not quite as accentuated and not statistically significant. In non-infected DHEA-treated animals no real change was noted in the number of PMNs.

FIGURE 6

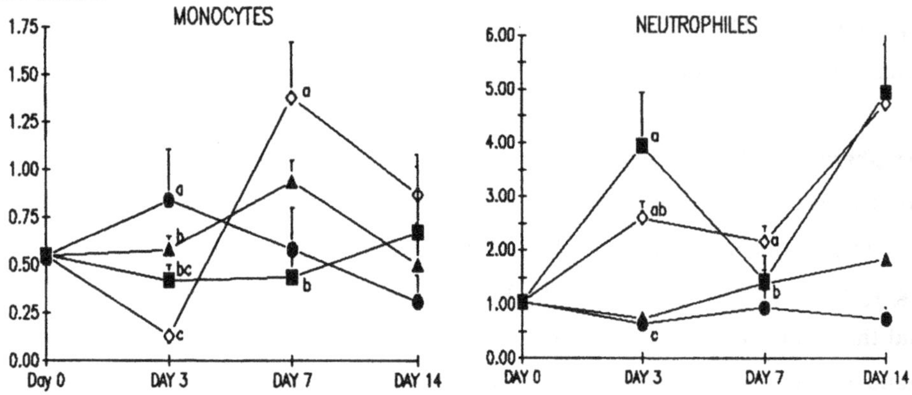

GROUP 1 ●—● CONTROL
2 ▲—▲ DHEA
3 ■—■ CVB4
4 ◇—◇ DHEA/CVB4

All values are mean cell count x 10^3/mm^3 blood ± S.E. Data points with different labels (a,b,c) are significantly different from each other ($p<0.05$) as determined by Tukey's studentized range test.

Role of Host Immunogenetics

The hairless mutation (hr/hr) in the inbred HRS/J mouse is associated with hereditary immunodeficiency and a deficiency in immunosurveillance against leukemogenesis (21,22,32). Experiments were done to test whether s.c. DHEA injection could affect the resistance of this strain to CVB4 infection. Inbred HRS/J hr/hr mice were injected s.c. with 1 g/kg DHEA and challenged i.p. with 10^5 pfu/animal of CVB4 one hour later. In contrast to the immunologically normal inbred C57BL/6J mice, DHEA did not protect this immunodeficient mutant from CVB4 lethality.

Mode of DHEA Administration

In contrast to protective s.c. DHEA injection our initial results show that DHEA given by the i.p. route was not associated with host protection from virus induced mortality or up-regulation of the immune response. We observed that s.c. injection of DHEA is associated with the formation of a local deposit leading to a relatively prolonged exposure to the lymphoid system. Long term feeding of 0.4% DHEA was also protective in the CVB4 model (Figure 1). It is, however, of particular interest to note that the magnitude and the range of protection against lethal virus infection associated with s.c. injection of DHEA was considerably greater than when DHEA was fed in the diet. Additionally, this is particularly interesting since the significant weight loss associated with dietary DHEA was not observed following subcutaneous injection of the steroid.

In Vitro Effect of DHEA

In vitro experiments were done to determine whether DHEA had any direct effect on CVB4 infectivity and replication. Cultured HeLa and VERO cells were incubated with either 2 μM, 20 μM or 50 μM DHEA and inoculated with 100 pfu of CVB4. No evidence of a reduction in the number of CVB4 plaque forming units could be detected at these concentrations of DHEA.

DISCUSSION

In general, the major steroid hormones of adrenocortical origin when administered at pharmacological doses have been regarded as immunosuppressive (33-37), particularly in viral infections (38-40). The administration of glucocorticoids results in higher viral tissue titers and increased symptomatology (21,41-44). In contrast, the results of this study demonstrate that DHEA prevents mortality induced with two distinct classes of viruses and a virulent streptococcus.

The protection against viral and bacterial induced mortality seen with a single s.c. injection of DHEA (but not DHEAS) does not appear to be achieved by a direct effect of the hormone on the microorganism but by a facilitation of the specific host immune response toward the particular infectious antigen. This supposition is supported by the observations that (a) DHEA failed to influence CVB4 replication in vitro, where immune mechanisms are not present; (b) DHEA was ineffective in the inbred HRS/J hr/hr mouse which is genetically immunodeficient; (c) Regulation and facilitation of the immune response by DHEA was seen in CVB4-infected mice with regard to the number of spleen IgM and IgG AFC; (d) Administration of DHEA alone was also associated with enlargement of the spleen germinal centers which suggests stimulation of the B lymphocyte dependent areas; (e) DHEA treatment of CVB4-infected animals resulted in a reduction in the "starry sky pattern", which is an indicator of cell killing, which was prominent in the spleens of CVB4-infected mice not treated with DHEA. Finally, (f) an increase in circulating mononuclear cells was observed in DHEA-treated/CVB4-infected mice which is consistent with the role of these cells in host defense against CVB4 infection (43).

While our studies do not reveal the specific effect(s) of DHEA on the immune system, there are suggestions from the work of Riley (45) that DHEA may interfere with the immunosuppressive action of glucocorticoids such as corticosterone. In Riley's studies, mice subjected to "rotation stress" experienced increased serum corticosteroid levels and developed thymic involution and reduced resistance to transplantable tumors. These involutional effects of stress were antagonized by the s.c. injection of 1 mg/animal of DHEA (46). In addition, DHEA also antagonized the effects of corticosterone injections on thymus involution. Viral infections results in an increase in glucocorticoid responses (47-52) and thymic involution as well as a generalized immunosuppression (53-56). Thus, it was reasonable to speculate that DHEA or its metabolites could act to protect the host by counteracting the glucocorticoid-mediated immune suppression. In this regard, a potent blocker of glucocorticoid synthesis, metyrapone, protects chickens against the lethal effects of Marek's disease, a herpesvirus-mediated lymphoproliferative disorder and also protects mice against murine sarcoma virus (57-59).

Recent observation from this laboratory indicate that the sparing effect of DHEA in these acute infection models is achieved by a reduction in the number of anti-viral cytotoxic T cells, leading to a reduced tissue pathology. Indeed, cytotoxic T lymphocytes have a major role in the pathogenesis of CVB4 infection while humoral immunity is protective (53-56). The opposite is seen in primary HSV infections (60,61); resistance is primarily mediated by T lymphocytes while antibody protection is not as significant. Our observation of the increased proliferation of the spleen germinal centers in DHEA-treated animals and the reduction in the viral killing of lymphocytes in DHEA-treated/CVB4-infected animals support this hypothesis. Furthermore, the alteration of circulating leukocytes and the elevation in monocytes seen in DHEA-treated/CVB4-infected mice suggest a modulatory effect of DHEA on the immune response at various stages of the infection. Monocytes have been reported to serve as effector cells in CVB4 infections (62), and it is possible that the changes in circulating monocyte levels reflects the action of DHEA on the distribution and generation of these cells.

In our studies, anti-viral effects were observed only when DHEA was given s.c. or p.o., indicating that the route of DHEA administration may be a critical factor in the up-regulation of the host immune response. As is evident the magnitude and the range of protection against lethal virus infection associated with subcutaneous injection of DHEA were considerably greater than when DHEA was fed in the diet. This suggests that the cutaneous layer may have an extremely important role in the host immune response. Recent reports show that the skin may, indeed, have unique immune functions (63,64). In fact, the skin is known to contain a population of cutaneous immune cells, which include the epidermal Langerhans cells and keratinocytes which produce an epidermal thymocyte-activating factor, similar to IL-1 (65). Interestingly, it has been suggested that the murine Thy-1+ dendritic epidermal cell has a role in immune surveillance (66,67) and antigen presentation (68). Consequently, it is possible that the increase in resistance following s.c. DHEA injection is associated with activation of specialized integumental immune functions.

Whatever the mechanism of DHEA's action in the acute viral models, our studies suggest that prolonged exposure to DHEA is also an important factor for obtaining the protective effect. Either prolonged feeding for 16 weeks or s.c. deposition of the hormone appear to be required for achieving antiviral protection. With regard to i.p. administration of the compound, a depot is not formed and, consequently, protection is not conferred to the host. Furthermore, injection of DHEA sulfate in the mouse, by either s.c. or i.p. routes does not yield anti-viral action, suggesting that the protective action of DHEA is through a pathway separate from sulfation. Additionally, the data supporting anti-glucocorticoid effects also suggest the mechanism is independent of the metabolic pathway for the sex steroids.

An unexpected finding in these studies was that the protection seen with DHEA was enhanced by increasing the virus dose in both infection models, Figures 2(a) and (b). These results suggest that a certain critical virus load is required to activate the protective mechanisms induced by the hormone. This phenomenon could be mediated by the need for a certain amount of viral antigen to trigger the pertinent DHEA-modulated immunological process. Another possibility is that a threshold amount of virus might be required to activate the adrenal cortex if the protective DHEA effect should prove to be mediated by counteraction of glucocorticoids or other steroid effects.

The protective effect of s.c. DHEA injection against intracranial HSV2 was obtained by injection of the hormone four hours prior to infection. Dose timing is critical however; if DHEA was injected within one hour of infection with HSV2 then no anti-viral effect was produced. Present results indicate that cutaneous DHEA is converted *in vivo* to androstenediol and it, in turn, is converted to androstenetriol which actually may mediate the facilitation of the host immune response towards the particular antigen or infectious agent, possibly by a cytokine activated pathway.

The protection against S. faecalis mediated by DHEA appears to be mediated by a different mechanism than the one observed with virus infections. This conclusion is based on the observation that DHEA's protective effect against bacterial mediated mortality is primarily seen within the first 24 hours following infection, Figure 3. We therefore have postulated that this protection may be mediated by an effect on the alternative complement pathway or possibly by an effect of DHEA against enterotoxin.

The results with dietary DHEA are also of particular interest, since they reveal that the effect of dietary DHEA on plasma neutral lipid levels will vary with the content of the lipid in the diet. It is apparent that DHEA's ability to lower the levels of plasma free cholesterol is negated by the addition of cholesterol and cholic acid to the diet. This is pertinent to virus mediated coxsackievirus pathology as cholesterol feeding enhances vascular pathology and viral lethality (26,27). Moreover, dietary cholesterol and DHEA interact to generate an additive effect on plasma free fatty acid levels, which was not evident with only dietary Cholesterol or DHEA alone. These result demonstrate that dietary DHEA may have a different effect on plasma lipid levels depending on the specific dietary regime used. This conclusion is also supported by the previous observation in this laboratory that certain interactions between dietary elements may take place that might cause significant changes in plasma lipids that can not be attributed to the individual elements alone (69). The significance of this to the protective action of dietary DHEA suggests that viral protection is not mediated through changes in blood lipid distributions.

In summary, our results show that the steroid hormone DHEA can be used effectively to mobilize the host cutaneous immune response for systemic protection against lethal infections both of viral or bacterial origin. Because the effects span such a wide variety of infectious pathogens we suggest that a broad stimulation of T_H, T_S, B-cell, and macrophages might be involved. Furthermore, due to its apparent role as a mediator on the host immune response and not on the infectious agent, we suspect that it could be used as a general facilitator of the immune response, when the host is challenged with an infectious agent.

ACKNOWLEDGEMENT:

The technical assistance of Ms. Karen Cameron and Ms. Christine K. Hogan is recognized. This work was supported in part by the Office of Naval Research (ONR) Contract N00014-85-K-0525, the National Foundation for Cancer Research, and The Enviroscience Foundation*. The authors are indebted to Dr. Jeannine Majde of ONR for her assistance and her vision.

* Mr. Shelley Krasnow, the founder and president of the environment foundation, recently passed away. His boundless curiosity, enthusiasm and willingness to test new frontiers will greatly be missed.

REFERENCES:

1. Windholz, M.E.. 1976. The Merck Index, Ninth Edition .

2. Diem, K. and C. Lentner. 1975. Geigy Scientific Tables (1975).

3. Tyrell, J.B. and P.H. Forsham. 1983. In: Basic and Clinical Endocrinology, (F.S. Greespan, P.H. Forsham, eds.) Lange Medical Publications. Los Altos California. pp. 258-294.

4. Barrett-Connor, E., K.T. Khaw, and S.S. Yen. 1984. New England J. Med. 315, 1519-1524.

5. Loria, R.M., L.B. Montgomery, L.A. Corey, and V.M. Chinchilli. 1984. Arch Virology. 81, 251-262.

6. Loria, R.M., L.B. Montgomery, N. Tutle-Fuller, H.M. Gregg, and V.M. Chinchilli. 1986. Diab. Res. & Clin. Prac. 2, 91-96.

7. Montgomery, L.B. and R.M. Loria. 1986. J. Med. Virol. 19, 255-268.

8. Markholst, H. and A. Lermark. 1987. In: Virus Infections and Diabetes Mellitus. (Y. Becker, ed.) Martinus Nijhoff Pub. Boston. pp. 111-124.

9. Bottazzo, G.F., R. Pujol-Borrell, and E.M. Gale. 1986. In: The Diabetes Annual, Vol 2, (K.G.M.M. Alberti, L.P. Krall, eds.) Elsevier Publ. Amsterdam. pp. 13-29.

10. Coleman, D.L., E.A. Leiter, and R.W. Schizer. 1984. Diabetologia 31, 830-833.

11. Coleman, D.L., R.W. Schizer, and E.A. Leiter. 1984. Diabetes 33, 26-32.

12. Coleman, D.L., E.H. Leiter, and N. Applezweig.1984. Endocrinologia 115, 238-248.

13. Coleman, D.L.. 1985. Endocrinologia 117, 2279-2283.

14. Loria, R.M., T.H. Inge, S.S. Cook, A.K. Szakal and W. Regelson. 1988. J. Med. Virology 26, 301-314.

15. Loria, R.M., D.A. Padgett, T.H. Inge, S.H. Cook, W. Regelson, J.R. Pascua, and H.P. Dalton. 1989. In: Symposium " Dehydroepiandrosterone Revisited " Sponsored by Roussel - Uclaf, Paris, France. January 19-20.

16. Gavan T.L. 1979. In: Clinical Diagnosis and Management by Laboratory Methods. Vol. 2 (Henry J.B. eds.) W.B.Saunders Comp. Philadelphia, London and Toronto, pp.1570-1571.

17. Berkovich, S. and M. Ressel. 1985. Proc. Soc. Exp. Biol. Med. 119, 690-694.

18. Berkovich, S. and M. Ressel. 1967. Archiv Fur die Gesamte Virusforschung 22, 246-251.

19. Mogensen, S.C.. 1977 Infection and Immunity 17, 268-273.

20. Baker, D.A., S.A. Plotkin. 1978. Proc.Soc. Exp. Biol. Med. 158, 131-134.

21. Yirrell, D.L., W.A. Blyth, and T.J. Hill. 1987. J. Gen. Virol. 68, 2461-2464.

22. Heiniger, H.J., H. Meier, N. Kaliss, M. Cherry, H.W. Chen, and R.D. Stoner. 1974. Cancer Res. 34, 201-211.

23. Johnson, D.A., L.D. Schultz, and H.G. Bedigian. 1982. Leukemia Res. 6, 711-720.

24. Loria, R.M., S. Kibrick, D. Downing, G. Madge, and L.C. Fillios. 1976. Nutrition Reports Internat. 12, 509-518.

25. Loria, R.M., S. Kibrick, and G. Madge. 1976. J. Inf. Dis. 133, 655-662.

26. Campbell, A.E., R.M. Loria, and G.E. Madge. 1978. Atherosclerosis. 31, 295-306.

27. Campbell, A.E., R.M. Loria, G.E. Madge, and A.M. Kaplan. 1982. Infect.Immun. 37, 307-317.

28. Moller, E., W. Bullock, and O. Makela. 1973. Eur. J. Immunol. 3, 172-179.

29. Cleary, M.P.. 1989. In: Hormones, Thermogenisis, and Obesity, (H. Lardy and F. Stratman, eds.). Elsevier, New York. pp. 365-376.

30. MacEwen, E.G., I.D. Kurzman, and A.L.M. Haffa. 1989. In: Hormones, Thermogenisis, and Obesity, (H. Lardy and F. Stratman, eds.). Elsevier, New York. pp. 399-404.

31. Sinkovics, J.G., R.J. Pienta, J.M. Trujillo, and M.J. Ahearn. 1969. J. Inf. Dis. 120, 250-254.

32. Sorger, K., M. Gessler, F.K. Hubner, H. Kohler, H. Olbing, W. Schulz, G.H. Thoenes, W. Thoenes. 1987. Clin. Nephrol. 27, 111-24.

33. Cupps, T.R. and A.S. Fauci. 1982. Immunolo. Rev. 65, 133-155.

34. Claman, H.N.. 1984. Clin. Immunol. Allergy 4, 317-329.

35. Grossman, C.J.. 1984 Endocrine Reviews 5, 435-455.

36. Goldien, A.. 1987. In: Basic and Clinical Pharmacology, Ed. 3 (B.G. Katzung, ed.). Appleton & Lange, Norwalk, Connecticut. pp. 449-460.

37. Parillo, J.E. and A.S. Fauci. 1979. Annual Reviews of Pharmacology and Toxicology 19, 179-201.

38. Kilbourne, E.D. and F.L. Horsfall. 1951. Proceedings Society of Experimental Biology and Medicine 77, 135-138.

39. Gatmaitan, B.G., J.L. Chason, and A.M. Lerner. 1970. J. Exp.Med. 131, 121-136.

40. Rytel, M.W. 1969. J. Infect. Dis. 120, 379-382.

41. Lynden, D.L. and S.A. Huber. 1984. Cellular Immunol. 87, 462-472.

42. Meek, E.S. and B. Golden. 1972. In: Infectious Diseases. (P.D. Hoeprich, eds.) Harper and Row, Publ. New York, pp. 1241-1242.

43. Hollinger, B.I.. 1985. In: Virology, (B.N. Fields, D.M. Knipe, R.M. Chanock, J.L. Melnick, B. Roizman, and R.R. Shope, eds.). Raven Press, New York pp. 1424.

44. Johnson, K.M.. 1985. In: Virology, (B.N. Fields, D.M. Knipe, R.M. Chanock, J.L. Melnick, B. Roizman, and R.R. Shope, eds.). Raven Press, New York pp. 1046.

45. Riley, V.. 1983. Science 212, 1100-1109.

46. Riley, V.. 1982. In: M. J. Murdoch Foundation Report, Pacific Northwest Research Foundation.

47. Smith, E.M., W.J. Meyer, and J.E. Blalock. 1982. Science 218, 1311-1312.

48. Dunn, A.J., M.L. Powell, and J.M. Gaskin. 1987. Science 238, 1423-1424.

49. Blalock, J.E.. 1987. Science 238, 1424-1425.

50. Hammond, J.B. and J.L Rosenberg. 1972. J. of Lab. Clin. Med. 79, 814-823.

51. Spackman, D. and V. Riley. 1974. Proc. Am. Assoc. Cancer Res. 15, 143.

52. Santisteban, G.A., V. Riley, and M.A. Fitzmaurice. 1972. Proc. Soc. Exp. Biol. Med. 139, 202-206.

53. Escobar, M.R. and P.D. Swenson. 1983. In: The Reticuloendothelial System. Vol 4 (N.R. Rose, B.V. Siegel, eds.) Plenum Press Publ. New York, pp. 201-253.

54. Woodruff, J.F. and J.J. Woodruff. 1975. In: Viral Immunology and Immunopathology. (A.L. Notkins, ed.). Academic Press, New York, pp. 393-418.

55. Rager-Zisman, B. and A.C. Allison. J. Gen. Virol. 19, 329-338.

56. Thong, H.Y., M.M. Vincent, S.A. Hensen, D.A. Fuccillo, M. Rola-Plesczynski, and J. Bellanti. 1975. Inf. & Immun. 12, 76-80.

57. Thompson, L., K.D. Elgert, W.B. Gross, and P.B. Siegel. 1980. Am. J. Vet. Research 41, 91-96.

58. Rettura, G., J. Seifter, M. Zisblatt, S. Levenson, N. Levine, and E. Seifter. 1973. J. National Cancer Institute 51, 1983-1985.

59. Spangelo, B.L., N.R.S. Hall, P.C. Ross, and A.L. Goldstein. 1987. Immunopharmacology 14, 11-20.

60. Russell, D.H., R. Kibler, L. Matrisian, D.F. Larson, B. Poulos, and B.E. Magun. 1985. J. Immunology 134, 3027-3031.

61. Lopez, C.. 1984. In: Immunobiology of Herpes Simplex Virus Infection. (B.T. Rouse, and C. Lopez, eds.) CRC Press, Boca Raton FL. pp. 45-69.

62. Woodruff, J.P.. 1979. J. Immunol. 123, 31-36.

63. Romerdahl, C.A. and M.L. Kripke. 1986. Cancer and Metastasis Review 5, 167-178.

64. Streilein, J.W. and R.E. Tigelaar. 1983. In: Photoimmunology. (Parrish, Kripke, Morison, eds.) Plenum Publishing, New York, pp. 95-130.

65. Sauder, D.N.. 1984. Lymphokine Research 3, 145-151.

66. Bergstresser, P.R., R.E. Tigelaar, J.H. Dees, and J.W. Streilein. 1983. J. Inv. Dermatol. 81, 286-288.

67. Tschachler, E., G. Schuler, J. Hutterer, H. Leibl, K. Wolff, and G. Stingl. 1983. J. Invest. Dermatol. 81, 282-285.

68. Sullivan, S., P.R. Bergstresser, R.E. Tigelaar, J.H. Dees, and J.W. Streilein. 1985. J. Invest. Dermatol. 84, 491-495.

69. Loria, R.M. and D.A. Padgett. 1990. Journal of Nutrition (submitted).

DHEA AND THYMUS INTEGRITY IN THE MOUSE

Vernon Riley[1], M.A. Fitzmaurice
Formerly with:
Pacific Northwest Research Foundation, Seattle, Washington

William Regelson

Department of Medicine, Virginia Commonwealth University/ Medical College of Virginia, Richmond, Virginia, 23219

Summary

This paper deals with the apparently opposing effects of stress or exogenous corticosterone (CSR), and exogenous dehydroepiandrosterone (DHEA) upon the integrity of the thymus in mice. DHEA's protective effects against corticosteroid or stress mediated thymic involution are demonstrable in young adult mice and depend upon the age of the mice subject to study. DHEA effects against thymic involution are not regenerative, but protective.

---

[1]The experiments described in this paper were designed by the late Vernon Riley and performed in his laboratory at the Pacific Northwest Research Foundation, Seattle, WA, shortly before he died. These data were originally presented in 1982 as a report to the Murdoch Foundation, which had provided support for these studies. Expert technical assistance was provided by H. Brent Solvason.

Dehydroepiandrosterone (DHEA)

Introduction: Stress, Adrenal Hormones, and Thymic Involution

Stress associated neuroendocrine alterations result in the release of corticosteroids, such as corticosterone(CSR) in the mouse, and cortisol in the human. Simultaneously, with stress, there is a lesser increased release of 17-ketosteroids, most notably dehydroepiandrosterone (DHEA) in the mouse.

The pattern of stress includes increased plasma CSR levels in young adult mice with an initial rapid peripheral lymphocytopenia followed by conspicuous thymic involution within 24 to 48 hours. In this regard, there is a large body of literature on immunoinhibition associated with stress or the exogenous administration of corticoids (1). Corticoids bind to specific cytoplasmic receptors which undergo an activation process and accumulate in the nucleus, where they modulate specific DNA sequences on receptor binding sites, resulting in the cytolysis of corticoid sensitive lymphoid cells. In young adult mice, a linear relationship was observed between the dose of exogenously administered CSR and resultant thymic involution (1). Various inhibitory effects of administered CSR upon the functions of T cell subsets involved in cell mediated immunity have also been observed (2).

Thymic involution is a well known age associated event and is observed whether endogenous CSR is stimulated via the pituitary adrenal axis by anxiety, stress, psychosocial circumstances or environmental factors (1-5). Alternatively, viral infections can also induce corticoid associated thymic involution via activation of the lymphoid adrenal axis (2,3,6). In relation to the above, although the endogenous CSR response of adult mice to various stressors does not diminish appreciably with increasing age (7), there is the well known dramatic natural decrease of endogenous production of DHEA with age (8). The deleterious effects of CSR upon thymus integrity and cell mediated immunity would be relatively unopposed in older animals if DHEA production is a key factor that modulates against thymic involution. For this reason, one must ask if the age related decline in DHEA has anything to do with age related thymic atrophy. Younger animals with higher DHEA levels would be more resistant to stress or CSR induced involution because of the opposing action of DHEA.

The gradual process of thymic involution is accelerated by long term chronic stress accompanied by chronically elevated levels of plasma CSR (1,9). The age and stress related involution of the thymus and the age associated decline in DHEA led us to consider that DHEA might provide a balance to corticosteroid or stress related effects, and it was the purpose of this study to determine if DHEA could block stress or corticosteroid mediated thymic involution.

Methods and Materials: Experimental Design

Thymic involution in young adult mice has long been employed as a qualitative bioassay for the presence of elevated circulating adrenal corticoids (6,9-11). A modification of this bioassay has been employed in these studies to demonstrate the protective effects of DHEA against corticoid induced thymolysis. In this regard, protective housing facilities and low stress handling procedures are essential for the maintenance of thymic integrity and the success of these experiments. CSR was administered subcutaneously or induced endogenously by anxiety inducing programmable rotation described in our previous papers (3).

DHEA was administered at various times relative to CSR administration or rotational stress. The mice were killed humanely by carbon dioxide 24 to 48 hours following experimental initiation and each thymus was carefully removed and weighed. Each thymus weight was then plotted as the percentile rank of the individual thymus within its experimental group. Statistical protective effects of DHEA against CSR-induced thymus involution were calculated employing the Chi-square test, the Wilcoxin sum of ranks test and Student's t test.

Experimental Mice

Significant protective effects of DHEA against CSR induced thymus involution were observed in male C57Bl, C3H/Bi, and BALB/C mice, and in female C57Bl mice. However, no systematic effort was made to compare the various mouse strains. Mice were obtained from the Central Animal Facility of the Fred Hutchinson Cancer Research Center, where they were bred within protective facilities. Uniform groups of at least 10 per group, in terms of age, sex and strain were employed for any given experiment, with their age range limited to within a 5 day span. Any abnormal appearing mice were discarded, and upon receipt, the mice were randomly segregated into groups of 5 or 10 in standard plastic cages. Male and female mice were kept in separate rooms to avoid sex or stress related elevated CSR responses (3,5,9). All mice were free from common pathogens and contamination with the LDH-virus to avoid interfering viral induced CSR elevations (2,6). Mice undergoing experimental procedures were kept in a separate room to avoid psychosocially induced CSR elevations caused by the communication of their distress to control mice (5).

Housing, Handling and Environment

Several basic discoveries relevant to corticosterone and thymolytic responses to anxiety-stress were made following the design of protective animal housing by Vernon Riley in 1967. The special barrier system employed in these studies consisted of ventilated enclosed shelves with a filtered laminar flow system to outside of the building allowing rapid exhaustion of released odors and aerosols (3-5,9,11, 19). When the plexiglas doors of these facilities were closed, each shelf became a separate, independent, and isolated unit within a larger protected facility. Standard 12 hour intervals of light and dark were controlled by automatic light switches, and the temperature was maintained at 24° C ($\pm$ 2). Standard plastic cages, 28 X 18 X 13 cm containing about 2 cm of San-I-Cel ground corncob bedding were employed. Food consisted of Wayne Lab Blox pellets, supplied ad libitum through the cage-top hopper. Clean water was constantly available through the use of pint glass bottles equipped with stainless steel drinking tubes. Cages and bedding were changed once a week with special handling procedures to minimize stressful disturbances. High sanitation standards were maintained by routine changing and sterilization of cages, drinking bottles and tubes.

Administration of CSR and DHEA

Both CSR and DHEA were administered subcutaneously (SQ) as insoluble water suspensions to give long lasting elevated concentrations of each compound in the plasma. When DHEA and CSR were given to the same mouse, they were injected SQ at distant sites from each other. DHEA was obtained from Porquiva, Mexico.

The Induction of Anxiety by Programmable Rotation

A programmable turntable was devised to induce anxiety in the absence of physical trauma or restraint. Entire groups of mice were rotated within their experimental cages to minimize handling. Intermittent rotation was administered at 45 rpm in 60 minute cycles (10 minutes of rotation followed by 50 minutes of rest). Reproducible anxiety induced elevations of plasma CSR were obtained following 20 minutes of such intermittent rotation (3). The reproducible effects of this mild anxiety producing stressor upon thymic involution, leukopenia, and various aspects of tumor development have been compared with those induced following various doses of natural or synthetic corticoids (3-5).

Thymus Weights

As mentioned previously, twenty four to forty eight hours following experimental initiation, mice were euthanized in a jar containing carbon dioxide. Thymuses were carefully removed, and quickly trimmed of surrounding tissues under continually moist conditions. They were then placed in groups of 5 in covered petri dishes lined with saline - moistened filter paper. Weighing was carried out as rapidly as possible. Each thymus was individually weighed to 0.1 mg accuracy.

Experimental Results

Protective effects of DHEA against CSR induced thymus involution:

Figures 1 and 2 contrast the results of two experiments in which protective effects of DHEA against CSR induced thymus involution were observed. Figure 1 illustrates an excellent protective effect of administered DHEA against stress induced thymus involution in young adult male mice. In this experiment, employing 13.5 week-old C57Bl mice, anxiety associated endogenous CSR was induced by programmed rotation. Under the circumstances of this experiment, 88% thymus protection was obtained by prior injection of DHEA, as contrasted to the dramatic 69% thymus involution induced by rotation in the absence of DHEA.

Figure 2 depicts the results of another experiment in which exogenous CSR was administered to partially simulate the physiological effects of anxiety associated stressors. Nine week old C3H/Bi male mice were employed. Although the calculated 82% thymus protection conferred by the administration of DHEA one hour prior to CSR injection was highly statistically significant, the plotted data illustrate that the thymic protective effects of DHEA are not always as complete as seen in Figure 1. Additional studies and analyses were undertaken to separate the various factors contributing to this experimental variation.

Fig. 3 shows the distinct difference between thymic weights in young versus old C57Bl male mice at 5 and 23 months expressed as a percentage decline from thymic weight at 5 months. While Figures 1 and 2 show rapid thymus involution as a response of young mice to an acute stressor, this figure demonstrates the end result of the normal gradual age-related involution of the thymus.

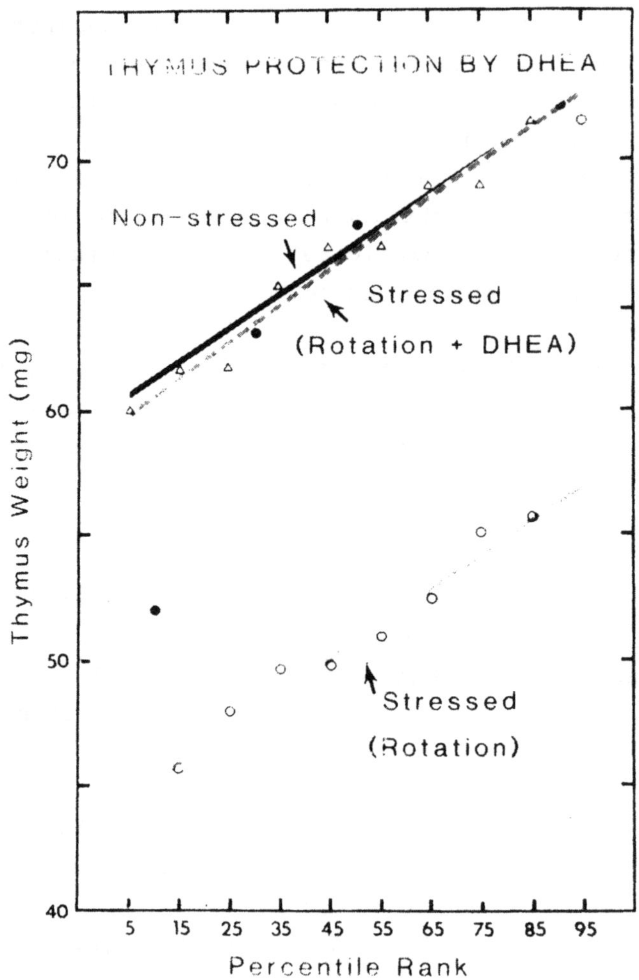

FIGURE 1

Protective effect of DHEA against thymus involution induced by rotational stress. The stressed mice were subjected to two rotational cycles in which the mice were rotated for ten minutes each hour at 45 rpm. DHEA (1 mg per mouse) was injected subcutaneously in the treated group one-half hour prior to the first rotation period. Thymuses were weighed 48 hours after the experiment was initiated. Each point represents the percentile rank of one mouse within its group. C57Bl male mice, 13.5 weeks of age, were employed. The protective effect of DHEA in this experiment was statistically significant ($P < .005$).

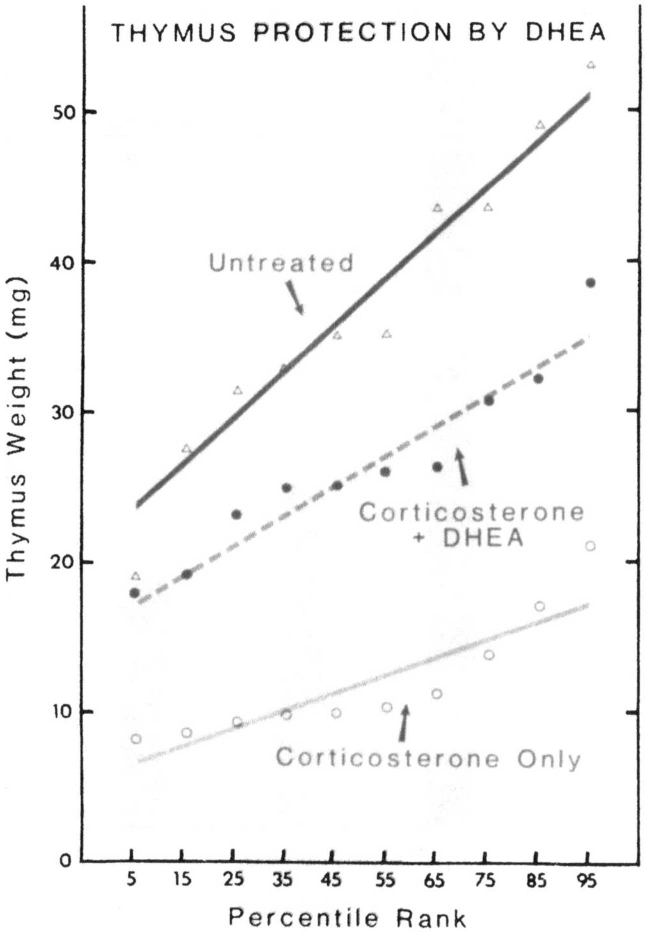

FIGURE 2

Statistically significant protective effect of DHEA (P <.002, Chi-square test) against thymus involution induced by corticosterone (1 mg/mouse). The DHEA treated mice received 1 mg of subcutaneous DHEA one hour prior to corticosterone administration. Each point represents the percentile rank of an individual mouse within its group. Each group consisted of ten male mice, nine weeks of age.

C3H/Bi mice were employed. Thymuses were weighed 40 hours following CSR administration.

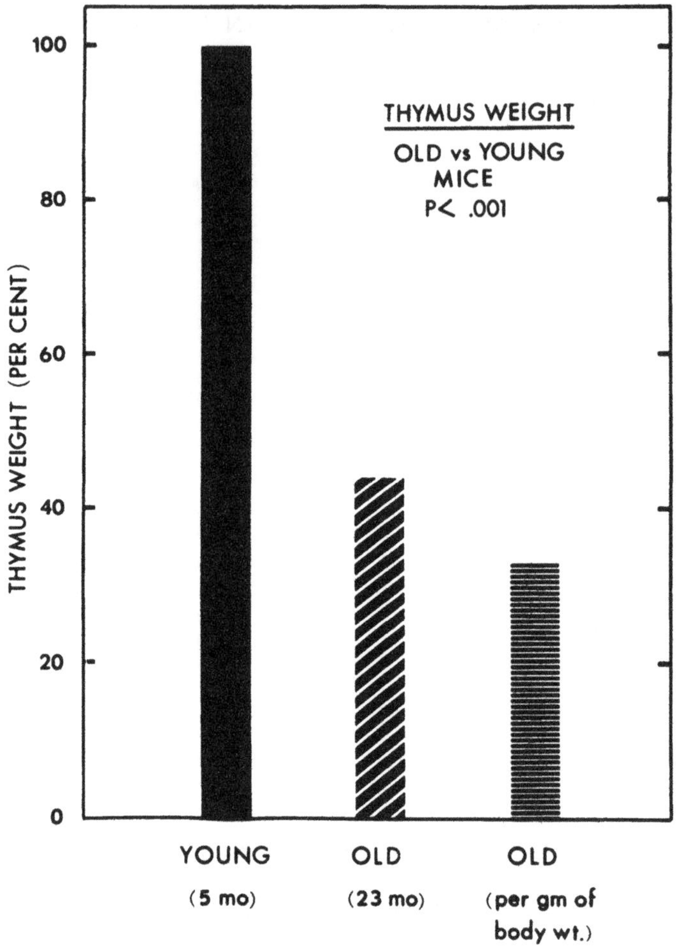

FIGURE 3

Young C57Bl male mice showing age related involution as a percentage decline from the 5 month thymic weight.

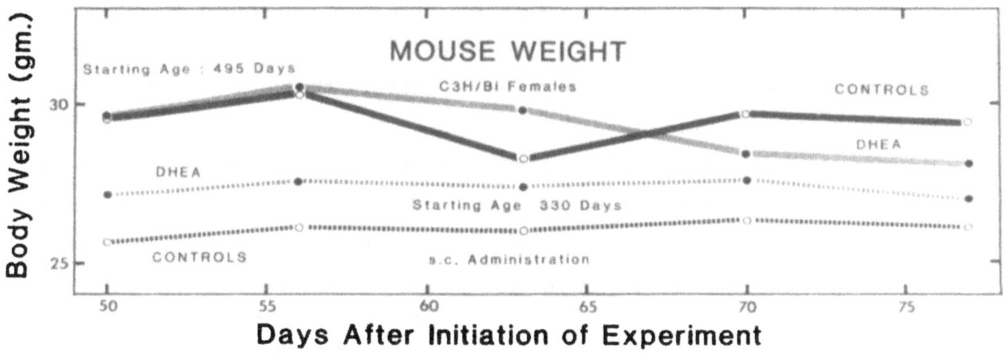

FIGURE 4

Changes in body weight on subcutaneous daily DHEA administration (1gm/mouse) beginning at 495 and 330 days.

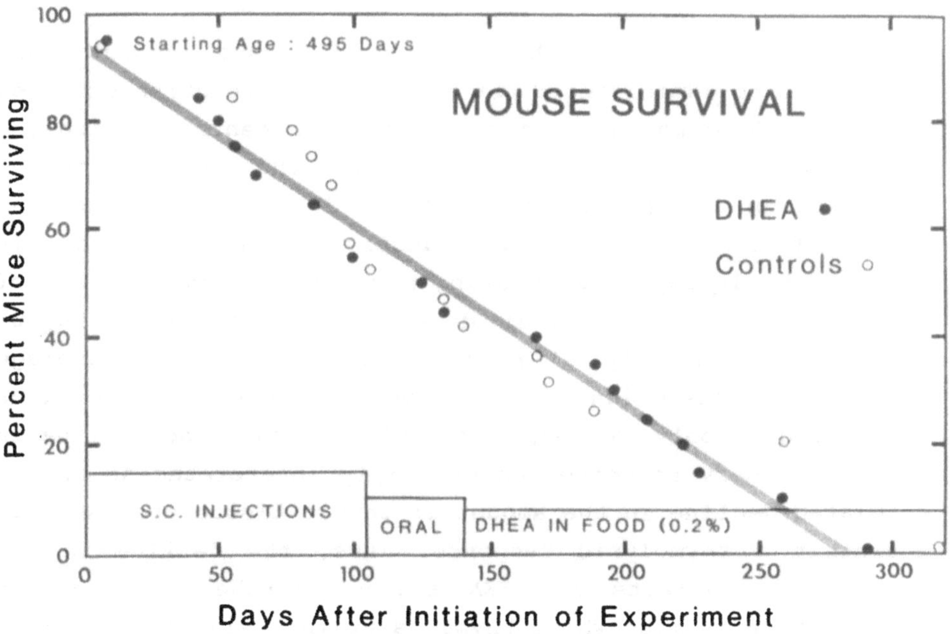

FIGURE 5

Shows the effects of DHEA administration on survival in C3H/Bi female mice as the percent mice surviving. No effect of DHEA is seen on survival of these older mice.

Effects of Initial Thymus Weight and Age in Young Adult Mice

Table 1 summarizes an analysis undertaken to examine the effects of mouse age and initial thymus weight upon the protective effects of administered DHEA against stress or CSR induced thymus involution. This shows that the best protection in young adult C3H/Bi or C57/Bl mice was obtained when the animals employed had initial thymus weights of 40 mg or more. In contrast, the slightly older mice in the lower group of experiments, with initial thymus weights of 30 mg or less, exhibited minimal thymus protection by DHEA. Thus, in adult mice having involuted thymuses as a function of age, or prior environmental stress, no significant protective effects of exogenous DHEA were observed.

Table 2 demonstrates the decline in thymus involutionary response to administered CSR as a general function of mouse age in C3H/Bi males. In a group of experiments in which the average mouse age was 7.4 weeks, the average thymus involution in response to a standard (1 mg/ 20 gm mouse) dose of CSR was 60 percent reduction in thymus weight. In contrast, in experiments using mice with an average age of 22.8 weeks, an identical dose of CSR induced a thymus involution of only 34 percent. Inasmuch as the average control thymus weights in the younger group was 51.6 mg as compared to only 18.0 mg in the older group, the decline in involutionary response can be simply interpreted as an age related decline in the amount of thymic tissue available to exhibit a thymolytic response. Thus, the decline in the protective effects of DHEA against further CSR induced involution in mice whose thymuses exhibit significant prior involution may simply be a reflection of a lack of cells available to respond to CSR, and hence, a lack of cells for DHEA to protect.

TABLE 1

THE INFLUENCE OF THYMUS WEIGHT UPON THE PROTECTIVE EFFECTS OF DHEA AGAINST CORTICOID INDUCED THYMUS INVOLUTION IN YOUNG ADULT MICE.

Experiments with Untreated Control Animals Having Thymus weights of 40 mg or more:

Experimental	% Thymus Protect. by DHEA	Untreated Control Thymus Wt. (mg)	Age of Mice (weekly)	Sex of Mice	Stressor
1	82	41	9	M	Corticosterone
2	85	40	11	M	Same
3	88	67	13.5	M	Rotation
4	74	52	17	F	Same
5	61	52	17	F	Same
MEAN	72.0*	50.4	13.5		

Experiments with Untreated Control Animals having Thymus Weights less than 30mg:

6	0	22	20	M	Corticosterone
7	40	14	20	M	same
8	0	17	20	M	same
9	22	16	30	M	same
10	0	16	26	M	same
11	22	21	28	M	same
MEAN	14.0*	17.7	24.0		

*The difference between the two groups is statistically significant ($P < .004$; Wilcoxin sum of ranks test).

TABLE 2

DECLINE IN CSR-INDUCED THYMUS INVOLUTION AS A FUNCTION OF MOUSE AGE

Experiments using mice from 4 to 10 weeks of age:

Thymus Involution Rank Number	% Thymus Involution*	Age of Mice (Weekly	Control Thymus Weight (mg)**
1	69	9	41
2	58	6	62
3	58	8	48
4	58	10	41
5	56	4	66
MEAN:	59.8	7.4	51.6

Experiments using mice 20 weeks of age and older:

Thymus Involution Rank Number	% Thymus Involution*	Age of Mice (Weekly)	Control Thymus Weight (mg)**
6	46	20	22
7	40	28	21
8	38	20	14
9	24	20	17
10	24	26	16
MEAN:	34.0	22.8	18.0

**The difference in percent thymus involution between the two groups is statistically significant (P <.005), employing the Wilcoxin sum of ranks test.

Thymus involution was induced by the administration of exogenous corticosterone (1 mg/mouse). The thymuses were weighed 48 hours following corticosterone administration.

Comparison of Young Adult and Pre-Pubertal Mice

Inasmuch as thymus weight and age are important correlates, further analysis seemed appropriate. The data shown in Table 3 in C3H/Bi mice are obviously related to the data illustrated by Tables 1 and 2. However, the item of special interest in Table 3 is that the youngest group of mice, from 4 to 7 weeks of age, although having larger average thymus weights than the mice who were 9 to 17 weeks old, exhibit very poor results in terms of protective effects of DHEA against CSR-induced thymus involution. Thus, in the case of very young, presumably pre-pubertal mice, the size of the thymus does not appear to be the only factor involved in the protective effects of DHEA against thymus involution. Since the data in Table 2 indicate that similar responses in terms of thymic involution following a standard dose of CSR were seen in mice aged 4-8 weeks and in those aged 9-10 weeks, a specific physiological phenomenon, involving the effects of DHEA upon stressed thymus tissue, appears to take place at or around the time of puberty.

The data of Table 3 indicate further, that, under the conditions of these experiments, optimal DHEA protection against thymus involution was observed in young adult mice aged 9 to 17 weeks, subsequent to puberty and prior to any significant degree of age or stress related thymic involution. Poor results were observed in prepubertal mice, and also in older mice with a significant degree of prior thymic involution.

TABLE 3

THE INFLUENCE OF AGE UPON THE PROTECTIVE EFFECTS OF DHEA AGAINST CORTICOID-INDUCED THYMUS INVOLUTION.

EXPERIMENTS USING MICE FROM 4 TO 7 WEEKS OF AGE:

Age in weeks	% Thymus Protection by DHEA	Untreated Controls Thymus Wt. (mg)	Sex of Mice	Stressor
4	14	42	M	Cortico-sterone
6	34	57	M	Cortico-sterone
6.8	6	71	F	Cortico-sterone

MEAN:
5.6($\pm$1.4) 18.0($\pm$14.4) 56.7($\pm$14.5)

EXPERIMENTS USING MICE FROM 9 TO 17 WEEKS OF AGE:

Age in weeks	% Thymus Protection by DHEA	Untreated Controls Thymus	Sex of Mice	Stressor
9	82	41	M	Cortico-sterone
11	85	40	M	Cortico-sterone
13.5	88	67	M	Cortico-sterone
17	74	52	F	Rotation
17	61	52	F	Rotation

MEAN:
13.5($\pm$3.6) 78.0*($\pm$10.8) 50.4($\pm$10.9)

EXPERIMENTS USING MICE FROM 20 TO 30 WEEKS OF AGE:

Age in weeks	% Thymus Protection by DHEA	Untreated Controls Thymus	Sex of Mice	Stressor
20	40	14	M	Cortico-sterone
20	0	22	M	Cortico-sterone
20	0	17	M	Cortico-sterone
26	0	16	M	Cortico-sterone
28	22	21	M	Cortico-sterone
30	22	16	M	Cortico-sterone
MEAN: 24(±4.5)	16.8(±17.0)	17.6(±3.1)		

*The DHEA protective effects observed in mice between 9 to 17 weeks of age are significantly different from the lesser protective effects observed in mice either younger ($P<.005$) or older ($P< .005$). The probability values were determined using Student's t test.

TABLE 4

COMPARATIVE PROTECTION OF THE THYMUS BY VARIOUS DHEA/CSR CONCENTRATION RATIOS[a]

Group Number	Experimental Treatment	Thymus[b] Weights (Average)[c] (mg) (SD)	Thymus Weight (%)	Thymus Weight Loss (%)	Relative Thymus Involut. (%)	Protection by DHEA (%)
1	Saline Controls	39 (±9)	100	0	0	----
2	DHEA Only	31 (±6)	79	21	30	(70)
3	DHEA/CSR (1 to 1)	26 (±6)	67	33	47	53
4	DHEA/CSR (1 to 1.5)	18 (±8)	46	54	78	22
5	DHEA/CSR (1 to 2)	14 (±4)	35	65	94	6
6	CSR Only	12 (±4)	31	69	100	0

a) Both DHEA and corticosterone (CSR) were injected as insoluble water suspensions to give long-lasting physiological levels of each compound in the blood plasma. The dose ratios were based upon the compound weights. However, since the molecular weights of the two compounds are similar, the weight ratios are essentially molar ratios. DHEA was injected approximately one hour prior to corticosterone injection.

b) The thymuses were weighed 48 hours following corticosterone injection.

c) Average thymus weights represent the means of ten mice per experimental group. C3H/Bi male mice, 9 weeks of age, were employed.

Timing Factors

While the CSR: DHEA molecular ratio is important for effective DHEA protective effects, sequential time factors are also relevant. In the experiment illustrated by Table 5, DHEA was administered to animals at various times prior to the injection of CSR. When DHEA was administered 4 hours prior to CSR, 88 percent thymus protection was obtained. The degree of this protection systematically decreased as the time interval between DHEA and CSR injections lessened. When both hormones were injected at different sites simultaneously the protection was only 60 percent. However, even when DHEA was administered one hour after CSR, a 33.7 percent protection was seen; this, however, was not statistically significant. This experiment shows that the prior presence of DHEA or its metabolites in the plasma greatly enhances its protective effects against CSR-induced thymic involution.

Effects of exogenous DHEA on body weight and survival

Figure 4 shows the effects of DHEA on body weight when given subcutaneously to groups of 10 mice beginning daily at 495 days and 330 days to C3H/Bi females weighed at the time periods indicated. It is only in the older mice that some moderate weight decline was seen. In these studies weight reduction does not seem to be a factor following DHEA administration.

Survival was studied in C3H/Bi females given DHEA subcutaneously, by oral intubation, and as 0.2% DHEA in food. There was no effect of DHEA administration on survival comparing treated and control mice. (Fig. 5).

Discussion/Conclusion:

In regard to the above experiments, the importance of the thymus to the development of the neuroendocrine and immune systems has been widely recognized. Neonatal thymectomy gives rise to altered neuroendocrine and immunological patterns, leading to an increase in autoimmune disorders with frequently lethal consequences (12-15). Thymectomy of young adult animals can have lethal consequences in malnourished or stressed animals, and can eventually lead to immune deficiencies, possibly related to a lack of thymic hormones, some of which are active in the promotion of cell mediated immunity (12-15). In the presence of an intact thymus, various neuroendocrine and immunological balances undergo developmental changes, followed by post-pubertal gradual age related alterations. In this regard, while intact adrenal function is essential for the optimal development of the pre-pubertal thymus, a gradual age related involution of the thymus commences at puberty (13,18) (See Fig. 3) concomitantly with an increased output of adrenal and gonadal steroids (10,17), and the onset of sexual behavioral changes and circadian or seasonal rhythms (17-20).

It thus becomes clear that endogenous changes in DHEA levels with adrenarche or aging could affect thymic response. The protective effects of DHEA against stress induced thymic involution could be subject to a number of biological factors that may influence the degree of thymus protection obtained by exogenous DHEA. This is interesting, because it also reflects the complexity of the natural homeostatic mechanisms that may be involved in maintaining an appropriate balance between the lympholytic effects of the adrenal corticoids on the one hand (stress) and the ability of DHEA or its metabolites to protect the organism against this potentially harmful action directed against important immunological elements of the host.

Our hypothesis that DHEA may have an opposing effect on thymic involution in response to stress or corticosterone has been borne out. Although DHEA administered alone exhibited a slight thymolytic effect, when administered prior to rotation induced stress or prior to CSR administration, a protective effect against thymus involution was observed. The pre-existence of a mature but non-involuted thymus was apparently essential to the demonstration of DHEA's protective effects, as adult mice exhibiting significant prior thymic involution due to advanced age or pre-existing stressful environmental conditions showed no protective or regenerative effects of DHEA upon the thymus.

In contrast to the protective effects seen in young adult mice with non-involuted thymuses, no protective effects of DHEA were observed if slightly younger, pre-pubertal mice were employed, even though the younger mice had larger thymuses prior to DHEA administration. Thus, exogenous DHEA apparently protects against the thymolytic effects of elevated endogenous or exogenous CSR within narrow, post-pubertal age limits. Our observations stress the importance of age as a factor in DHEA response. Neonatal, young, mature, and older mice may respond to DHEA in entirely different ways.

In regard to DHEA effects, we saw no significant effects of DHEA on mouse survival when given via various routes to older mice at 495 days of age. There were also no dramatic weight loss effects of DHEA seen on subcutaneous administration of DHEA (Fig. 4) beginning at 330 and 495 days of age although further study via different routes and at different ages and sex has to be done to determine if trends in weight are meaningful.

Recently, Crudo (22) has reported the down regulation of hepatic glucocorticoid receptors induced by subcutaneous DHEA in Sprague Dawley rats. Kalimi et al (in this text) report a similar DHEA-induced down regulation of hepatic glucocorticoid receptors. In Crudo's study, it did not relate to receptor affinity but to a decline in available receptors. A similar phenomenon may underlie the block to thymic involution induced by DHEA.

Whether the administration of DHEA might be protective in young adult mice against CSR-induced thymic involution in older mice which have been shielded from extensive thymus involution was not determined in these studies. The route of DHEA administration may be critical as SQ dosage is vital to the action of DHEA stimulating host resistance to viral and bacterial infection (Loria et al. this volume). In addition, different patterns and routes of DHEA administration may be important to receptor availability, governed by down regulation or circadian rhythms. Experiments involving various DHEA:CSR ratios and sequential times of administration provide suggestive evidence that DHEA may compete with CSR for receptor binding sites on certain sensitive murine thymocytes (21).

In conclusion, it is our observation that DHEA can exert protective action against thymic involution, depending on age related and stress related factors.

TABLE 5

RELATIVE PROTECTIVE EFFECTS OF DHEA UPON CORTICOSTERONE INDUCED THYMUS INVOLUTION AS A FUNCTION OF THE RELATIVE TIME OF INJECTION OF CORTICOSTERONE AND DHEA[1]

Group No.	Experimental Treatment	Thymus Wts[2] (mg)	Thymus Wt. (%)	Thymus Wt Loss (%)	Relative Thymus Involution (%)	Thymus Protection by DHEA (%)
1	Controls (Saline)	40.0(±6.9)	100	0	0	---
2	DHEA-4 hrs[3] (CSR, 0hr)	38.4(±8.1)	96.0	4.0	12.0	88.0[4]
3	DHEA-2 hrs (CSR, 0hr)	35.9(±10.9)	89.8	10.2	30.7	69.3
4	DHEA-1 hr (CSR, 0hr)	34.5(±6.9)	86.2	13.8	41.6	58.4
5	DHEA 0 hr (simultaneous)	34.8(±8.7)	87.0	13.0	39.2	60.8
6	DHEA +1 (CSR, 0hr)	31.2(±6.0)	78.0	22.0	66.3	33.7
7	CSR only	26.7(±3.9)	66.8	33.2	100	0

1. The weight ratio of DHEA to CSR was 0.835 to 1.
2. The thymuses were weighed approximately 40 hours following CSR injection
3. Both CSR and DHEA were injected subcutaneously as insoluble water suspensions to give long-acting physiological levels of both compounds in the plasma.
4. The difference in thymus weights between Group 2 (DHEA injected 4 hours prior rto CSR) and Group 7 (CSR only) is statistically significant ($P<.01$, Student's T test.

(BALB/C male mice, 11 weeks of age were employed)

References

1. Spackman, D.H., M.A. Fitzmaurice, and V. Riley. 1985. In: Perspective on Behavioral Medicine, Vol. 2 (R. Williams, ed). Academic Press, San Diego. pp. 149-182.

2. Fitzmaurice, M.A. 1988. Inter. J. of Neuroscience. 39, 307-324.

3. Riley, V. 1981. Science, 212, 1100-1109.

4. Riley, V., M.A. Fitzmaurice, D.H. Spackman. 1981. In: Psychoneuroimmunology (R. Ader, ed.). Academic Press, New York. pp. 31-102.

5. Riley, V., M.A. Fitzmaurice, D.H. Spackman. 1981. In: Perspectives on Behavioral Medicine, Vol.1 (S. Weiss, J. Herd and B. Fox, eds.) Academic Press, New York, pp.371-400.

6. Fitzmaurice, M.A., V. Riley, and G.A. Santisteban. 1972. Pathologie-biologie. 20, 743-750.

7. Landfield, P.W., J.C. Waymire, and G. Lynch. 1978. Science 202, 1098-1101.

8. Orentreich, N., J.L. Brind, R.L. Rizer, J.H. Vogelman, 1984. J. Clin. Endocrinol. Metab. 59, 551-555.

9. Riley, V., M.A. Fitzmaurice, and D.H. Spackman. 1982. In: Biological mediators of behavior and disease: neoplasia. (S. Levy, ed.). Elsevier Press, New York. pp 175-218.

10. Santisteban, G.A., and Dougherty. 1954. Endocrinology. 54, 130-146.

11. Santisteban, G.A., V. Riley, and M.A. Fitzmaurice. 1972. Proc. Society for Exp. Biol. and Med. 139, 202-206.

12. Bach, J.F., M. Dardenne, J.M. Pleau, and M.A. Bach. 1975. Annals of the New York Academy of Science. 249, 186-210.

13. Luckey, T.D., 1973. In: Thymic Hormones, (T.D. Luckey, ed.) University Park Press, Baltimore, MD., pp. 275-314.

14. Trainin, N., Y. Levo, and V. Rotter. 1974. Europ.J. Immunol. 4, 634-637.

15. Trainin, N., M. Small, and Y. Kimhi. 1973. In: Thymic Hormones. (T.D. Luckey, ed.). University Park Press, Baltimore, Md., pp. 135-158.

16. Smith, H., I.-M Chen, R. Kubo, and K.S.K. Tung. 1989. Science, 245, 749-752.

17. Santisteban, G.A., C. Guslander, and K. Wilhight. 1969. Anatomical Record, 163, 2.

18. Metcalf, D. 1966. Recent results in cancer research. Vol. 5, The Thymus, Springer-Verlag, New York.

19. Riley, V. 1972. Proceedings of the 23rd Annual Session of the American Association for Laboratory Animal Science, no. 22A.

20. Krieger, D.T. 1972. Science, 221, 1301-1304.

21. Turnell, R.W., N. Kaiser, R.J. Milholland, and F. Rosen. 1974. J. Biological Chem. 248, No. 4, 1133-1138.

22. Crudo, D. 1985. Endocrine Soc. 71st Annual Mtg. Seattle, Washington, Abstract # 1455.

Effect of Dehydroepiandrosterone in Lymphocytes and Macrophages Infected with Human Immunodeficiency Viruses

R. F. Schinazi, B. F. H. Eriksson, B. H. Arnold

Veterans Affairs Medical Center, Decatur, Georgia 30033, and Laboratory of Biochemical Pharmacology, Department of Pediatrics, Emory University School of Medicine, Atlanta, Georgia 30322

P. Lekas, M. S. McGrath

AIDS Activities Division, Department of Medicine, UCSF San Francisco General Hospital, San Francisco, California 94120

Introduction

Dehydroepiandrosterone (DHEA) is one of the most abundant naturally occurring adrenal steroids in humans. Several beneficial biological attributes of DHEA have been claimed (1,2). For example, it has been reported to have immunoregulatory, anti-diabetic, anti-cancer and anti-stress activity (2,3-6). DHEA has also been reported to prevent obesity without suppressing food consumption, to prolong life, and to depress the mutagenic effects of carcinogens (2,3,7,8).

This compound is a potent non-competitive inhibitor of mammalian glucose-6-phosphate dehydrogenase (G-6-PD), a rate-limiting enzyme which leads to the production of NADPH for biosynthetic reductions in the cytosol and the production of ribose-5-phosphate as a precursor for nucleotide synthesis (5,6,9). In contrast, DHEA-sulfate (DHEA-S), a main metabolite of DHEA, is essentially inactive as an inhibitor of G-6-PD (5,9). The anticarcinogenic and cell differentiation inhibitory action of DHEA may be related to inhibition of G-6-PD (3).

Henderson et al. (10) have demonstrated that DHEA and 16α-bromoepiandrosterone (Epi-Br) markedly inhibited Epstein-Barr virus-induced transformation of human lymphocytes at concentrations of 50-100 μM. Recent observations have found that sera from persons with AIDS had subnormal levels of DHEA (11,12) and that elevated levels were present in high risk seronegative homosexual men, suggesting that DHEA may reduce the risk of progression of disease in virus-infected individuals (J. Sonnabend, D. Seaman, and T. Fleischer; M. Jacobson, and M. S. McGrath, personal communication). An alternative explanation for the low levels of DHEA may be the recent finding that adrenocortical lesions leading to adrenocortical insufficiency are common in individuals infected with HIV-1(13). Uncontrolled clinical trials in Europe using an oral formulation of this drug have suggested that DHEA can maintain the reduction of persistent viral p24 antigenemia produced by 3'-azido-3'-deoxythymidine (AZT) (Dr. J. Lange, personal communication).

Nestler et al. (14) recently demonstrated that high doses (1.6 g per day) of oral DHEA administered to normal individuals for 28 days caused no untoward effects. In the obese individuals, DHEA reduced body fat, increased muscle mass and reduced serum low density lipoprotein cholesterol levels, properties that could be beneficial in individuals with AIDS and AIDS related complex.

Since the spectrum of activity of DHEA and structural analogues against human retroviruses has not been reported, several studies were initiated which were aimed at: (a) delineating the activity of DHEA and analogues in human peripheral blood mononuclear cells (PBMC) and CEM cells infected with human immunodeficiency virus type 1 (HIV-1); (b) determining the capacity of DHEA to inhibit virus replication when treatment was delayed or withdrawn at different times after infection; (c) establishing whether the antiviral activity observed was due to a direct inactivation of HIV-1; (d) determining whether DHEA would inhibit HIV-1 replication in normal human macrophages; (e) assessing the HIV-1 reverse transcriptase inhibition potential of these drugs; and (f) evaluating the interaction of DHEA and AZT in HIV-1 infected and uninfected human PBMC. These basic virological studies are especially timely since FDA-authorized clinical trials with a copolymer adsorbate of DHEA (EL10 - Elan Corporation, plc) in symptomatic HIV-infected individuals have begun at the University of California, San Francisco and the Community Research Initiative, New York.

Materials and Methods

Compounds. DHEA and Epi-Br were provided by Elan Corporation, Athlone, Ireland. Fresh stock solutions (2 mM) of these steroids in DMSO (0.5%) were prepared prior to the *in vitro* assays. DHEA-S was obtained from Sigma (St. Louis, MO.). AZT and 3'-azido-3'-deoxythymidine-5'-triphosphate (AZT-TP) were synthesized and purified in our laboratory according to published methods (15-17). HPA-23 (21-tungsto-9-antimoniate) was obtained from C. L. Hill (Emory University, Atlanta, GA.). Phosphonoformate was obtained from Astra Alab AB, Södertälje, Sweden.

Cells. Human PBMC from healthy HIV-1 seronegative and hepatitis B virus seronegative donors were isolated by Ficoll-Hypaque discontinuous gradient centrifugation at 1,000 x g for 30 minutes, washed twice in phosphate-buffered saline (pH 7.2; PBS), and pelleted at 300 x g for 10 minutes. Before infection, the cells were stimulated by phytohemagglutinin (PHA) at a concentration of 6 μg/ml for 3 days in RPMI 1640 medium supplemented with 15% heat-inactivated fetal calf serum, 1.5 mM L-glutamine, penicillin (100 U/ml), streptomycin (100 μg/ml), and 4 mM sodium bicarbonate buffer. CEM (CEM-CCRF) cells are a T-lymphoblastoid cell line obtained from the American Type Culture Collection, Rockville, MD. These cells were maintained in RPMI 1640 medium supplemented with 10% heat-inactivated fetal calf serum, penicillin (100 U/ml), and streptomycin (100 μg/ml). The normal levels of DHEA and DHEA-S in the media used which contained fetal calf serum was less than 2 ng/ml as determined by radioimmunoassay (SmithKline Laboratories, Ltd, Atlanta, GA).

Viruses. HIV-1 (strain LAV-1) was obtained from Dr. P. Feorino (Centers for Disease Control, Atlanta, GA). The virus was propagated in human PBMC using RPMI 1640 medium as described (18,19) without PHA or fungizone and supplemented with 7% v/v interleukin-2 (Advanced Biotechnologies, Silver Spring, MD), 7 μg/ml DEAE-dextran (Pharmacia, Uppsala, Sweden), and 370 U/ml of antibody to human leucocyte (α) interferon (ICN, Lisle, IL.). Virus obtained from cell-free culture supernatant was titrated and stored in aliquots at -80°C until use. A low passage HIV-1 isolate, designated HIV-DV, shown previously to be effective at infecting macrophages from normal donors (20), was utilized for the experiments with macrophages.

Inhibition of virus replication in human PBMC. The procedures for the antiviral assays in human PBMC have been published previously (19). Briefly, uninfected PHA-stimulated human PBMC were infected with 63,000 disintegrations per minute (dpm) of RT activity per 10^7 cells per 10 ml of medium. Thirty minutes later, the drugs at twice their final concentrations were added to the cultures. Uninfected and untreated PBMC at equivalent cell densities were grown in parallel as controls. The cultures were maintained in a humidified 5% CO_2-95% air incubator at 37°C for 6 days after infection, at which point all cultures were sampled for supernatant RT activity since RT levels peaked (> 10^6 dpm/ml) in infected untreated control flasks at this time.

Since virus yield in PBMC is not significant until 4 days after infection, experiments were designed to determine if the antiviral activity of DHEA persisted when drug was removed or added within the first 4 days. For the drug withdrawal experiments, infected PBMC were treated with 60 and 100 µM DHEA, corresponding to approximately the 85% and 95% effective antiviral concentrations. On specified days after infection, the cells were washed with PBS, and the cultures re-established in medium without drug. Virus was harvested after 6 days and quantitated by a reverse transcriptase assay, as described below.

Delayed treatment experiments were performed to determine if the activity of DHEA was reduced if the drug was added at different times after infection. Infected PBMC were exposed to the drugs on days 0, 1, 2, 3, and 4 after infection. The volume of the drug did not increase the final volume by more than 0.1 ml (1%). The virus inocula, concentration of the compounds used, and the harvesting and quantitation of virus were identical to the drug withdrawal experiment.

Experiments in CEM cells were conducted as described for the PBMC, except that the RPMI 1640 medium contained only 10% heat-inactivated fetal calf serum and antibiotics. Virus was harvested 14 days after infection when RT activity peaked.

RT activity assay. The virus in clarified supernatant was concentrated by centrifugation at 40,000 rpm for 30 minutes using a rotor (70.1 Ti; Beckman Instruments, Inc., Fullerton, CA) and then resuspended in virus disrupting buffer. The amount of virus present was determined by a RT assay performed in 96-well microdilution plates, as described previously (19,21).

In vitro macrophage HIV-1 infection assay. Monocytes/macrophages were isolated, as previously described (20), from buffy coats of blood obtained from Irwin Memorial Blood Bank, San Francisco, CA. The cells were placed in Teflon culture vessels (Savillex, Minnetonka, MN) in RPMI-1640 supplemented with 10% AB-positive (blood group) human serum at a density of 5 x 10^5 cells/ml. After 7-20 days in culture (a time when lymphocyte contamination was minimal), macrophages were exposed to HIV-DV at room temperature for one hour at a multiplicity of infection approximating one (one $TCID_{50}$ unit/cell). Unbound virus was removed by washing with undiluted fetal calf serum. Cells were then resuspended and 10^5 cells/well added to a 96 well microdilution plate in the absence or presence of various DHEA dilutions in duplicate. On day 9 after acute infection, supernatants were harvested and HIV-1 p24 antigen was quantitated using the Abbott HTLV-III EIA (19). Percent inhibition of p24 in DHEA treated cells compared with untreated, infected control cells was calculated for all experiments. The lower limit of sensitivity for the EIA was 30 pg/ml of p24.

Cytotoxicity assay and effect on protein synthesis. (a) PBMC and CEM cells. The drugs were first evaluated for their potential toxic effects on uninfected PHA-stimulated human PBMC. Flasks (25 cm^2) were seeded so that the final cell concentration was 2 x 10^5 cells/ml. The cells were cultured with and without drug for 6 days at which time aliquots were counted for cell growth and viability, as assessed by the trypan blue dye-exclusion method using a hemacytometer, as described previously (22,23). The median inhibitory concentration (IC_{50}) was calculated using the median effect method (24,25). The cytotoxicity of the drugs in CEM cells was determined 4 days after addition of the drugs to the cultures using the same procedure.

(b) Macrophages. Since these cells were not stimulated, the effect of DHEA on cell proliferation cannot be measured by standard growth methods. Therefore, protein synthesis in normal uninfected macrophages was measured. Drug-treated and untreated cells were cultivated for 9 days as described above. The cells were washed, and then 100 µl [^{3}H]-leucine (10 µCi/well, specific activity 50.0 Ci/mmol, NEN Research Products, Boston, MA) in leucine deficient RPMI 1640 containing 2% human serum was added. After a 16 hour incubation, the cells were washed, TCA precipitated, and the amount of radioactivity determined. The radioactivity in the untreated cells was consistently in the range of 400,000 to 700,000 counts per minute per 10^5 cells.

Virus inactivation studies. HIV-1 (equivalent to 200,000 dpm RT activity) was added to 2 ml of medium containing 0, 80, 100, and 200 µM of DHEA in sterile 10 ml Oak Ridge ultracentrifuge tubes (Nalgene Co.; Rochester, NY) and incubated in triplicate, for 2 hours at 37°C, in an atmosphere containing 5% CO_2-95% air. Following incubation, 3 ml of cold medium was added to each tube prior to ultracentrifugation at 40,000 rpm for 30 minutes in a Beckman 70.1 Ti rotor at 4°C. Supernatants were discarded, pellets resuspended in 1 ml fresh medium, diluted, and titrated in PBMC in fresh medium (26). Tubes with medium containing 0.5% DMSO, and handled as above, were included in order to assess the effect of DMSO on HIV-1 inactivation. After 6 days in culture, the virus in the supernatant was concentrated by centrifugation and quantitated (26) by an RT assay.

Evaluation of compounds on purified retroviral RT. HIV-1 RT was isolated from detergent disrupted virions obtained from the cell-free supernatant of infected PHA-stimulated PBMC. The enzyme was purified by passing the extract through ion-exchange chromatography columns, as described previously (16,27). The reactions were started by the addition of 10 µl of partially purified RT, incubated at 37°C for 60 minutes, and processed as previously described (16,27).

Median-effect method. Dose-effect relationships were analyzed by the median-effect equation (25). Synergy, summation (additivity), and antagonism of drug effects were quantitatively analyzed by the multiple-drug effect analysis developed by Chou and Talalay (25). The slope (m) and EC_{50} (or IC_{50}) values were obtained by using a computer program developed by Chou and Chou (24). This analysis generated the combination effect as depicted below: when the combination index (C.I.) = 1, summation is indicated; when C.I. < 1, synergy is indicated; and when C.I. > 1, antagonism is indicated. C.I. values were determined for a mutually non-exclusive interaction since this analysis provides more conservative results and is appropriate when the median-effect (Chou) plots for the drugs alone and in combination are not parallel. Additional details on the use of this method have been described by Schinazi et al. (38).

Results

Cell culture studies. The antiviral effects of DHEA, DHEA-S, and Epi-Br were evaluated in HIV-1 infected PBMC; the amount of virus or viral antigens produced in the supernatant of infected cultures 6 days after infection were evaluated by a reverse transcriptase assay or enzyme immunoassay. DHEA was found to be at least 4,000-fold less active than AZT (Table 1). Using a reverse transcriptase assay, DHEA, DHEA-S, and Epi-Br were active against HIV-1 replication with EC_{50} values of 15.1, 82.3 and 3.7 µM, respectively (Fig. 1). The antiviral activity was more variable for DHEA-S (EC_{50} range of 25-85 µM) than for the other steroid analogues. In CEM cells infected with HIV-1, the EC_{50} values for DHEA, DHEA-S, and Epi-Br were 40.3, 19.8, and 1.1 µM (Fig. 2).

Additional studies were performed with DHEA in human PBMC infected on Day 0 to determine if and how long the anti-HIV-1 activity remained after removal of the compound and how late treatment could be initiated after infection and still produce an antiviral effect. When the drug was removed from infected cultures on different days after infection, a significant antiviral activity of DHEA was apparent only if the drug was present in the media for at least 3 days.

TABLE 1. Effect of AZT and DHEA Alone and in Combination in Human PBMC Infected with HIV-1.

Treatment	Ratio	Concn (μM)	% inhibition[a]	
			RTA[b]	EIA[c]
AZT		0.0001	11.4	-4.0
		0.001	31.8	14.0
		0.01	83.1	62.7
		0.1	98.5	95.1
DHEA		1	1.7	6.4
		4	38.7	9.9
		8	27.6	26.6
		16	51.0	25.0
		32	60.6	56.3
		64	87.7	76.8
		128	96.7	92.6
AZT/DHEA	1:1,000	0.00025/0.25	21.6	2.9
		0.001/1	19.0	3.2
		0.002/2	21.4	15.8
		0.004/4	54.4	26.8
		0.008/8	75.1	50.3
		0.016/16	91.3	81.3
		0.032/32	98.1	95.0
AZT/DHEA	1:4,000	0.00025/1	28.6	9.9
		0.001/4	40.7	17.9
		0.002/8	60.6	43.2
		0.004/16	64.2	41.5
		0.008/32	77.9	64.0
		0.016/64	96.6	91.8
		0.032/128	99.3	96.9

[a] The yield of virus or virus products present in the supernatant 6 days after infection was determined by a reverse transcriptase assay (RTA) and enzyme immunoassay (EIA), as described in Materials and Methods.

[b] Mean counts for the blank, uninfected control, and infected untreated cultures in the original supernatant were 977 ± 108, 3,837, and $1.03 \pm 0.15 \times 10^6$ dpm (± SD), respectively.

[c] EIA was performed on supernatant using a commercial Abbott antigen kit. Samples were treated with TX-100 and stored at -80°C prior to testing. The ng HIV p24/ml of infected control supernatant was 316 ± 20.5. The blank and uninfected control had an value ≤ 0.6 ng p24/ml

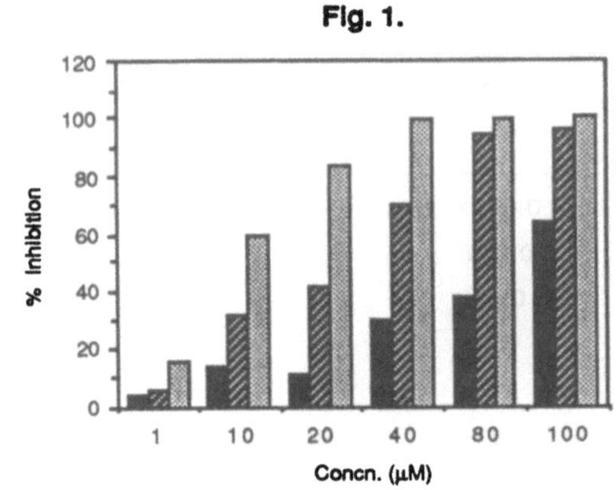

FIG. 1. Effect of DHEA (striped bars), DHEA-S (solid bars), and Epi-Br (dotted bars) on the replication of HIV-1 in human PBMC. Six days after infection, the virus infected untreated control had a mean (± S.D.) value of $1.39 \pm 0.06 \times 10^6$ dpm/ml of reverse transcriptase activity. The mean (± S.D.) values for the blank and uninfected control were 1,058 ± 286 dpm and 5,470 dpm, respectively. Percent inhibition was calculated for values in infected treated cells as compared with untreated, infected control cells.

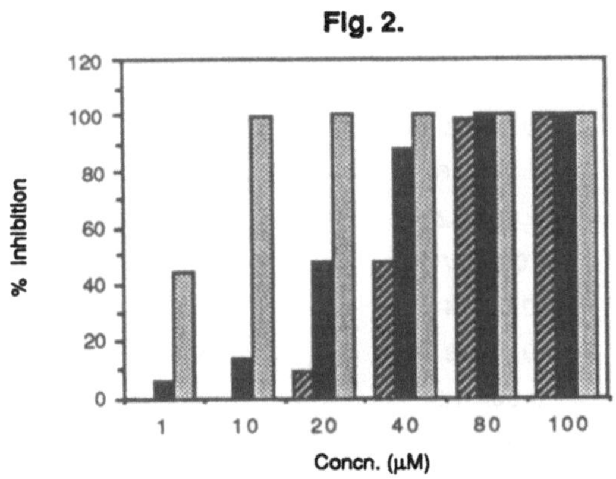

FIG. 2. Effect of DHEA (striped bars), DHEA-S (solid bars), and Epi-Br (dotted bars) on the replication of HIV-1 in CEM cells. Virus was harvested from supernatants 14 days after infection. The mean reverse transcriptase activity of triplicate uninfected control flasks was 361,932 dpm/ml.

The antiviral activity of DHEA could not be reversed if drug was removed on Days 3-5 after infection (Fig. 3). Although this steroid was still effective up to Day 3 after infection, late treatment was not as effective as early treatment (Fig. 4). These results indicated that the anti-HIV-1 activity of DHEA occurs, at least in part, at an early stage after infection.

Experiments were also conducted to determine if DHEA could neutralize the infectivity of virus particles. Virus, which had been incubated for 2 h in the presence of the drugs, was concentrated by centrifugation, diluted and then co-cultivated with fresh mitogen-stimulated cells for 6 days. The virus obtained from the supernatant was concentrated and the amount of RT determined.

The mean reciprocal dilution at which 50% of the cultures were RT positive (> 10,000 dpm/ml) were similar for all the drug concentrations tested. The results indicated that even at 200 μM, DHEA was unable to neutralize HIV-1. Thus, the antiretroviral activity of DHEA was not due to a direct virucidal effect.

Fig. 3.

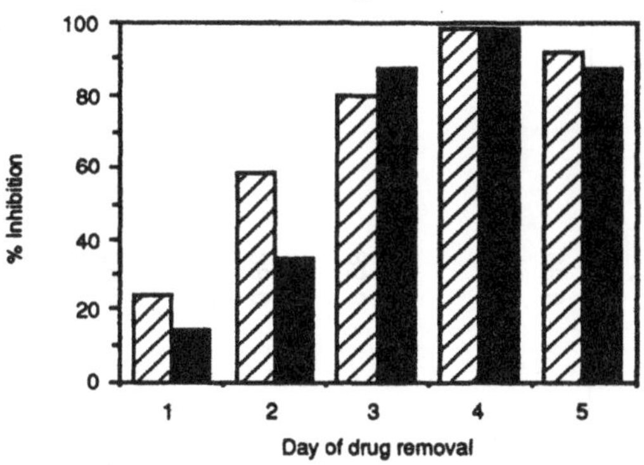

FIG. 3. Effect of DHEA on HIV-1 replication when cultures were infected on day 0 and incubated for various periods of time in the presence of drug. Virus yield (reverse transcriptase activity) was measured 6 days after infection and the inhibition was calculated relative to untreated cells. Striped bars 60 μM; solid bar 100 μM.

DHEA also decreased p24 antigen levels indicating a reduction of infection in newly exposed normal peripheral blood derived macrophages to HIV. Figure 5 shows that high concentrations of DHEA were most effective. DHEA could not be tested at concentrations greater than 100 μM in this system because of its poor solubility. Supernatants from the infected, untreated cultures contained highly variable levels of HIV-1 p24 antigen (control range 1-70 ng/10^5 cells) dependent on donor macrophage susceptibility to infection with HIV-1. The EC_{50} in this assay system using a MOI of one was between 10 μM and 100 μM with marked donor variability in both susceptibility to infection with HIV-1 and in antiviral effects of DHEA. Under similar conditions, the EC_{50} for AZT was about 0.004 μM (S. Crowe, personal communication).

TABLE 2. Median effective concentration and combination index (C.I.) values for AZT and DHEA alone and in combination in HIV-1 infected human peripheral blood mononuclear cells.

Treatment (drug ratio)	Assay used	Parameter[a]			C.I. at F_a[b] of		
		m	EC_{50} (μM)	r	0.50	0.70	0.90
AZT	RTA	0.91	0.0014	0.99			
	EIA	1.04	0.0059	0.99			
DHEA	RTA	1.53	16.98	0.98			
	EIA	1.31	23.65	0.98			
AZT/DHEA	RTA	1.60	3.42	0.98	3.12	2.14	1.21
(1:1,000)	EIA	1.74	6.64	0.99	1.73	1.25	0.76
AZT/DHEA	RTA	0.92	4.60	0.92	1.31	1.51	2.05
(1:4,000)	EIA	1.16	11.24	0.96	1.18	1.18	1.19

[a] m is the slope, EC_{50} is the median effective concentration, and r is the correlation coefficient, as determined from the median-effect plot.

[b] C.I. < 1 indicates synergy; C.I. > 1 incates antagonism; C.I. = 1 indicates additivity. F_a is a component of the median effect equation referring to the fraction of the system affected (e.g., 0.50 means the C.I. at a 50% reduction of RT activity). C.I. values were determined for a mutually non-exclusive interaction.

Fig. 4.

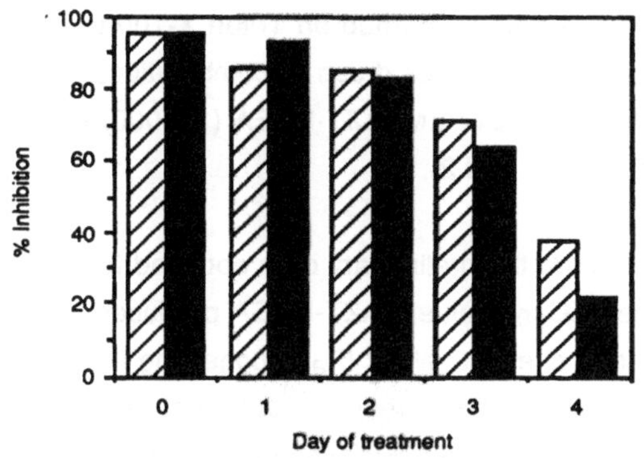

FIG. 4. Effect of DHEA on HIV-1 replication when treatment was delayed. Striped bars: 60 µM; solid bar: 100 µM.

DHEA did not inhibit [^{3}H]-leucine incorporation into macromolecules when tested up to 100 µM. At the concentrations used in all the antiviral or cytotoxicity assays (< 0.5%), DMSO resulted in no antiviral activity in PBMC or CEM cells (data not shown).

FIG.5.

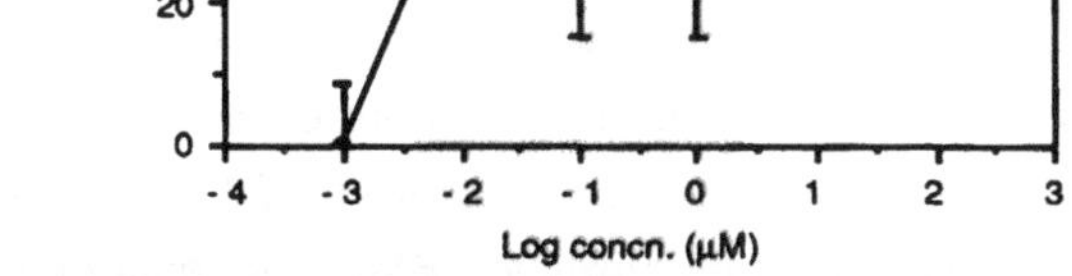

FIG. 5. Effect of DHEA on normal human macrophages infected with HIV-1. Viral p24 antigen levels in supernatants from treated and untreated infected cells were determined 9 days after acute infection (see Materials and Methods Section). Each point represents data from six separate macrophage donors in duplicate (error bar = 1 S.D.)

Combination studies. The interaction of 1:1,000 or 1:4,000 molar ratios of AZT (range 0.25-32 nM) and DHEA (1-128 µM) on the replication of HIV-1 in human PBMC, was determined by reverse transcriptase assay (RTA) of disrupted virions obtained from supernatant of cells that were exposed to virus and the drugs alone and in combination. An HIV-1 p24 enzyme immunoassay was also performed on Triton X-100-treated supernatant to confirm the activity. The EC_{50} for the various drugs or combinations as determined by an RTA was: DHEA = 17 µM, AZT = 0.0014 µM, AZT-DHEA (1:1,000) = 3.42 µM; and AZT-DHEA (1:4,000) = 4.60 µM (Tables 1 and 2).

The drug ratios were selected from the preliminary data obtained for the single drugs and was based on the ratio of their EC_{50}s-values. AZT-DHEA produced a combination index greater than 1 over a wide range of effect values (F_a), suggesting antagonism. Analyses of the EIA data indicated similar results. The correlation coefficient (r^2) for the 25 pairs of data points shown in Table 1 was 0.93. The EIA showed a complete inhibition of p24 production at 128 µM of DHEA. The EC_{50} value for the EIA data was about 2-fold greater than that obtained when virus yield was assessed by RT assay.

Cell cytotoxicity. To assess whether the viral inhibition resulted from direct cytotoxicity effects, cells were exposed to various concentrations of the steroids and the cells were monitored for growth and viability for 6 days. As seen in Table 3, when uninfected mitogen-stimulated PBMC were exposed to 10 and 100 µM DHEA or DHEA-S, no significant effect on cell growth was noted (cell viability was > 95%). However, whereas the IC_{50} for DHEA was determined to be 482 µM, Epi-Br caused significant cell death at 100 µM (Table 3).

In CEM cells exposed to the drugs for 4 days, Epi-Br was cytotoxic at concentrations as low as 10 µM (IC_{50} = 2.6 µM). The combination of DHEA and AZT at the highest ratios tested for the antiviral assays did not markedly affect the growth of uninfected PBMC. Only at concentrations greater than 500 µM DHEA was significant toxicity apparent (≥ 46% inhibition). Similar results were obtained when the cells were exposed to the drugs for 12 days. For example, the EC_{50} for DHEA was reduced from 482 µM to 384 µM which indicates that this compound is essentially non-toxic to PBMC even after prolonged *in vitro* drug exposure (data not shown).

Inhibition of purified HIV-1 RT. Studies were performed to determine if DHEA or DHEA-S interfered with the activity of partially purified HIV-1 RT in a cell-free system. As shown in Table 4, none of the steroids inhibited HIV-1 RT when tested up to 100 µM. In contrast, the EC_{50}s-values for phosphonoformate, 3'-azido-3'-deoxythymidine-5'-triphosphate, and HPA-23 were 0.16, 0.005, and about 60 µM, respectively.

TABLE 3. Effect of Steroids and AZT on the Growth of Uninfected Mitogen Stimulated Human Peripheral Blood mononuclear Cells and CEM Cells.

Treatment	Concn. (μM)	% Growth inhibition of:	
		PBMC[a]	CEM[b]
DHEA	10	15.7	14.3
	100	19.7	31.4
	500	51.0	
	1,000	53.5	37.1
DHEA-S	10	0	0
	100	3.2	0
	1,000	51.7	14.3
Epi-Br	1	0	20.0
	10	24.3	77.1
	100	64.4	81.1
AZT	1	0	0
	10	0	0
	100	14.0	0
AZT/DHEA (ratio 1:4,000)	0.016/64	0	
	0.032/128	14.9	
	0.064/256	27.7	
	0.128/512	55.9	
	0.256/1,024	49.0	
AZT/DHEA (ratio 1:1,000)	0.016/16	0	
	0.032/32	9.8	
	0.064/64	12.1	
	0.128/128	20.0	
	0.256/256	26.0	
	0.512/512	49.0	

[a] PBMC were counted after drug exposure for 6 days by the trypan blue exclusion method. The results represent the mean of two separate experiments with different donor cells. Untreated cultures had a mean of 9.5×10^5 or 1.07×10^6 cells/ml.

[b] CEM cells were counted after drug exposure for 4 days. Untreated cultures had 1.8×10^6 cells/ml.

Discussion

Several antiretroviral agents are currently at different stages of clinical development (28-30). The most prominent members of that group are the nucleoside analogues which include AZT, 3'-azido-2',3'-dideoxyuridine (AzddU, CS-87), 2',3'-dideoxyinosine (ddI) and 2',3'-didehydro-3'-deoxythymidine (D4T) (30). The development of non-nucleosides, such as DHEA, are particularly important since they probably have different mechanisms of action, making them particularly suitable against potentially drug-resistant variants, (31) or as good alternative drugs alone or in combination for the therapy of retroviral infections (32).

Our studies indicated that DHEA was a modest selective inhibitor of HIV-1 replication in primary human lymphocytes. In these cells, the drug appeared to be more potent than DHEA-S, but was less potent than Epi-Br and AZT. The selective antiviral activity of DHEA was confirmed in CEM cells, a continuous lymphoblastoid cell line. However, the potency of DHEA in primary lymphocytes was about twice as great as that found in CEM cells. The concentrations of DHEA or DHEA-S necessary for antiviral activity in these cells were markedly greater than the normal serum concentrations present in humans (5-10 nM and 5-10 µM, respectively) (11,18). Due to the observed toxic effects of Epi-Br in both PBMC and CEM cells, this compound was not considered for further evaluation. DHEA was not virucidal and did not inhibit purified HIV-1 reverse transcriptase. The mechanism(s) involved in the antiviral activity of DHEA and its sulfated form must be on sites other than the HIV-1 RT. It is conceivable that a metabolite of DHEA could inhibit this viral enzyme. The drug seemed to be more protective when cells were treated early after virus infection than when added late. DHEA was also a modest inhibitor of HIV-1 replication in human macrophages. Although significant donor macrophage variability in response to DHEA was seen, DHEA levels approaching those present *in vivo* (10 nM) continued to be associated with significant measurable *in vitro* antiviral activity (Fig. 5); the no-effect dose was 1 nM. A recent report indicated that DHEA was ineffective *in vitro* against coxsackie-virus B4 and herpes simplex virus type 2, but it had a marked protective activity against lethal infection with both viruses in mice after a single prophylactic dose (33). The effects of DHEA were associated with an 80% elevation in the number of spleen IgM antibody forming cells (33). These results suggest that, in addition to its antiviral effects, DHEA may also have a significant role as an up-modulator of immune response *in vivo*. It is also possible that the antiviral activity of DHEA may be related to its corticosteroid antagonist properties (34; W. Regelson, personal communication). Certain corticosteroids were shown by Klein et al. to enhance lysis of HIV infected cells (35). Since infected cells appear to be more vulnerable to the lytic action of corticosteroids, then DHEA may provide a protective effect.

TABLE 4. Inhibition of Purified HIV-1 Reverse Transcriptase by DHEA, DHEA-S, AZT-TP, PFA and HPA-23.

Compound	Concentration (μM)	Percent inhibition relative to control [a]	EC_{50} (μM)
DHEA	1	-2	>100
	10	-4	
	100	0	
DHEA-S	1	4	>100
	10	0	
	100	0	
HPA-23	1	-2	~ 60
	10	-6	
	100	96	
AZT-TP	0.001	21	0.005
	0.005	51	
	0.01	65	
	0.05	87	
PFA	0.05	22	0.16
	0.1	40	
	0.2	55	

[a] The assay was performed without bovine serum albumin (normally added to stabilize the enzyme), and with 1 μM [^{3}H]dTTP as substrate and 3.1 μg/ml of $(rA)_n$-$(dT)_{12-18}$ as synthetic template-primer. Uninhibited enzyme activity gave 26,580 ± 1,910 cpm (mean value ± SD of 8 determinations). This was equivalent to an incorporation of 3.6 pmoles of dTMP into newly synthesized DNA per hour for a 100 μl reaction mixture. EC_{50} values were determined from the median-effect plots.

When used alone, certain steroids can exacerbate viral infections (36,37). Therefore, it was important to assess the interaction of DHEA and AZT since such a combination may be considered for the treatment of retroviral infections. Studies on the combination of AZT and DHEA at two different ratios (1:1,000, and 1:4,000) indicated mostly antagonistic anti-HIV-1 effects between these two drugs in infected cells, irrespective of the assay used to quantify virus yield or viral products. These *in vitro* results revealed that the combination of AZT and DHEA may decrease the efficacy of AZT or exacerbate virus replication at certain concentrations and ratios of the drugs. No increased toxicity was noted in the uninfected cells, suggesting that the antagonism observed was not secondary to toxic effects. An inhibition of the pentose phosphate pathway by DHEA with a subsequent lowering of the cellular NADPH pool has been postulated (9). This effect would decrease *de novo* synthesis of purine and pyrimidine nucleotides which may affect viral nucleic acid synthesis to a greater extent. However, a recent report by Ennas et al. (9) indicated that DHEA inhibited radiolabeled thymidine uptake of mitogen-stimulated human lymphocytes from both G-6-PD and G-6-PD-deficient individuals in a dose-dependent and reversible fashion, suggesting that the inhibition of thymidine uptake is G-6-PD independent. This inhibitory effect was found even when DHEA was added to cells in the last hour of culture, simultaneously with the addition of the labeled thymidine. Based on this report, the activity of AZT would be expected to be greater in the combination experiments, since lower intracellular levels of thymidine would be expected in the presence of DHEA and thymidine prevents the antiviral activity of AZT (23,30). However, it is also possible that DHEA inhibits cellular uptake of AZT which would reduce its antiviral potency (9,38). Studies in erythrocytes indicate that AZT permeates cell membranes by nonfacilitated diffusion and not by the nucleoside transporter (38). The effect of DHEA on the transport of natural and modified nucleoside into cells has not been explored. Additional studies of the effect of DHEA on the *de novo* purine and pyrimidine pathways, and on the decreased uptake of nucleosides by DHEA-treated cells, may also explain the interference noted on the activity of AZT in the presence of DHEA. The interaction of DHEA with other anti-HIV-1 agents is not known and is beyond the scope of our studies.

The combination of DHEA and AZT also merits additional studies in animal models for retroviral infections since there is, at present, no way to predict whether the results observed *in vitro* will occur *in vivo*. Like AZT, DHEA is known to be partly metabolized to the glucuronide form (2,30). A potential competition between these two drugs may produce an elevation of the serum concentration of the drugs which may be beneficial provided the combination does not also increase bone-marrow suppression (23). Additional studies on the interaction of these drugs with the enzyme responsible for these bioconversions, presumably the liver's UDP glucuronyl transferase, are warranted.

The observed HIV-1 antiviral effect, however modest, together with the reported immune up-regulatory effects in experimental animals suggest that further laboratory and clinical studies to elucidate a possible beneficial role of DHEA in infected individuals are justified.

Summary

The ability of dehydroepiandrosterone (DHEA) and 3'-azido-3'-deoxythymidine (AZT) to inhibit the replication of human immunodeficiency virus type 1 (HIV-1) was examined in human peripheral blood mononuclear cells. Reverse transcriptase activity associated with virus and the amount of HIV-1 p24 antigens in the supernatant were used to assess the antiviral activity. Using the former assay, the EC_{50} for DHEA and AZT were 17 μM and 0.0014 μM, respectively. Results obtained by an enzyme immunoassay (EIA) were similar. DHEA-sulfate was markedly less active than DHEA. In contrast, 16α-bromoepiandrosterone was more potent and also more toxic than DHEA. The specific antiviral activity of DHEA was confirmed in CEM cells, a continuous lymphoblastoid cell line. In primary lymphocytes, the antiviral activity of DHEA could not be reversed when the drug was removed from infected cultures 3-5 days after infection. Although this steroid was still effective when added up to 3 days after infection, late treatment was not as effective as early treatment. DHEA did not have a direct virucidal effect on infectious virus. In contrast to the 5'-triphosphate of AZT and other known antiretroviral agents, DHEA did not inhibit HIV-1 reverse transcriptase enzymatic activity when tested up to 100 μM, suggesting that the mechanism(s) involved in its antiviral activity must be on sites other than this viral enzyme. Acute infection of normal human macrophages was also inhibited by DHEA at concentrations between 10 μM and 100 μM. At these concentrations DHEA was not toxic to macrophages as assessed by macromolecular synthesis. Multiple-drug effect analyses were used to quantitatively determine the interaction of AZT and DHEA in human PBMC infected with HIV-1 at a ratio of 1:1,000 and 1:4,000. Analyses of the cell culture reverse transcriptase and EIA data indicated mostly antagonistic interactions. At therapeutic levels, no apparent toxicity to uninfected cells was observed for the drugs alone or in combination. The mechanism for the antiviral activity of DHEA and its interaction with AZT in infected human lymphocytes is unclear, and may be related to the inhibition of nucleoside transport and regulation of *de novo* synthesis of nucleotides.

Acknowledgements

The technical assistance of V. Saalmann, D. Cannon, and B. Oswald is gratefully acknowledged. Supported by Public Health Service Grants AI 26055 and AI 25899 from the National Institutes of Health, the Department of Veterans Affairs, and by Elan Corporation, plc.

References

1. Kent S. 1982. DHEA: "miracle drug?" Geriatrics. 37:157.

2. Sonka J. 1976. Dehydroepiandrosterone: metabolic effects. Acta Univ Carol Med University Karlova, Prague, Czechoslovakia.

3. Gordon G.B., L.M. Shantz, and P. Talalay. 1987. Modulation of growth, differentiation and carcinogenesis by dehydroepiandrosterone. Advan. Enz. Regul. 26:355-382.

4. Miller D.S., N.N. Teng, and S.C. Ballon. 1986. Epithelial ovarian carcinoma in patients with intersex disorders: the role of pituitary gonadotropins in ovarian tumorigenesis. Gynecol. Oncol. 24:299-308.

5. Pashko L.L., A.G. Schwartz, M. Abou-Gharbia, and D. Swern. 1981. Inhibition of DNA synthesis in mouse epidermis and breast epithelium by dehydroepiandrosterone and related steroids. Carcinogenesis. 2:717-721.

6. Schwartz A.G., L. Pashko, and J.M. Whitcomb. 1986. Inhibition of tumor development by dehydroepiandrosterone and related steroids. Toxicol. Pathol. 14:357-362.

7. Barrett-Conner E., K.T. Khaw, and S.S.C. Yen. 1986. A prospective study of dehydroepiandrosterone sulfate, mortality, and cardiovascular disease. N. Engl. J. Med. 315:1519-1524.

8. Pashko L.L., R.J. Rovito, J.R. Williams, E.L. Sobel, and A.G. Schwartz. 1984. Dehydroepiandrosterone (DHEA) and 3β-methylandrost-5-en-17-one: inhibitors of 7,12-dimethylbenz[a]anthracene (DMBA)-initiated and 12-O-tetradecanoylphorbol-13-acetate (TPA)-promoted skin papilloma formation in mice. Carcinogenesis. 5:463-466.

9. Ennas M.G., S. Laconi, S. Dessai, G. Milia, M.R. Murru, and P.E. Manconi. 1987. Influence of dehydroepiandrosterone on G-6-PD activity and ^{3}H-thymidine uptake of human lymphocytes *in vitro*. Toxicol. Pathol. 15:241-244.

10. Henderson E,. A. Schwartz, L. Pashko, M. Abou-Gharbia, and D. Swern. 1981. Dehydroepiandrosterone and 16α-bromo-epiandrosterone: inhibitors of Epstein-Barr virus-induced transformation of human lymphocytes. Carcinogenesis. 2:683-686.

11. Bricaire F., B. Rouveix, and R. Pignal. 1988. Taux des hormones sexuelles et du cortisol chez les malades atteints de syndrome d'immunodéficit acquis. La Presse Médicale 7:1540.

12. Merril C.R., M.Z.G. Harrington, and T. Sunderland. 1989. Plasma dehydroepiandrosterone levels in HIV infection. J. Amer. Med. Assoc. 261:1149.

13. Bricaire F., C. Marche, D. Zoubi, B. Regnier, and A.G. Saimot. 1988. Adrenocortical lesions and AIDS. Lancet 1:881.

14. Nestler J.E., C.O. Barlascini, J.N. Clore, and W.G.Blackard. 1988. Dehydroepiandrosterone reduces serum low density lipoprotein levels and body fat but does not alter insulin sensitivity in normal men. J. Clin. Endocrinol. Metab. 66:57-61.

15. Lin T.S., and W.H. Prusoff. 1978. Synthesis and biological activity of several amino analogues of thymidine. J. Med. Chem. 21:109-112.

16. Schinazi R.F., B.F.H. Eriksson, and S.H. Hughes. 1989. Comparison of inhibitory activity of various antiretroviral agents against particle derived and recombinant HIV-1 reverse transcriptase. Antimicrob. Agents Chemother. 33:115-117.

17. Yoshikawa M., T. Kato, and T. Takenishi. 1969. Studies of phosphorylation. III. Selective phosphorylation of unprotected nucleosides. Bull. Chem. Soc. Jap. 42:3505-3508.

18. McDougal J.S., S.P. Cort, M.S. Kennedy, C.D. Cabridilla, P.M. Feorino, D.P. Francis, D. Hicks, V.S. Kalyanaramen, and L.S. Martin. 1985. Immunoassay for the detection and quantitation of infectious human retrovirus, lymphadenopathy-associated virus (LAV). J. Immun. Meth. 76:171-183.

19. Schinazi R.F., D.L. Cannon, B.H. Arnold, and D. Martino-Saltzman. 1988. Combinations of isoprinosine and 3'-azido-3'-deoxythymidine in lymphocytes infected with human immunodeficiency virus type 1. Antimicrob. Agents Chemother. 32:1784-1787.

20. Crowe S., J. Mills, and M.S. McGrath. 1987. Quantitative immunocytofluorographic analysis of CD4 surface antigen expression and HIV infection of human peripheral blood monocyte/macrophages. AIDS Res. Human Retro. 3:135-145.

21. Spira T.J., L.H. Bozeman, R.C. Holman, D.T. Warfield, S.K. Phillips, and P.M. Feorino. 1987. Micromethod for assaying the reverse transcriptase of human T-cell lymphotropic virus type III/lymphadenopathy-associated virus. J. Clin. Microbiol. 25:97-99.

22. Schinazi R.F., J. Peters, C.C. Williams, D. Chance, and A.J. Nahmias. 1982. Effect of combination of acyclovir, and vidarabine or its 5'-monophosphate on herpes simplex viruses in cell culture and in mice. Antimicrob. Agents Chemother. 22: 499-507.

23. Sommadossi J.-P., R. Carlisle, R.F. Schinazi, and Z. Zhou. 1988. Enhancement of the antiretroviral activity of 3'-azido-3'-deoxythymidine by selective rescue of human bone marrow progenitor cells using metabolic modulation with uridine. Antimicrob. Agents Chemother. 32:997-1001.

24. Chou J., and T.-C. Chou. 1985. Dose-effect analysis with microcomputers: Quantitation of ED_{50}, LD_{50}, synergism, antagonism, low-dose risk, receptor binding and enzyme kinetics. A computer software for Apple II Series and IBM-PC and Instruction Manual. Elsevier-Biosoft, Elsevier Science Publishers, Cambridge, U.K.

25. Chou T.-C., and P. Talalay. 1984. Quantitative analysis of dose-effect relationships: the combined effects of multiple drugs or enzyme inhibitors. Adv. Enz. Regul. 22:27-55.

26. Martin L.S., J.S. McDougal, and S.L. Loskoski. 1985. Disinfection and inactivation of the human T lymphotropic virus type III/lymphadenopathy-associated virus. J. Inf. Dis. 152:400-403.

27. Eriksson B., L. Vrang, H. Bazin, J. Chattopadhyaya, and B. Öberg. 1977. Different patterns of inhibition of avian myeloblastosis virus reverse transcriptase activity by 3'-azido-3'-deoxythymidine 5'-triphosphate and its *threo* isomer. Antimicrob. Agents Chemother. 31:600-604.

28. Fischl M.A., D.D. Richman, M.H. Grieco, M.S. Gottlieb, P.A. Volberding, O.L. Laskin, J.M. Leedom, J.E. Groopman, D. Mildvan, R.T. Schooley, D.D. Jackson, D.T. Durack, D. King, and the AZT Collaborative Working Group. 1987. The efficacy of azidothymidine (AZT) in the treatment of patients with AIDS and AIDS-related complex. A double-blind, placebo-controlled trial. N. Engl. J. Med. 317:185-191.

29. Hirsch M.S. 1988. AIDS commentary. Azidothymidine. J .Infect. Dis. 157:427-431.

30. Schinazi R.F. 1988. Strategies and targets for anti-human immunodeficiency virus type 1 chemotherapy. In: *AIDS in children, adolescents and heterosexual adults: an interdisciplinary approach to prevention.* R.F. Schinazi, and A.J. Nahmias, (eds), Elsevier North Holland, New York pp.126-143.

31. Larder B.A,. G. Darby, and D.D. Richman . 1989. HIV with reduced sensitivity to zidovudine (AZT) isolated during prolonged therapy. Science. 243:1731-1734.

32. Eriksson B.F.H., and R.F. Schinazi. 1989. Combinations of 3'-azido-3'-deoxythymidine (Zidovudine) and phosphonoformate (Foscarnet) against human immunodeficiency virus type 1 and cytomegalovirus *in vitro.* Antimicrob. Agents Chemother. 33:663-669.

33. Loria R.M., T.H. Inge, S.S. Cook, A.K. Szakal, and W. Regelson. 1988. Protection against acute lethal viral infections with the native steroid dehydroepiandrosterone (DHEA). J. Med. Virol. 26:301-314.

34. Regelson W., R. Loria, and M. Kalimi. 1988. Hormonal intervention:"buffer hormones" or "state dependency" - the role of dehydroepiandrosterone (DHEA), thyroid hormone, estrogen, and hypophysectomy in aging. Annals N. Y. Acad. Sci. 521:260-273.

36. Klein A., B. Bruser, J.B. Robinson, P.H. Pinkerton, and A. Malkin. 1987. Effect of non-viral fraction of acquired immunodeficiency syndrome plasma on the vulnerability of lymphocytes to cortisol. J. Endocrinol. 112:259-264.

37. Schinazi R.F., J. Peters, D. Chance, and A.J. Nahmias. 1982. Studies *in vitro* and *in vivo* of combinations of antivirals and anti-inflammatory agents in relation to the treatment of herpes simplex viruses. In: *Current Chemotherapy and Immunotherapy.* P. Periti, and G.G. Grassi, (eds.), The American Society for Microbiology, Washington, D.C., Volume II, pp.1085-1087.

38. Yirrell D.L., W.A. Blyth, and T.J. Hill. 1987. The influence of androgen paralysis in mice following intravenous inoculation of herpes simplex virus. J. Gen. Virol. 68:2461-2464.

39. Zimmerman T.P., W.B. Mahony, and K.L. Prus. 1987. 3'-Azido-3'-deoxythymidine. An unusual nucleoside analogue that permeates the membrane of human erythrocytes and lymphocytes by nonfacilitated diffusion. J. Biol. Chem. 262:5748-5754.

40. Richman D.D., M.A. Fischl, M.H. Grieco, M.S. Gottlieb, P.A. Volberding, O.L. Laskin, J.L. Leedom, J.E. Groopman, D. Mildvan, M.S. Hirsch, G.G. Jackson, D.T. Durack, and S. Nusinoff-Lehrman. 1987. The toxicity of azidothymidine (AZT) in the treatment of patients with AIDS and AIDS-related complex. A double-blind, placebo-controlled trial. N. Engl. J. Med. 317:192-197.

41. Schinazi R.F., T.-C. Chou, R.T. Scott, X. Yao, and A.J. Nahmias. 1986. Delayed treatment with combinations of antiviral drugs in mice infected with herpes simplex virus and application of the median effect method of analysis. Antimicrob. Agents Chemother. 30:491-498.

DEHYDROEPIANDROSTERONE (DHEA) AND DIABETIC SYNDROMES IN MICE

D.L. Coleman
The Jackson Laboratory, Bar Harbor, Maine 04609

Introduction

DHEA and its sulfate derivative are major adrenal secretory products in humans, yet their biological function remains unknown. DHEA has been shown to have wide-ranging beneficial effects in a diverse variety of disease conditions in animal models (1-5, this volume). This chapter will summarize our findings regarding the marked therapeutic effects of DHEA in mutant mice with diabetic-obesity syndromes of varying degrees of severity (4,6). DHEA and DHEA sulfate are both converted to a wide variety of compounds, both inert and metabolically active (androgens and estrogens), which makes it difficult to assess whether the action of DHEA is due to DHEA itself, to its metabolites, or a combination of both. The hypoglycemic action of DHEA was only seen when fed, suggesting that some metabolite or metabolites produced *in vivo* in the liver or during passage through the gut wall or in the gut lumen by microbial action were the active agents. Of the many metabolites of DHEA tested, only the 5β reduced compounds (etiocholanolones) and estrogens were effective (6). The etiocholanolones, although 4 times more active than DHEA, still had to be fed to be effective, suggesting their conversion to more active metabolites. In contrast, estradiol was markedly more effective against diabetes symptoms, but not the obesity, than any compound by all routes of administration. Conversion of etiocholanolones to estrogen seems unlikely because the 5β saturated ring should preclude aromatization and uterotropic effects were not observed when female mice were treated with etiocholanolones. Although

Dehydroepiandrosterone (DHEA)

many mechanisms of action have been suggested for DHEA, none adequately explain the beneficial effects observed in diabetes or obesity syndromes.

Results

Diabetes-obesity conditions

Obese (*ob*) and diabetes (*db*) are two autosomal recessive mutations in the mouse that cause similar, if not identical, syndromes when maintained on the same inbred background (7). Either mutation when maintained on the C57BL/KsJ (BKs) inbred background produces obesity and life-shortening diabetes associated with atrophy of the β cells of the islets of Langerhorn. In contrast, either mutation maintained on the C57BL/6J (B6) inbred background produces a more severe obesity associated with mild and transient hyperglycemia that is more severe in males. The initial stages of the syndrome produced by each mutation on each background are identical. These include hyperinsulinemia associated with mild hypoglycemia occurring at 10-14 days of age and prior to any hyperphagia. After weaning, hyperphagia occurs, and hyperglycemia becomes pronounced in spite of progressive elevations in plasma insulin concentrations. The hyperglycemia is either transient or sustained depending on the ultimate islet response (atrophy, BKs or hypertrophy, B6). Viable yellow (A^{vy}) is a dominant mutation at the agouti locus that interferes with eumelanin synthesis resulting in a mouse with yellow pelage. Associated with this coat color mutation is an increased susceptibility to mammary tumor development in females and hepatomas in males (8). Additionally, these mutants have a mild glucose intolerance associated with mild hyperinsulinemia and a slowly developing, but severe, obesity (9).

DHEA (0.4%), fed to BKs-*db/db* mutants reduced blood sugar concentrations to normal, prevented the islet atrophy typical of chow fed mutants and sustained plasma and pancreatic insulin at much higher concentrations than that seen in chow fed mutants (Table 1). No effect on weight gain was observed in treated mutants even though food consumption was increased 20-30%. DHEA (0.2%) was somewhat less effective than 0.4% (Table 1) with respect to

Table 1: Effect of DHEA on Diabetes-Obesity Syndromes in Mice

Treatment	Food Consumption, g/day	Weight Gain, g/16 wk	Blood Sugar, mg/dl	Plasma IRI, µU/ml	Pancreatic IRI, U/g
Bks - ++:					
none	3.8 ± 0.2[1]	11.4	131 ± 2.0	42.4 ± 2.0	3.84 ± 0.63
DHEA (0.4%)	4.0 ± 0.2	9.6	101 ± 1.4	51.4 ± 3.3	3.89 ± 0.13
BLKs-*db*/*db*:					
none	4.2 ± 0.2	26.1	448 ± 30	205 ± 28.9	1.06 ± 0.25
DHEA 0.2%	N.D.	N.D.	194 ± 28*	731 ± 215*	3.76 ± 0.85*
DHEA 0.4%	6.2 ± 0.1*	27.9	156 ± 19*	2374 ± 1883*	10.40 ± 0.08*
B6 - ++:					
none	3.8 ± 0.2	12.4	140 ± 4.0	37.3 ± 5.4	2.67 ± 0.17
DHEA (0.4%)	3.0 ± 0.1	11.0	118 ± 3.5	37.4 ± 8.9	2.29 ± 0.12
B6 - *db*/*db*:					
none	4.8 ± 0.2	30.1	140 ± 15	2302 ± 454	8.50 ± 1.9
DHEA (0.1%)	N.D.	N.D.	126 ± 13	>4000*	9.48 ± 1.8
DHEA (0.4%)	2.8 ± 0.2	20.4*	111 ± 9.2	1414 ± 345*	8.97 ± 1.5
C3H - *A*/*A*:					
none	4.4 ± 0.2	25.0	137 ± 9.5	152 ± 37.0	N.D.
DHEA (0.4%)	5.8 ± 0.2	11.2*	105 ± 3.0	41 ± 1.8	N.D.
C3H A^{vy}/A:					
none	5.3 ± 0.3	34.0	230 ± 3.1	495 ± 6.2	N.D.
DHEA (0.4%)	6.4 ± 0.4	15.3*	124 ± 3.7*	82 ± 8.7*	N.D.

[1] Figures represent mean ± SEM from groups of 4 - 6 male mice fed powdered chow ± DHEA starting at weaning and continuing for 16 weeks. N.D., not determined. IRI., immunoreactive insulin

* Significantly different from mice fed chow alone ($p < .05$).

prevention of hyperglycemia and maintenance of higher than normal plasma and pancreatic insulin concentrations. In contrast, treatment of B6-*db/db* mutants prevented the transitory increase in blood sugar concentration and decreased the rate of weight gain (Table 1). Plasma insulin, although still elevated, was less than that typical of untreated B6 mutants. Pancreatic insulin remained higher than normal. Treatment of B6 mutants with 0.1% of DHEA seemed equally as effective as 0.4% with regard to all parameters studied. Histological examination of the pancreas in treated mutants maintained on either inbred background revealed large hyperactive islets with enlarged sinusoids and no signs of islet atrophy.

Groups of normal mice (BK++ or B6++) fed DHEA (0.4%) gained slightly less weight than chow fed controls and at autopsy an obvious decrease in the size of the fat pads was observed. This probably reflected a decrease in body fat although body composition studies were not undertaken. Uteri and seminal vesicles were enlarged, indicative of either estrogen and androgen stimulation depending on the sex. Other than these findings, all tissues appeared normal on gross examination and no signs of toxicity were observed in any treated group.

Treated C3H-A^{vy}/A mutants responded to treatment in a similar fashion to that seen with B6-*db/db* mutants except the responses were more rapid and readily apparent. Body weight gain was completely normalized in spite of an increase in food consumption. A significant decrease in the plasma insulin concentration to values typical of those values found in normal mice of most inbred strains occurred concomitant with a drop in the blood sugar concentration to normal (Table 1). Normal C3H-*A/A* mice fed chow also exhibit mild obesity and when treated with DHEA exhibit a decreased rate of weight gain associated with a drop in plasma insulin concentration and an obvious decrease in the total mass of body fat reserves. Again, no toxic effects were observed in treated mice of this strain.

Lipogenesis

Our studies (10) and those of others (11) suggested that one or more futile cycles might produce the decrease in metabolic efficiency observed with DHEA treatment. Lipogenesis and lipolysis are reciprocal with DHEA pathways that utilize considerable energy and if the rates of both pathways are increased on DHEA treatment, it is conceivable that considerable amounts of energy could be wasted. We found that the rate of incorporation of radiolabeled acetate into triglyceride was increased 5 fold in DHEA-treated normal mice. However the already elevated lipogenesis rate in chow fed mutants was not increased further by DHEA treatment and no increase in the lipolysis rate could be demonstrated in either normal or mutant mice even though a decrease in lipid stores associated with increased lipogenesis would suggest some parallel increase in lipolysis rate (10).

Gluconeogenesis

One well-documented abnormality associated with the obese and diabetic mutants maintained on either inbred background is an increased rate of gluconeogenesis in spite of excessive circulating insulin (7). Activities of the key gluconeogenic enzymes remain elevated and unregulated by insulin, suggesting total insulin resistance with respect to this pathway (10). Treatment of mutants with DHEA reduced both the overall rate of lactate conversion to glucose and the activity of the enzyme, phosphoenolpyruvate carboxykinase, to normal within 24 hours after treatment, suggesting a rapid restoration of insulin sensitivity. This increased sensitivity to insulin also may be involved in the maintenance of normal blood sugar concentration in spite of reduced plasma insulin concentrations in treated B6-*db*/*db* and C3H-A^{vy}/A mutants.

Corticosteroids

Both corticosteroids and DHEA are synthesized from a common precursor (pregnenolone) and a major increase in the synthesis of one compound may be reflected as a decrease in the synthesis of the other. Mutant (*ob/ob* and *db/db*) mice maintained on any background exhibit hypercorticism and this has been suggested to be a cause of, or at least contribute to, the obesity-diabetes syndrome. Adrenalectomy has been shown to either reduce the severity (12) or even normalize it (13). The magnitude of the adrenalectomy effect has been shown to be dependent on the type of diet fed (14,15). Our studies with adrenalectomized mice have failed to produce the effects seen by others. Instead the rate of weight gain and degree of hyperglycemia remained similar to that of intact mutants (4). However, DHEA treatment was more effective in the adrenalectomized BKs mutants when compared with intact BKs mutants with normalization of blood sugar being produced with 0.1% rather than 0.4% DHEA (4).

Discussion

DHEA is a steroid that when fed in pharmacological doses has beneficial effects in a variety of obese states in rodents. Although many potential mechanisms of action have been put forward, none readily explains the effects seen with diabetic mice. Most of the known effects of DHEA such as its inhibition of glucose 6-phosphate dehydrogenase limiting the production of reduced pyridine nucleotides (16) and its thermogenic effect as demonstrated in rats (17) have been ruled out in these mutants (18, and Coleman, unpublished observations).

We and most other investigators have found the best effects are seen when DHEA is fed. This suggests that DHEA is converted in the body or in the gut, possibly by micoorganisms, into more active metabolites. DHEA can be converted to many metabolites most of which involve oxidation via a cytochrome p450 system. Some of these metabolites may be metabolic effectors in their own right acting directly on specific pathways or interacting in many pathways to produce many subtle changes that in turn produce a variety of beneficial effects including the large decrease in metabolic efficiency seen in treated obese mutants. Any action on specific pathways may directly relate to their beneficial effects in different abnormalities.

Hypercorticism occurring secondary to obesity is a prominent feature of both the *db* and *ob* mutation that plays a major role in exacerbating the condition. Replacement therapy with exogenous DHEA might tend to normalize the corticosterone to DHEA balance thereby permitting more normal metabolism. Our observation that adrenalectomy potentiates the action of DHEA suggests that the restoration of a more normal milieu could be responsible for some of the effects of DHEA.

The anti-obesity effects of DHEA are seen in spite of normal or increased food intake. Many of the effects of DHEA on tumor development, auto-immune diseases, aging and diabetes-obesity syndromes can also be produced by food restriction in rodents (19). The effects of adrenalectomy on diabetes-obesity mutants has also been shown to be diet dependent (14,15). These observations suggest a possible interaction of DHEA treatment with both the quality and quantity of the diet fed. Systematic studies of these interactions must be undertaken and understood in order to better understand the action of DHEA.

The activation of a futile cycle or cycles by DHEA would explain some of the decreased metabolic efficiency observed in diabetic mutants. Although a variety of potential cycles have been investigated, all are deficient in one or more aspects with respect to the beneficial effects produced in mutant mice. One consistent effect of treatment observed with obesity mutants is the potentiation of the action of insulin that allows normalization of glucose homeostasis and the rate of gluconeogenesis with a lower circulating insulin requirement than that seen prior to treatment. This decreased need for insulin coupled with the decreased glucose stress resulting from the normalized gluconeogenic rate could be sufficient to prevent islet atrophy, thereby preventing severe diabetes and life shortening in BKs mutants, and to reduce the hyperplasia and hypertrophy of β-cells, decrease plasma insulin and slow obesity development in B6 mutants. Since insulin is a major anabolic hormone involved in the control of many functions, especially lipogenesis and cell replication, this decrease in insulin concentration could be responsible for both the anti-tumor and anti-obesity effects of DHEA. Additionally, lowering the plasma insulin concentration could permit normal action of the counter-regulatory hormones (glucagon, catecholamines), thereby restoring normal metabolic cycling and increasing energy expenditure. Conclusive studies defining the mechanisms of action of DHEA must await the identification of any active metabolites, and the establishment of their action, if any, as biological effectors in multiple metabolic pathways.

Acknowledgements

The research discussed in this review article was supported by grant DK 14461 from the NIDDK and a grant from the Kroc Foundation. The National Institute of Diabetes, Digestive and Kidney Diseases is not responsible for the contents of this manuscript nor do the contents necessarily represent the official view of that agency. The Jackson Laboratory is fully accredited by the American Association of Laboratory Care.

References

1. Gordon, C.B., L.M. Shantz and P. Talalay. 1987. Adv. Enz. Regulation. *26,* 355.

2. Yen, T.T., J.A. Allan, D.V. Pearson, J.M. Acton and M.M. Greenberg. 1977. Lipids *12,* 409.

3. Cleary, M.P. 1989. In: Hormones, Thermogenesis and Obesity (H. Lardy and F. Strutman, eds). Elsevier, New York. pp.365-376.

4. Coleman, D.L. 1989. In: Hormones, Thermogenesis and Obesity (H. Lardy and F. Stratman, eds). Elsevier, New York. pp.377-383.

5. Schwartz, A.G., J.M. Whitcomb, D.K. Fairman, L.A. Hastings, M.L. Lewbart and L.L. Pashko. 1989 In: Hormones, Thermogenesis and Obesity (H. Lardy and F. Stratman, eds). Elsevier, New York. pp.441-453.

6. Coleman, D.L., E.H. Leiter and N. Applezweig. 1984. Endocrinology *115,* 239.

7. Coleman, D.L. 1978. Diabetologia *14,* 141.

8. Heston, W.E. and G. Vlahakis. 1961. J. Natl. Cancer Inst. *26,* 969.

9. Coleman, D.L. 1985. Endocrinology *117,* 2279.

10. Coleman, D.L. 1988. In: Pathogenesis and New Approaches to the Study of Noninsulin-dependent Diabetes (A.Y. Chang and A.R. Diani, eds.). Alan R. Liss, Inc., New York. pp. 161-175.

11. Cleary, M.P., J. Billheimer, A. Finan, J.L. Sarlin and A.G. Schwartz. 1986. Horm. Met. Res. *16 suppl. 1,* 43.

12. Dubuc, P.U. and H.S. Wilden. 1985. Intl. J. Obesity *10,* 91.

13. Saito, M. and G. Bray. 1985. Am. J. Physiol. *246,* 1220.

14. Smith, C.K. and D.R. Romsos. 1986. Am. J. Physiol. *249,* R13.

15. Grogan, C.K., H.-K. Kim and D.R. Romsos. 1987. J. Nutr. *117,* 1115.

16. Marks, P.H. and J. Banks. 1972. Proc. Natl. Acad. Sci. *46,* 493.

17. Lardy, H., C.-Y. Su, M. Kneer and S. Wiegus. 1989. In: Hormones, Thermogenesis and Obesity (H. Lardy and F. Stratman, eds.). Elsevier, New York. pp. 415-426.

18. Coleman, D.L., R.W. Schwizer and E.H. Leiter. 1984. Diabetes *33,* 26.

19. Weindruch, R., R.L. Walford, S. Fligiel and D. Guthrie. 1986. J. Nutr. *116,* 641.

REGULATION OF DEHYDROEPIANDROSTERONE METABOLISM BY INSULIN, AND METABOLIC EFFECTS OF DEHYDROEPIANDROSTERONE IN MAN

John E. Nestler, John N. Clore, and William G. Blackard
Division of Endocrinology and Metabolism, Medical College of Virginia/Virginia Commonwealth University, Richmond, VA 23298

Introduction

The aging process is accompanied by the development of atherosclerosis, obesity, and glucose intolerance. These sequelae contribute substantially to the morbidity and mortality of the elderly, yet the mechanisms of their pathogenesis remain poorly understood.

Several characteristics of the adrenal steroid dehydroepiandrosterone (3β-hydroxy-5-androsten-17-one; DHEA) strongly suggest that it may play a role in the aging process. DHEA is the most abundantly produced adrenal steroid, and serum concentrations of its sulfate ester, DHEA-sulfate (DHEA-S), are approximately 20-fold higher than those of any other circulating steroid hormone (1). Peak serum levels of DHEA and DHEA-S occur at age 25 years (1,2). Levels decrease progressively thereafter, diminishing by 95% by age 85-90 years (1,2), as the incidence of atherosclerosis, obesity, and diabetes rise. This suggests that higher levels of DHEA or DHEA-S may be protective against the development of these degenerative processes. The mechanism(s) of the age-related decline in serum DHEA and DHEA-S levels is not known, and similar declines in other adrenal androgens or glucocorticoids have not been described (3).

In addition, over the past two decades numerous studies have

Dehydroepiandrosterone (DHEA)

been conducted examining the effects of DHEA in animals. Many of these studies have shown that DHEA administration 1) improves serum lipids and retards the development of atherosclerosis (4-8), 2) results in lower body weight without affecting appetite or food intake (5,9-31), 3) prevents the development of diabetes in genetically diabetic (-db/db) or obese (-ob/ob) mice (15-19), and 4) increases tissue sensitivity to insulin in aged normal mice (17). A more detailed discussion of these animal studies can be found in the other chapters of this book.

Studies examining the effects of DHEA in man, however, are limited. Therefore, we recently conducted several studies designed to test the hypothesis that the progressive decline in serum levels of DHEA and DHEA-S with aging plays a role in the development of the degenerative processes which accompany the aging process.

Results

I. Effects of hyperinsulinemia on DHEA and DHEA-S metabolism

Our interest in DHEA was initially stimulated by our finding of an unexpected decline in serum DHEA-S levels in women during sustained hyperinsulinemia (32). In a study designed to assess the effects of acute hyperinsulinemia on serum androgens, five normal women were studied with the hyperinsulinemic-euglycemic clamp technique. Each woman received a 0.1 U/kg insulin bolus dose, followed by a 10 mU/kg·min insulin infusion for 12 h. A mean (±SE) serum insulin concentration of 1832 ± 292 µU/ml was achieved, while the serum glucose concentration was clamped at 116 ± 5 mg/dl.

Serum testosterone and progesterone levels remained relatively constant during the insulin infusion. In contrast, serum

DHEA-S concentrations decreased progressively, declining by 39% at 12 h (144 ± 29 vs 234 ± 35 µg/dl at zero time; P<0.01) (Fig. 1). This steady decline in serum DHEA-S levels was shown not to be due to diurnal variation, suppression of ACTH release, or alterations in serum prolactin levels (33).

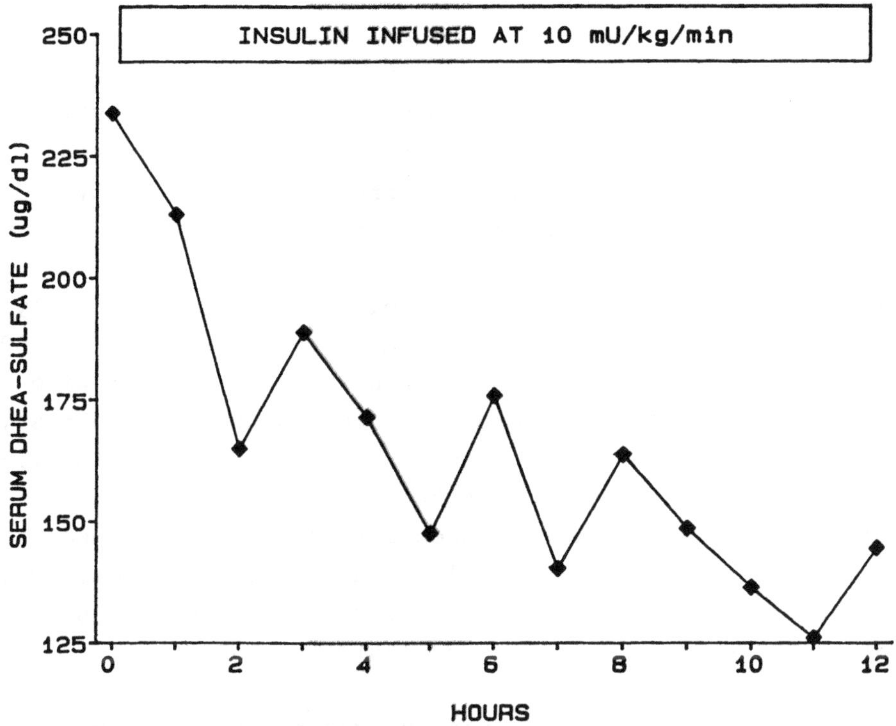

Fig. 1. Mean Serum DHEA-S Levels in Five Women during a Hyperinsulinemic-Euglycemic Clamp.

To determine whether this fall in serum DHEA-S levels might have been due to insulin-stimulated 1) hydrolysis of DHEA-S to DHEA, 2) conversion of DHEA/DHEA-S to androstenedione, and/or 3) urinary excretion of these steroids, 10 additional men were studied by the hyperinsulinemic-euglycemic clamp technique and during a control saline infusion (34). Each man received a

0.1 U/kg insulin bolus dose, followed by a 10 mU/kg·min insulin infusion for 4 h. An average insulin level of 1726 ± 36 µU/ml was achieved, while the serum glucose level was maintained at 90.5 ± 2.3 mg/dl.

During hyperinsulinemia, serum DHEA-S levels again declined progressively, and were only 79.1 ± 3.2% of the baseline value by the end of the study (hour 6) ($P<0.05$). Serum DHEA levels also fell progressively, and were 52.9 ± 10.2% of the baseline value by the end of the study ($P<0.05$). Since serum DHEA-S and DHEA levels fell concurrently, this finding suggests that hyperinsulinemia does not reduce serum DHEA-S levels by increasing the hydrolysis of DHEA-S to DHEA.

The percent fall in serum DHEA levels correlated positively with the percent fall in serum DHEA-S levels ($r=0.44$; $P<0.02$). This positive correlation is striking given the known wide intradiem variability of serum DHEA levels and, in contrast, the relative lack of fluctuation of serum DHEA-S levels (1,2). This correlation suggests that the decline in serum DHEA concentrations may have been secondary to an insulin-mediated reduction in serum DHEA-S levels. Thus, hyperinsulinemia may decrease serum DHEA levels indirectly by reducing the storage pool of DHEA-S available for conversion to DHEA.

DHEA is converted by the enzyme 3β-hydroxysteroid dehydrogenase to androstenedione, and enhanced conversion of DHEA to androstenedione by insulin could theoretically explain the fall in serum DHEA/DHEA-S levels during hyperinsulinemia. If this were so, one might expect serum androstenedione levels to rise as serum DHEA/DHEA-S levels fell. In this study, however, serum androstenedione levels also fell progressively during experimentally-induced hyperinsulinemia, and were only 48.1 ± 3.3% of the baseline value at the end of the study ($P<0.05$). This suggests that the decline in serum DHEA and DHEA-S levels was not due to insulin-stimulated conversion of DHEA/DHEA-S to androstenedione.

Notably, no change in serum DHEA-S, DHEA or androstenedione levels occurred in paired control studies, during which 0.45% saline was infused at rates matched exactly to the rates of the dextrose and insulin infusions during the hyperinsulinemic clamp studies.

Finally, we also considered the possibility that hyperinsulinemia could lower serum DHEA and DHEA-S levels by increasing urinary excretion of these steroids. Indeed, despite decreasing serum DHEA and DHEA-S levels during hyperinsulinemia, urinary DHEA-S and DHEA-glucuronide excretions were increased by 50% ($P<0.05$) and 86% ($P=0.05$), respectively, when compared to urinary excretion of these steroids during the control studies. In contrast, urinary excretion of unconjugated DHEA was unchanged. Quantitatively, however, increased urinary excretion of conjugated DHEA during hyperinsulinemia accounted for only ~5% of the concomitant fall in serum DHEA-S concentrations.

These observations indicate that hyperinsulinemia reduces serum DHEA-S, DHEA, and androstenedione levels in men, and suggest that neither insulin-stimulated hydrolysis of DHEA-S nor increased conversion of DHEA/DHEA-S to androstenedione are likely mechanisms for the fall in serum DHEA-S levels. Moreover, although hyperinsulinemia increases urinary excretion of conjugated DHEA, the contribution of this mechanism to the insulin-induced fall in serum DHEA-S levels is negligible. Thus, it is likely that insulin, in addition to its stimulation of urinary steroid excretion, must either decrease adrenal DHEA-S production and/or increase DHEA-S catabolism. A possible mechanism whereby hyperinsulinemia might specifically decrease adrenal androgen production, without affecting adrenal production of glucocorticoids or mineralocorticoids, is inhibition of 17,20-desmolase activity. This possibility has not yet been evaluated.

II. Metabolic effects of DHEA administration to normal men

This decline in serum DHEA and DHEA-S levels in response to hyperinsulinemia was particularly intriguing in view of recent reports which demonstrated that DHEA administration prevented the development of diabetes mellitus in genetically diabetic (-db/db) or obese (-ob/ob) mice (15-19) and increased tissue sensitivity to insulin in aged normal mice (17). We speculated that in response to insulinopenia the body might raise DHEA and DHEA-S levels in order to increase its sensitivity to insulin, whereas during hyperinsulinemia serum DHEA and DHEA-S would be reduced (32). In this manner DHEA and/or DHEA-S would act as an endogenous regulator of tissue sensitivity to insulin.

In addition, DHEA had been reported to exert beneficial effects on lipids (4-6) and obesity (9-31) in animals. A study in man had demonstrated an inverse correlation between fetal serum DHEA-S and LDL levels (35), while a more recent study demonstrated an inverse correlation between DHEA-S levels and death from cardiovascular disease in adult men (36). It seemed plausible that DHEA might exert some of its beneficial effects on lipids and obesity by altering tissue sensitivity to insulin.

Therefore, to test the hypotheses that DHEA administration would increase tissue sensitivity to insulin, decrease serum cholesterol levels, and decrease body fat content, we conducted a prospective, randomized, double-blind study in which 5 normal men ingested oral DHEA (1600 mg/day) and 5 normal men ingested placebo for 28 days (37). The DHEA and placebo groups were matched for age and weight. Before and after DHEA or placebo administration, tissue sensitivity to insulin, serum lipid levels, and anthropometric parameters were determined.

To assess compliance, serum DHEA-S levels were checked weekly. Mean serum DHEA-S levels did not change in the placebo group, but rose 2.5- to 3.5-fold in the DHEA group (Table 1).

Table 1. Serum DHEA-S Concentrations in Study Subjects before and during Oral Placebo or DHEA (1600 mg/day) Administration.

	Serum DHEA-S (μg/dl)	
Day	Placebo group (N=5)	DHEA group (N=5)
0	365 ± 56	403 ± 71
7	395 ± 98	1218 ± 281
14	404 ± 82	1463 ± 460
21	383 ± 84	1126 ± 343
28	320 ± 64	1431 ± 460

Serum androstenedione concentrations were similar in the placebo and DHEA groups on day 0 (1576 ± 71 vs 1225 ± 165 pg/ml, P=NS), and serum androstenedione did not change in the placebo group during the study. In the DHEA group, however, the mean serum androstenedione concentration increased 2-fold from 1225 ± 165 pg/ml to 2473 ± 333 pg/ml ($P<0.004$). Serum concentrations of estrone, estradiol, total testosterone, free testosterone, and sex hormone-binding globulin did not change in either the placebo or DHEA group.

Serum concentrations of high-density lipoprotein (HDL) cholesterol, very low density lipoprotein (VLDL) cholesterol, and triglycerides did not change in either the placebo or DHEA group. In the DHEA group, however, the mean serum total cholesterol concentration fell by 7.0% from 186 ± 8 mg/dl to

173 ± 11 mg/dl (P<0.05), whereas no change occurred in the placebo group. In the DHEA group, this fall in serum total cholesterol was due almost entirely to a 7.5% fall in serum low-density lipoprotein (LDL) cholesterol concentrations from 124 ± 4 mg/dl to 115 ± 5 mg/dl (P<0.01) (Fig. 2). Again, no change in serum LDL cholesterol levels occurred in the placebo group (Fig. 2).

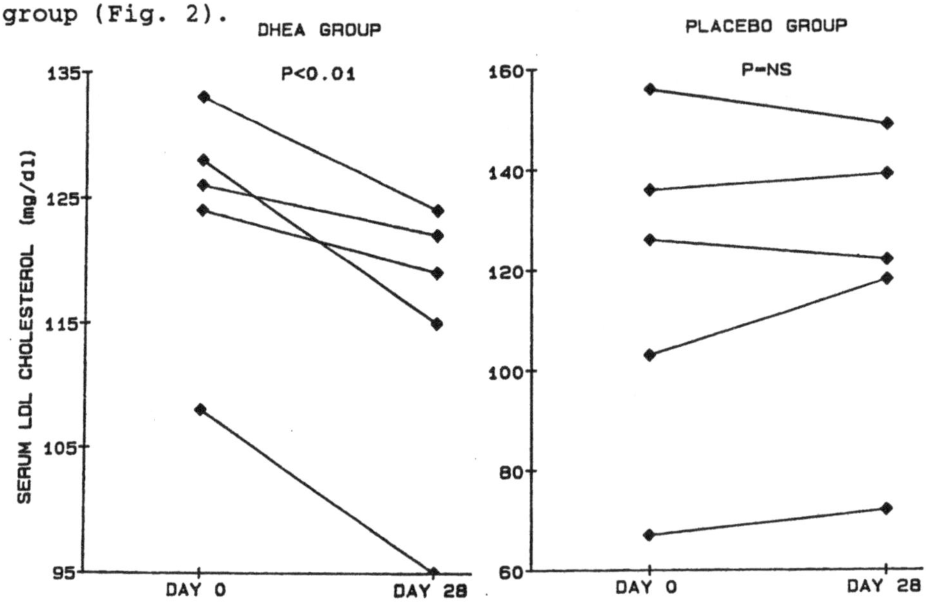

Fig. 2. Serum LDL Cholesterol Levels in Study Subjects before and after Oral Placebo or DHEA (1600 mg/day) Administration.

In addition, serum apolipoprotein B (apoB) levels fell in all 5 men who ingested DHEA, with a fall of 24.0% in the mean value (from 1229 ± 108 µg/ml to 934 ± 56 µg/ml; P<0.01) (Fig. 3) (38). In contrast, in the placebo group serum apoB rose in 2 men and fell in 3 men (Fig. 3), and no significant change in the mean value occurred. It should be noted, however, that in contrast to the LDL cholesterol results, the magnitude of apoB reduction between the DHEA and placebo groups was not statistically different.

These results suggest that DHEA may exert anti-atherogenic

effects by lowering both serum LDL cholesterol and apoB levels. In many studies serum apoB levels were found to be better predictors of the severity of coronary artery disease than levels of total cholesterol or LDL cholesterol (39-43), and an increased ratio of LDL apoB to LDL cholesterol appears to be present in patients with documented coronary disease and in children of individuals with premature myocardial infarctions and hyperapolipoproteinemia (44,45). The 3-fold greater fall in serum apoB than serum LDL cholesterol (24.0% versus 7.5%) which we observed in the DHEA group may explain the observation that DHEA can exert anti-atherogenic effects in some animal models despite no apparent reduction in serum cholesterol levels (6,7).

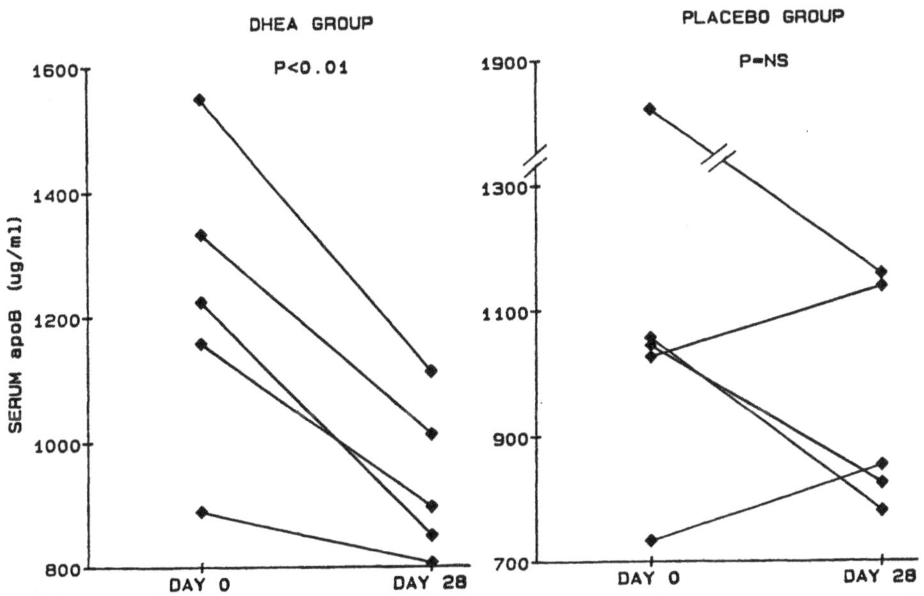

Fig. 3. Serum apoB Levels in Study Subjects before and after Oral Placebo or DHEA (1600 mg/day) Administration.

No statistically significant change in anthropometric measurements occurred in either the placebo or DHEA group (Table 2). DHEA administration, however, was associated with a reduction in percent body fat in four of the five men, as assessed by hydrostatic weighing (Table 2), with a mean decrease of 31%. The single man in the DHEA group who did not lose body fat was the leanest subject in the group (body mass index = 20.0 kg/m^2) and weighed about 90% of ideal body weight. These results suggest that the reduction in body fat mass by DHEA may in part depend upon preexisting adiposity, since the maximal reduction in body fat occurred in the most obese man, whereas the single man who did not lose body fat was the leanest. Since weights did not change in the DHEA group, it is likely that the reduction in body fat mass was coupled with an increase in muscle mass.

Table 2. Anthropometric Data in Study Subjects before and after Oral Placebo or DHEA (1600 mg/day) Administration.

	Placebo group (N=5)	
	Day 0	Day 28
Weight (kg):	75.7 ± 7.0	74.3 ± 7.0
%Body fat (%):	16.9 ± 3.5	16.7 ± 2.2
	DHEA group (N=5)	
	Day 0	Day 28
Weight (kg):	77.5 ± 3.4	78.2 ± 3.5
%Body fat (%):	15.9 ± 3.7	10.9 ± 1.2

Fasting serum glucose levels on day 0 and day 28 were similar in both the placebo group (89.6 ± 3.3 vs 98.0 ± 2.4 mg/dl; P=NS) and the DHEA group (87.8 ± 2.4 vs 91.6 ± 2.0 mg/dl; P=NS). Fasting serum insulin levels on day 0 and day 28 were also similar in both groups (placebo group, 16.6 ± 1.6 vs 12.7 ± 1.8 µU/ml [P=NS]; DHEA group, 14.6 ± 2.4 vs 12.6 ± 1.1 µU/ml [P=NS]).

Tissue sensitivity to insulin was assessed by the hyperinsulinemic-euglycemic clamp technique (46) during a 2-h infusion of insulin at a rate of 1.0 mU/kg·min. In the placebo group, tissue sensitivities to insulin on day 0 and day 28 were similar (7.1 ± 1.0 vs 7.9 ± 1.6 mg glucose/kg·min; P=NS) (Fig. 4). Likewise, tissue sensitivities to insulin were similar on day 0 and day 28 in the DHEA group (6.6 ± 0.9 vs 6.3 ± 1.0 mg glucose/kg·min; P=NS) (Fig. 4). The day 0 and day 28 steady-state insulin levels (during the 20- to 120-min time period of the clamp) in the placebo group (68.9 ± 5.4 vs 60.4 ± 3.2 µU/ml; P=NS) and DHEA group (64.4 ± 9.0 vs 70.3 ± 8.4 µU/ml; P=NS) were comparable.

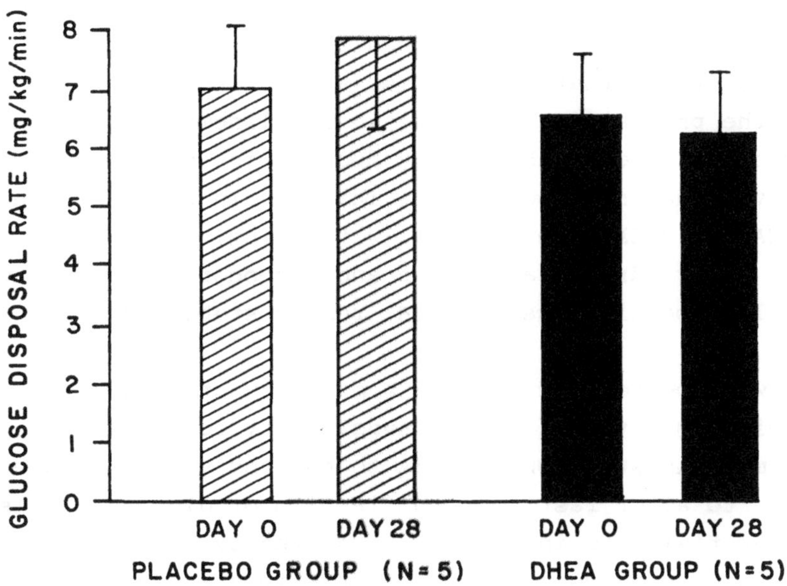

Fig. 4. Glucose Disposal Rates during Clamp Studies.

Contrary to our working hypothesis, DHEA administration had no effect on tissue sensitivity to insulin, unlike androgen therapy, which results in decreased tissue sensitivity to insulin (47). It should be emphasized, however, that these results do not exclude the possibility that DHEA administration might augment insulin sensitivity if administered for a longer duration or to a group of individuals with preexisting insulin resistance (e.g., obese Type II diabetic patients).

Summary

In summary, acute hyperinsulinemia results in a reduction in serum DHEA and DHEA-S levels in man, but the mechanism(s) subserving these reductions remains unknown. Since suprapharmacologic serum insulin levels were attained in our studies, the physiologic relevance of our findings is unclear. Nonetheless, it should be noted that other investigators, in a preliminary communication, have reported a similar decline in serum DHEA-S levels during physiologic elevations of serum insulin levels (48). Although speculative, the increased insulin resistance that accompanies aging -- and, consequently, the progressive increase in insulinemia -- may in some way contribute to the steady decline in serum DHEA and DHEA-S levels as one grows older. This may in part explain why serum DHEA and DHEA-S levels fall with aging, while serum levels of other adrenal androgens and glucocorticoids, as well as adrenal responsiveness to ACTH, remain stable (3).

Although we hypothesized that DHEA might function as a physiologic regulator of tissue sensitivity to insulin, short-term administration of DHEA to healthy young men does not appear to alter fasting serum insulin or glucose levels or tissue sensitivity to insulin (as measured by the insulin-glucose clamp technique). These observations argue against any effect of DHEA on glucose homeostasis in man, although studies in elderly individuals (whose basal DHEA/DHEA-S levels are low) or in individuals with diabetes mellitus (who are resistant to insulin) still need to be conducted.

Of our observations, the one that has engendered the most public interest is the association of DHEA administration with a reduction in body fat mass. This is not surprising, since obesity is such a common disorder and effective therapies remain elusive. We caution, however, that the results we observed were neither uniform nor statistically significant, and await confirmation. We suspect that even if DHEA does possess anti-obesity properties, this may be difficult to demonstrate conclusively, and further studies of ours in young individuals suggest that the intersubject variability in body fat mass reduction after DHEA administration may be great (unpublished observations). Some individuals may be obese precisely because they are resistant to DHEA's actions, and this resistance may not be surmountable by the administration of exogenous DHEA. Furthermore, although some individuals may be obese because their serum DHEA levels are low, it is likely that many individuals are obese for reasons unrelated to DHEA, and these individuals may well not respond to DHEA administration with a reduction in fat mass. A group of subjects in whom the effects of DHEA on fat mass might most easily be detected is the elderly, whose basal levels are low. It is probably this group that should be targeted for future human studies of DHEA's anti-obesity effects.

We believe that the most promising therapeutic effect of DHEA lies in its hypolipidemic and, possibly, anti-atherogenic properties. The reductions in serum LDL cholesterol and apoB levels which we observed are particularly striking, since young normocholesterolemic men were studied. Basal serum DHEA and DHEA-S concentrations in these subjects were presumably at peak levels. Animal studies now show a protective effect of DHEA against the development of aortic atherosclerosis (6,7) without a significant change in serum lipids, and a recently completed study in a heart-transplanted rabbit model of accelerated atherosclerosis demonstrated a 72% reduction in the development of coronary luminal stenosis in rabbits fed DHEA (8). Although a hypolipidemic effect of DHEA may have contributed to DHEA's protection against the development of coronary atherosclerosis in this study, we should remain open to the possibility that DHEA may exert anti-atherogenic effects through other mechanisms as well (e.g., interpolation into cellular membranes, alterations in the physico-chemical properties of lipoproteins, suppression of superoxide radical generation, etc.).

It is highly likely that a steroid produced as abundantly as DHEA, and whose metabolism is regulated so distinctly from other steroids, serves an important biologic function. Although numerous animal studies suggest such functions, their applicability to human physiology is uncertain. Over the past decade a renewed interest in DHEA has emerged, and this will hopefully result in an increasing number of human studies designed to delineate DHEA's metabolic effects in man and the possible role of DHEA in the aging process.

Acknowledgement

We thank Drs. Cornelius Barlascini, Keith Usiskin, and Devin Welty, and the staff of the General Clinical Research Center of the Medical College of Virginia for their assistance in performing the studies. We also thank Dr. Jerome Strauss III of the University of Pennsylvania for some of the steroid measurements, and Drs. Yadon Arad and Henry Ginsberg of Columbia University for the apoB determinations. These studies were supported by NIH grants DK18903 and RR00065, a Fellowship Grant from the Virginia Affiliate of the American Diabetes Association (K.S.U.), and a Feasibility Grant from the American Diabetes Association (J.E.N.).

References

1. Orentreich, N., J.L. Brind, R.L. Rizer and J.H. Vogelman. 1984. J. Clin. Endocrinol. Metab. 59,551-555.

2. Migeon, C.J., A.R. Keller, B. Lawrence and T.H. Shepard II. 1957. J. Clin. Endocrinol. Metab. 17,1051-1062.

3. Parker, L.N. 1989. Adrenal androgens in clinical medicine. Academic Press, San Diego. pp. 118-134.

4. Ben-David, M., S. Dikstein, G. Bismuth and F.G. Sulman. 1967. Proc. Soc. Exp. Biol. Med. 125,1136-1140.

5. Kurzman, I.D., E.G. MacEwen and A.L.M. Haffa. Int. J. Obesity (in press)

6. Gordon, G.B., D.E. Bush and H.F. Weisman. 1988. J. Clin. Invest. 82,712-720.

7. Arad, Y., J. Badimon, L. Badimon, W.C. Hembree and H.N. Ginsberg. 1989. Arteriosclerosis 9,159-166.

8. Eich, D.M., D.E. Johnson, J.E. Nestler, K. Daijin, M.L. Hess and A.S. Wechsler. Clin. Res. (in press)

9. Yen, T.Y., J.A. Allan, D.V. Pearson and J.M. Acton. 1977. Lipids 12,409-413.

10. Schwartz, A.G. 1979. Cancer Res. 39,1129-1132.

11. Schwartz, A.G., G.C. Hard, L.L. Pashko, M. Abou-Gharbia and D. Swern. 1981. Nutrition and Cancer 3,46-53.

12. Nyce, J.W., P.N. Magee, G.C. Hard and A.G. Schwartz. 1984. Carcinogenesis (Lond) 5,57-62.

13. Pashko, L.L. and A.G. Schwartz. 1983. J. Gerontol. 38, 8-12.

14. Cleary, M.P., R. Seidenstat, R.H. Tannen and A.G. Schwartz. 1982. Proc. Soc. Exp. Biol. Med. 171,276-284.

15. Coleman, D.L., E.H. Leiter and R.W. Schwizer. 1982. Diabetes 31,830-833.

16. Coleman, D.L., E.H. Leiter and N. Applezweig. 1984. Endocrinology 115,239-243.

17. Coleman, D.L., R.W. Schwizer and E.H. Leiter. 1984. Diabetes 33,26-32.

18. Coleman, D.L. 1985. Endocrinology 117,2279-2283.

19. Coleman, D.L. 1989. In: Proceedings of the Eighteenth Steenbock Symposium "Hormones, Thermogenesis, and Obesity," (H.A. Lardy and F. Stratman, eds.), Elsevier Science Publishers, New York. pp. 377-384.

20. Weindruch, R., G. McFeeters and R.L. Walford. 1984. Exp. Geront. 19,297-304.

21. Lucas, J.A., S.A. Ahmed, L.M. Casey and P.C. MacDonald. 1985. J. Clin. Invest. 75,2091-2093.

22. Tannen, R.H. and A.G. Schwartz. 1982. Fedn. Proc. 42,463.

23. Cleary, M.P., J. Billheimer and A.G. Schwartz. 1982. Gerontol. 22,136.

24. Cleary, M.P., A. Shepherd and B. Jenks. 1984. J. Nutr. 114,1242-1251.

25. Shepherd, A. and M.P. Cleary. 1984. Am. J. Physiol. 246,E123-E128.

26. Cleary, M.P., N. Fox, B. Lazin and J.T. Billheimer. 1985. Nutr. Res. 5,1247-1257.

27. Cleary, M.P., A. Shepherd, J. Zisk and A. Schwartz. 1983. Nutrition and Behavior 1,127-136.

28. Tagliaferro, A.R. and J.R. Davis. 1983. Fedn. Proc. 43,326.

29. Casazza, J.P., W.T. Schaffer and R.L. Veech. 1986. J. Nutr. 116,304-310.

30. Pashko, L.L., D.K. Fairman and A.G. Schwartz. 1986. J. Gerontol. 41,433-438.

31. Cleary, M.P., S.S. Hood, C. Chando, C.T. Hansen and J.T. Billheimer. 1984. Nutr. Res. 4,485-494.

32. Nestler, J.E., J.N. Clore, J.F. Strauss III and W.G. Blackard. 1987. J. Clin. Endocrinol. Metab. 64,180-184.

33. Schiebinger, R.J., G.P. Chrousos, G.B. Cutler and D.L. Loriaux. 1986. J. Clin. Endocrinol. Metab. 62,202-209.

34. Nestler, J.E., K.S. Usiskin, C.O. Barlascini, D.F. Welty, J.N. Clore and W.G. Blackard. 1989. J. Clin. Endocrinol. Metab. 69,1040-1046.

35. Parker Jr, C.R., E.R. Simpson, D.W. Bilheimer, K. Leveno, B.R. Carr and P.C. Macdonald. 1980. Science 208,512-514.

36. Barrett-Connor, E., K-T. Khaw and S.S.C. Yen. 1986. N. Engl. J. Med. 315,1519-1524.

37. Nestler, J.E., C.O. Barlascini, J.N. Clore and W.G. Blackard. 1988. J. Clin. Endocrinol. Metab. 66,57-61.

38. Nestler, J.E., C.O. Barlascini, J.N. Clore, W.G. Blackard, H.N. Ginsberg and Y. Arad. 1989. Clin. Res. 37,22A.

39. Sniderman, A., S. Shapiro, D. Marpole, B. Skinner, B. Teng and P.O. Kwiterovich Jr. 1980. Proc. Natl. Acad. Sci. USA 77, 604-608.

40. De Backer, G., M. Rosseneu and J.P. Deslypere. 1982. Atherosclerosis 42,197-203.

41. Schmidt, S.B., A.G. Wasserman, R.A. Muesing, S.E. Schlesselman, J.C. Larosa and A.M. Ross. 1985. Am. J. Cardiol. 55,1459-1462.

42. Kwiterovich Jr, P.O. and A.D. Sniderman. 1983. Prev. Med. 12,815-834.

43. Brunzell, J.D., A.D. Sniderman, J.J. Albers and P.O. Kwiterovich Jr. 1984. Arteriosclerosis 4,79-83.

44. Sniderman, A., B. Teng, J. Genest, K. Cianflone, S. Wacholder and P.O. Kwiterovich Jr. 1985. Am. J. Cardiol. 55,291-295.

45. Freedman, D.S., S.R. Srinivasan, C.L Shear, F.A. Franklin, L.S. Webber and G.S. Berenson. 1986. N. Eng. J. Med. 315, 721-726.

46. Rizza, R.A., L.J. Mandarino and J.E. Gerich. 1981. Am. J. Physiol. 240,E630-E639.

47. Mooradian, A.D., J.E. Morley and S.G. Korenman. 1987. Endocr. Rev. 8,1-28.

48. Diamond, M.P., D. Grainger, A.L. Gill, K. Zych, G. Knab and R.A. DeFronzo. 1989. In: Scientific Program and Abstracts for the 36th Annual Meeting of the Society for Gynecologic Investigation. p. 293.

THE ROLE OF DHEA IN OBESITY

Margot P. Cleary
The Hormel Institute, University of Minnesota
Austin, Minnesota 55912

Introduction

Interest in the possible role of dehydroepiandrosterone (DHEA) in the regulation of body weight dates back almost 30 years. In 1964, Šonka and coworkers (1) reported that they could detect no DHEA in urine samples of 27 of 32 elderly obese diabetic individuals. The remaining subjects had normal values. A second report indicated that obese people tended to excrete less DHEA in their urine than did normal weight subjects (2). Several other studies on metabolism of DHEA and DHEA-sulfate in obese subjects were also published (3-5). These authors speculated that lower levels or the absence of DHEA may have been associated with increased fat biosynthesis and activation of insulin due to a lack of inhibition of the enzyme, glucose-6-phosphate dehydrogenase (G6PD). Shortly before the above studies were published, Marks and Banks (6) had reported that DHEA inhibited this important enzyme of the pentose phosphate shunt. This finding was substantiated by other investigators who showed that the inhibition of G6PD by DHEA was unique to mammalian forms of the enzyme (7,8-12). Despite this interest in a possible interrelation of DHEA, G6PD activity and obesity, it was not until the latter half of the 1970's that studies in rodents began to appear that indicated possible pharmacological effects attributable to DHEA treatment, particularly with respect to changes in body weight.

Dehydroepiandrosterone (DHEA)

DHEA Treatment in Mice

In 1977, Yen et al. (13) reported that long-term treatment of lean and obese ($A^{vy/a}$) obese mice with DHEA [administered by intubation (500 mg/kg, 3x weekly)] lowered body weight gain in both male and female mice. Body weights of obese treated mice reached levels similar to those of lean mice. It was also reported that 0.2% DHEA in the diet resulted in lower body weight in obese male and female mice, but intubation with 150 mg/kg or intraperitoneal injection at 10 mg/kg of DHEA was only effective in obese female mice. A lower incorporation of ^3H_2O into total liver lipid was also reported in mice that were treated with DHEA. The most striking aspect of this work was that this effect of DHEA on body weight occurred independently of an alteration in food intake.

Other investigators have confirmed that DHEA administration results in lower body weight in mice (14-19). In these studies, DHEA was administered either by intubation or was included in the food. Lower fat pad weights were found to accompany changes in body weight in C3H DHEA-treated mice when compared to control mice (15). In obese diabetic mice treated with DHEA, lower blood sugar and insulin levels were reported (19). In some DHEA-treated diabetic mice strains, body weight changes were not found despite lower serum glucose and insulin levels (20-22). This appears to, in some cases, be attributable to elevations of food intake but in others to something inherent in the background strain of the mice. As in the earlier human studies cited above, there was speculation that the mechanism of this antiobesity action of DHEA was the inhibition of G6PD leading to lowered availability of NADPH and lowered rates of fatty acid synthesis. However, only one study presented any enzyme data (18). These authors did not interpret their findings to be supportive of this hypothesis, although they did report lowered erythrocyte G6PD activity in some treated mice.

Initial Studies in Rats

The first studies of DHEA administration to rats were undertaken to evaluate the effects of DHEA on G6PD activity. Willmer and Foster (23) reported that daily subcutaneous injections of DHEA at levels of either 2.5, 5 or 10 mg to rats that were fasted for 48 hours and then refed a high glucose diet for 48 hours resulted in lower hepatic G6PD activity relative to non-treated rats. Similar results were reported several years later by Tepperman et al. (24). In another study, lower liver G6PD activity was measured 2 hours after the rats received intraperitoneal injection of 200 mg/kg DHEA but not after 4 hours (25). Due to the short duration of these studies, there was no evaluation as to whether DHEA administration had any effect on body weight or fat.

Preliminary studies by Cleary et al. (26) were undertaken to address the question of whether DHEA treatment could alter body weight in rats. Young growing Sprague-Dawley rats intubated with DHEA (450 mg/kg, 3x weekly) from 21 to 79 days of age did not have changes in body weight. However, intraperitoneal injections at a dose of 50 mg/kg 3x weekly from 21-84 or 140-189 days of age resulted in significantly lower body weight in high-fat fed rats and in chow-fed Sprague-Dawley rats from the older group. No changes in body weights were found in the younger chow-fed rats injected with DHEA but these rats had significantly increased their food intake. Whether DHEA would lower body weight when included in the diet was also evaluated. When mixed into a powdered chow-type diet at 0.3% from 21 to 98 days of age and then at 0.4% from 99-161 days of age, DHEA treatment did not alter body weight gain in young rats. When DHEA was included in a high fat diet at these same doses, body weights of the treated rats were significantly lower than those of control rats despite an increase in food intake. In the rats treated with DHEA in the diet, liver G6PD activity was not affected in chow-fed DHEA rats but was lower in high fat-fed DHEA rats compared to control rats. Marked elevations of hepatic malic enzyme activity were found in both chow and high fat-fed rats treated with

DHEA. This confirmed an earlier observation of Tepperman et al. (24) that showed elevation of malic enzyme activity in response to short-term DHEA treatment. In general, DHEA treatment resulted in lower body weight in fatter, older rats while the younger, leaner rats did not respond to treatment.

In an additional study in young adult Sprague-Dawley male rats, the level of DHEA in the diet was raised to 0.6% (26). After 20 weeks, weight gain of the treated rats was 44% less than that of the control rats, while food intakes were essentially identical. Most of our subsequent studies have used this level and mode of administration for DHEA. Other researchers have also found that long-term DHEA treatment results in lower body weight in Sprague-Dawley rats (17,27,28).

Effect of DHEA Treatment on Body Weight in Obese Zucker Rats

The genetically obese Zucker rat has been used extensively as a model of juvenile-onset obesity. This is a homozygous recessive trait identified as the "fa" gene (29). Homozygous dominant and heterozygous rats are lean. These obese rats have hyperplastic-hypertrophic adipose depots (30), are hyperlipidemic (31,32) and with age develop marked insulin resistance (33,34). This type of obesity has been shown to be resistant to changes in body fat in response to food restriction. This results whether food restriction is initiated postweaning (35-37) or preweaning (38) and has been shown to occur even when combined pre- and postweaning food restriction is used (39).

The yellow obese mouse described earlier has a genetic obesity which is maturity-onset and results in hypertrophic adipose depots (i.e, bigger fat cells). These mice responded quite dramatically to DHEA treatment (13). However, Yen et al. (13) claimed that there was no effect of DHEA treatment on ob/ob mice with early onset hyperplastic-hypertrophic obesity similar to that of the obese

Zucker rats. It was therefore of interest to determine how obese Zucker rats would respond to DHEA administration.

Six-week-old lean and obese female Zucker rats were fed 0.6% DHEA in their diets for 15 weeks (40). Both lean and obese DHEA-treated rats had lower final body weights and rates of weight gain when compared to control rats (Figure 1A and B). Cumulative food intake in lean DHEA rats was similar to that of lean control rats (Figure 1C). Obese DHEA rats did have 20% lower food intake compared to obese control rats, but this was more than accounted for by the 40% lower body weight of the obese DHEA rats. When food efficiency ratios (body weight gain divided by food intake) were calculated, the lean and obese DHEA rats had similar values that were significantly lower than values calculated for the control rats (Figure 1D). This is an indication of the inefficient utilization of calories by DHEA-treated rats. An additional study using a pair-fed control group substantiated that decreased food intake does not account for the lower body weight of the DHEA-treated obese rats (41).

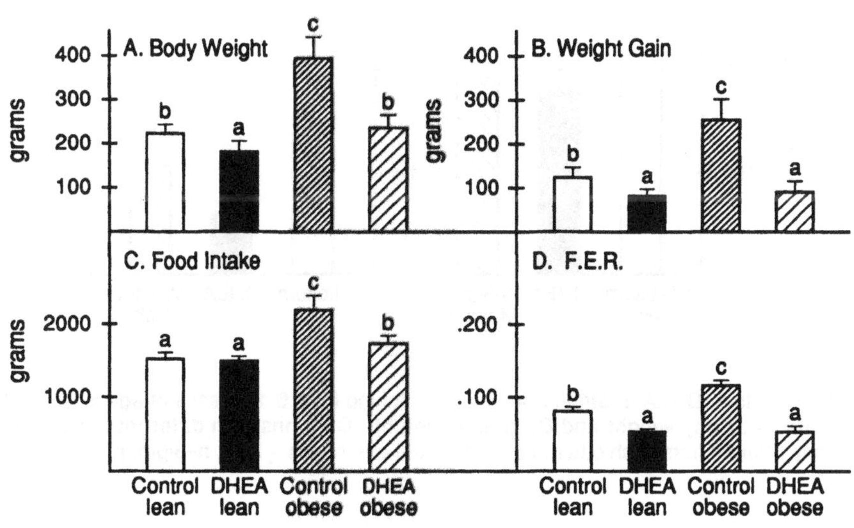

Figure 1. Effects of DHEA treatment from 6-21 weeks of age in lean and obese female Zucker rats. A, body weight; B, weight gain; C, food intake; D, F.E.R. (food efficiency ratio). Columns with different superscripts are significantly different from each other at $p<0.05$. Data are means $\pm$ S.D., n=11 for lean groups and n=7 for obese groups. Adapted from (40).

These changes in body weight were shown to be accompanied by less body fat and lower fat cell size and number in the DHEA-treated rats (40). Fat free mass and organ growth were preserved in DHEA-treated obese rats (40,41). This is a different response than that found in food-restricted obese rats who preserve body fat at the expense of lean tissue. Recently, we have shown that when obese rats were food restricted to the degree necessary to weight-match them to DHEA-treated obese rats (Figure 2A), fat pad weights of the weight-matched rats were substantially greater than those of the DHEA-treated rats (Figure 2B). In fact, the fat pad weights of the weight-matched rats were similar to those of ad libitum-fed rats. Thus, the results of both pair-feeding and weight-matching studies indicate that DHEA treatment is mediated through a mechanism separate from a decreased caloric intake.

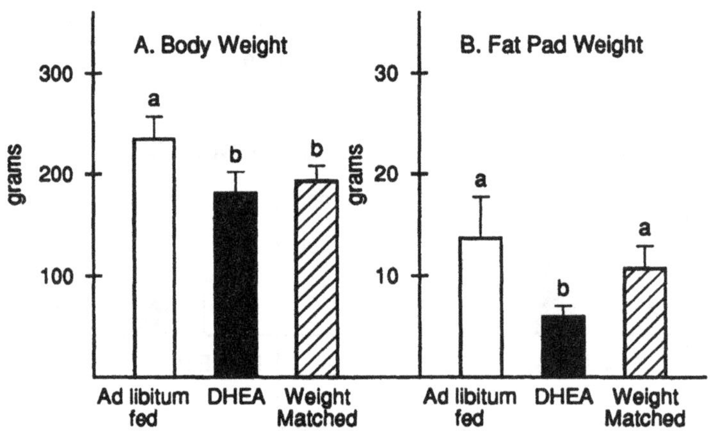

Figure 2. Effects of DHEA treatment or weight matching from 5-10 weeks of age in obese female Zucker rats. A, Body weight and B, fat pad weight. Columns with different superscripts are significantly different from each other at $p<0.05$. Data are means ± S.D., n=5/group.

Another group of obese rats was treated with DHEA as above from 6-21 weeks of age and then returned to a chow diet until 36 weeks of age. Following return to chow feeding, weight gain was greater in the formerly DHEA-treated rats than in never treated rats (40). However, body weights and fat pad weights of the formerly DHEA-treated rats did not reach levels of never treated rats as is characteristic of food-restricted rats following refeeding.

The effects of administering DHEA to adult obese Zucker rats were also investigated (42). Lean and obese female rats 8 months of age were treated for 6 months with DHEA at a level of either 0.6% or 1.0%. The control lean rats gained weight relative to their initial body weights (Table 1). Lean rats fed the 0.6% DHEA diet maintained their initial body weight, and the 1.0% DHEA lean group lost weight. The obese control rats gained weight while the two DHEA groups lost weight. Final body weights and fat pad weights were lower in both lean and obese DHEA-treated groups compared to their respective control group at the termination of the experiment. However, due to the overwhelming effects in the obese rats, differences for lean rats did not reach statistical significance.

It is unclear precisely how DHEA treatment results in lower body fat content. There could be either an inhibition of lipid deposition or an enhancement of lipid mobilization. Activities of adipose tissue enzymes involved in lipid metabolism were found to be lower in DHEA-treated obese Zucker rats compared to control obese rats (43). These enzymes included G6PD, fatty acid synthetase, malic enzyme, and lipoprotein lipase. These findings, however, could be due to the lower serum insulin levels that are associated with DHEA treatment (see next section). Lipolysis has not been found to be different in DHEA-treated versus control rats (44).

Table 1. Effect of DHEA Treatment from 8-14 Months of Age on Body Weight and Fat Pad Weight in Lean and Obese Female Zucker Rats[1,2]

	Initial body weight (g)	Final body weight (g)	Fat pad weight (g)
Control-lean	228±18[a,5]	240±13[a,*]	5.80±1.99[a]
0.6%[3]-lean	230±18[a]	226±26[a]	3.77±1.18[a]
1.0%[4]-lean	226±20[a]	202±11[a,**]	2.49±0.57[a]
Control-obese	461±38[b]	626±90[d,*]	73.40±15.97[c]
0.6%[3]-obese	454±42[b]	374±29[c,**]	29.40±6.18[b]
1.0%[4]-obese	470±52[b]	316±61[b,**]	25.16±4.79[b]

[1]Adapted from data presented in (42). [2]Data are mean ± S.D. (n=6/group). [3]DHEA included in the diet at a level of 0.6%. [4]DHEA included in the diet at a level of 1.0%. [5]Values within a column with different superscript letters are significantly different at $p<0.05$. [*]Indicates final body weight significantly greater than initial body weight. [**]Indicates final body weight significantly less than initial body weight.

Due to numerous metabolic effects of DHEA treatment on liver metabolism (see below), the lowering of serum insulin and the general consensus that overall energy utilization is altered, it appeared that the decreased adipose tissue mass found in treated rats was a consequence of an indirect response to DHEA. However, a recent study showed that addition of DHEA to the medium can interfere with the conversion of confluent cultured 3T3-L1 preadipocytes to lipid-filled adipocytes (45). This has been attributed to inhibition of G6PD (46) and therefore leaves open the possibility that DHEA treatment might have a specific effect on adipose tissue.

DHEA Treatment and Serum Insulin

The obese Zucker rat develops hyperinsulinemia shortly after weaning and eventually becomes insulin resistant. DHEA treatment of these rats has been shown to consistently result in lower serum insulin levels (42,43,47-50). Lower serum insulin levels have sometimes been measured in lean DHEA-treated Zucker rats. Recently, we have found lower serum insulin levels in DHEA-treated Sprague-Dawley rats (51). This occurred whether or not the rats were hyperinsulinemic and/or obese as a result of consuming a condensed milk-corn oil diet.

Decreased amounts of body fat and smaller fat cell size in conjunction with lower serum insulin led to the hypothesis that peripheral insulin resistance was improved in DHEA-treated obese rats. However, following 22 weeks of DHEA treatment (6-28 weeks of age), glucose metabolism in isolated adipocytes of obese Zucker rats was not improved in either the absence or presence of insulin (49). In fact, rates of conversion of glucose to CO_2 and fatty acids were significantly less in adipocytes from DHEA-treated obese rats than in those from either ad libitum-fed or pair-fed obese rats. In a more recent study (47), no differences in glucose metabolism were found in either isolated fat cells or muscle strips obtained from DHEA-treated obese rats after 5 weeks of treatment when compared to tissue from either control or weight-matched obese rats.

Pancreatic insulin content was found to be lower after long-term DHEA administration in both lean and obese rats (50). After 5 weeks of treatment, serum insulin was lower but not pancreatic insulin in obese rats (47). We have also recently shown (52) that after 5 weeks of treatment there were no differences in response to glucose tolerance tests between DHEA-treated and control obese rats. In light of these studies and Nestler's et al. (53) recent findings of no differences in glucose disposal in DHEA-treated compared to non-treated human subjects, it appears that DHEA treatment may either primarily effect secretion and/or synthesis of pancreatic insulin or enhance liver clearance of insulin. These possibilities are currently under investigation.

DHEA and Liver Metabolism

DHEA treatment has been shown to increase liver weight in rats on either an absolute and/or relative to body weight basis (26-28,40,41,54,55). Lower liver lipid (55-57) and glycogen levels (28) have been found in DHEA-treated rats, while hepatic DNA, RNA and protein content have been reported to be higher (40,41). There also appears to be an increase in protein associated with the hepatic mitochondrial fraction (52,58). Stereologic examination of electron micrographs also suggests an increase in mitochondrial number following DHEA treatment (M. P. Cleary, unpublished).

Activity of a number of hepatic enzymes of glycogen-glucose metabolism have been reported by Mayer et al. (28) to be lower after long-term DHEA treatment. These included glycogen phosphorylase a, glycogen phosphorylase a + b, hexokinase, glucokinase, pyruvate kinase and fructose 1,6-bisphosphatase as well as the previously discussed G6PD. Pyruvate carboxylase has also been reported to be lower in DHEA-treated rats (54). Activities of other liver enzymes are reported to be elevated by DHEA treatment; these include glucose-6-phosphatase (28), α-glucosidase (28), mitochondrial and cytosolic glycerol-3-phosphate dehydrogenase (59), cytosolic lactate dehydrogenase (59), isocitrate dehydrogenase (54) and peroxisomal and microsomal carnitine acetyl transferase (51).

As previously noted, malic enzyme activity has been found to be elevated in livers of DHEA-treated rats by a number of investigators (24,26,42,43,48,54). Recently, this increase has been shown to be due to stimulation of the rate of transcription of the gene for malic enzyme in the liver (60). This appeared to be specific to the liver, as no alterations of malic enzyme activity or transcription rates were found in brain, heart, kidney and testes (60).

Lower activities of several hepatic lipogenic enzymes, i.e., fatty acid synthetase and G6PD, have been measured following DHEA treatment in obese Zucker rats (43). However, as indicated earlier for Sprague-Dawley rats, malic enzyme activity was elevated in livers of both lean and obese Zucker rats treated with DHEA. In general, these findings and other observations in the literature [such as no effects on intermediates of the pentose pathway (54), no differences in the [$NADP^+$/NADPH] ratio (54,61) and no differences in the fate of radiolabeled glucose (44,62) in livers of DHEA-treated versus control rats] appear to question whether inhibition of G6PD can explain DHEA's antiobesity action. An evaluation of existing literature indicates that in long-term studies when DHEA treatment results in lower hepatic G6PD activity either lower body weight and/or body fat is also found (63). It is felt that G6PD inhibition is associated with DHEA's anticancer properties (64,65).

The activity of the enzyme, long-chain fatty acyl-Coenzyme A (LCFA-CoA) hydrolase, was found to be elevated in response to DHEA treatment (48). This led us to hypothesize that a futile cycle of deacylation/reacylation of fatty acids was induced by DHEA treatment. Recently, this was investigated in more detail (58). Following 3 days of DHEA treatment by intraperitoneal injection, obese Zucker rats had elevated LCFA-CoA hydrolase and synthetase activities indicating that such a cycle may, in fact, be induced. These findings were accompanied by higher levels of LCFA-CoA and CoA in livers of DHEA-treated rats. In addition, although no change in liver weight was present, higher yield of mitochondrial protein was obtained and elevated rates of state 3 mitochondrial respiration were measured using both succinate and glutamate-malate as substrates. This was found when respiration was expressed per milligram protein, per gram liver and per liver. These observations suggested a combination of enhanced ATP synthesis and induction of this futile cycle as a mechanism to explain DHEA's antiobesity action.

More recently completed work suggests that the elevation in state 3 mitochondrial respiration occurs within 3 hours of treatment prior to the elevation of LCFA-CoA hydrolase activity or increases in liver weight and mitochondrial protein content (66). Initially, i.e., after 3 and 12 hours of treatment, only glutamate-malate supported respiration was higher. After 1 and 3 days, both succinate and glutamate-malate supported state 3 respiration were elevated in mitochondria from treated rats. At these four time points, there was little change in state 4 respiration. When expressed per milligram protein, respiration rates return to "normal" after 7 days of treatment (58,65) and with longer treatment are actually lower than control values (51,52). However, due to the increases in mitochondrial protein, expression of respiration rates per gram or per liver results in higher values in DHEA-treated rats. A summary of these findings for DHEA-treated obese Zucker rats are presented in Table 2.

How these affects of DHEA on mitochondria are mediated is unknown. After both 1 and 7 days of treatment, a number of changes in fatty acid composition of mitochondrial phosphatidylcholine, phosphatidylethanolamine and cardiolipin were detected (66). The trend was towards a decrease in ω3 fatty acids and an increase in ω6 fatty acids with no change in total lipid content. Whether these alterations of fatty acid composition play a role in altering mitochondrial metabolism in DHEA-treated rats remains to be determined. That the ω6/ω3 ratio may play a role in mitochondrial respiration is supported by two recent studies. In these experiments, when menhaden oil was fed to rats, lower rates of mitochondrial respiration were reported (67,68). Presumably, this diet manipulation would elevate the ω3 fatty acids and lower the ω6 fatty acids, resulting in the opposite findings from those of DHEA-treated rats.

Table 2. Summary of the Effect of DHEA Treatment on Rates of Hepatic Mitochondrial Respiration, Mitochondrial Protein and Liver Weight in Obese Zucker Rats

	Length of treatment in days (mode of administration)						
	1/8 (intubation)[1]	1/2 (diet)[2]	1 (diet)[2]	3 (i.p.)[3]	7 (diet)[2]	7 (i.p.)[3]	35 (diet)[4]
State 3 Respiration							
per mg protein	↑[5](GM)[6]	↑(GM)	↑	↑	-	-	↓(GM)
per gram	↑(GM)	↑	↑	↑	↑	↑	↑
State 4 Respiration							
per mg protein	-	-	-	-	-	-	↓(GM)
per gram	-	↑	↑(GM)	-	↑	↑(S)[7]	↑
Mitochondrial protein							
per gram	-	↑	-	↑	↑	↑	↑
Liver weight	-	-	-	-	↑	↑	↑

[1]20 mg DHEA by intubation (65). [2]0.6% DHEA in diet (65). [3]50 mg/kg DHEA day 1 and 25 mg thereafter (58). [4]0.6% DHEA in diet (47). [5]↑ indicates significantly greater; ↓ indicates significantly lower; and - indicates no difference than control. [6](GM) indicates only found with glutamate-malate as substrate and succinate unchanged; [7](S) indicates only found with succinate as substrate and glutamate-malate unchanged; otherwise includes both glutamate-malate and succinate.

In order to study the mechanism of DHEA action on mitochondrial respiration in more detail, additional studies with isolated mitochondria were undertaken (69). However, the addition of DHEA to isolated mitochondria resulted in instantaneous inhibition of respiration with all substrates used except succinate. This was shown to be true for mitochondria from liver, adrenal gland, heart, brain, kidney and brown adipose tissue. These results confirmed findings presented in an earlier report by Šonka et al. (70). In general, these findings provided little help in understanding the in vivo response of DHEA.

DHEA Treatment in Rats with Diet-induced Obesity (DIO)

Most studies of DHEA treatment were done using either Sprague-Dawley rats or obese Zucker rats, although several other rat strains have also been used. Effects of DHEA administration in BHE rats have been undertaken in one long-term (71) and several short-term studies (56,57,72). When fed a high sucrose diet, BHE rats manifest numerous abnormalities of lipid and carbohydrate metabolism and have been reported to develop atherosclerosis, although they are not obese (73). In the one long-term study, markedly lower body weights were found in the DHEA-treated BHE rats compared to control BHE rats (71). In Wistar rats fed a semipurified diet with a high sucrose content, 2 weeks of DHEA treatment resulted in significantly lower body weight gain (55).

In addition to being of genetic origin, obesity can be the result of surgical or electrical damage to various parts of the brain. One such obese model produced from knife cuts of the medial hypothalamus has been used in conjunction with DHEA treatment (74). Body weight gain was lower in both DHEA-treated rats with knife cuts and in DHEA-treated rats with sham operations when compared to either non-treated knife cut or non-treated sham operated rats. Lower food intake was reported for the DHEA-treated rats and this was provided as an explanation for the compromised weight gain. However, no body fat measurements or calculations of energy utilization were made making it difficult to evaluate the results adequately.

A variety of dietary manipulations have been employed to induce obesity in rats. However, no DHEA treatment studies have been undertaken in rats made obese by such techniques except for one recent study (51). This diet manipulation involved feeding Sprague-Dawley rats commercial rodent food mixed with condensed milk and corn oil. These diet-induced obese (DIO) rats have been reported to have many changes in lipid and carbohydrate metabolism similar to those of obese Zucker rats and BHE rats (75-78). There are several interesting aspects of feeding this diet mixture to Sprague-Dawley rats. When fed to young animals, all rats became obese, without increasing their energy intake (77). In rats weighing over 400 g upon introduction of the diet, two populations of rats resulted (75,76). Based on assessment of body weight, approximately 50% of the rats gained more weight than chow-fed rats and were considered to be obese. The remaining rats showed resistance to the dietary manipulation by consuming less calories than either chow-fed rats or rats that became obese after consuming the condensed milk diet. Body weights of these diet-resistant (DR) rats were similar to those of chow-fed rats. DHEA treatment was investigated in both younger and older rats with this type of obesity (51).

In the first experiment, the condensed milk diet was fed beginning at 10 weeks of age when rats weighed ~300 g. After 8 weeks, the rats fed this diet weighed 15% more than chow-fed rats. DHEA at a level of 0.6% was then added to the diet of half of the DIO rats. Within 2 weeks, the DHEA-treated DIO rats lost weight and weighed the same as chow-fed rats. Their weight remained similar to those of chow-fed rats for the remaining 4 weeks of treatment. Final body weights are shown in Table 3. At the termination of the experiment, epididymal fat pad weights and cellularity were similar in DIO-DHEA and chow-fed rats, and these values were significantly lower than those of DIO rats (Table 3). Similar results were found for retroperitoneal fat pads (data not shown). Serum glucose was the same in all three groups while insulin was lower in DIO-DHEA rats compared to the other two groups (Table 3). Serum triacylglycerol and cholesterol levels were elevated in DIO rats compared to chow rats and were lowered by DHEA treatment to levels comparable to those of chow-fed rats (Table 3).

Table 3. Results of DHEA Treatment in Young Diet-Induced Obese Rats[1,2]

	Chow[3] (n=5)	DIO[4] (n=5)	DIO-DHEA[5] (n=7)
Final body weight (g)	513±56[a,6]	630±59[b]	517±50[a]
Epididymal fat pad weight (g)	8.54±2.57[a]	18.10±5.84[b]	8.41±2.38[a]
Fat cell number x 10^6	18.32±5.13[a]	26.53±7.01[b]	19.26±2.06[a]
Fat cell size (μg lipid/cell)	0.399±0.166[a]	0.561±0.127[b]	0.362±0.109[a]
Glucose (mg/dl)	144±14[a]	152±5[a]	132±18[a]
Insulin (μμ/ml)	76±23[b]	65±26[b]	42±20[a]
Cholesterol (mg/dl)	65±9[a]	86±14[b,*]	73±14[a,b,*]
Triacylglycerol (mg/dl)	139±54[a]	335±163[b]	115±59[a]

[1]From data presented in (51). [2]Data are means ± S.D. [3]Fed powdered chow-type diet from 10-23 weeks of age. [4]Fed chow-condensed milk-corn oil diet from 10-23 weeks of age. [5]Fed chow-condensed milk-corn oil diet from 10-17 weeks of age, then 0.6% DHEA added for remainder of experiment. [6]Across a row, values with different superscripts are significantly different from each other at $p<0.05$. [*]$p<0.1$ between these two groups.

A second study was conducted using rats weighing ~460 g at the start of the experiment (51) with the expectation that both DIO and DR rats could be produced by feeding the condensed milk diet. After 8 weeks on the condensed milk diet, DIO and DR groups were able to be identified based on body weight. DHEA was then fed for 2 weeks to one-half of the rats in these two groups as well as to one-half of a group of chow-fed rats. Body weights were lower in all groups treated with DHEA (Figure 3A). Food intake was not lower in DHEA-treated rats but was lower in DR rats compared to the other two groups, i.e., DIO or chow (Figure 3B). Consuming the condensed milk diet resulted in significant increases in epididymal and retroperitoneal fat pad weights and fat cell numbers in the DIO rats compared to the rats in the other two groups. Just 2 weeks of DHEA treatment tended to lower these values in DIO-DHEA rats when compared to DIO-control rats. There was little effect of DHEA treatment on adipose tissue cellularity in chow-DHEA or DR-DHEA groups compared to their respective control group.

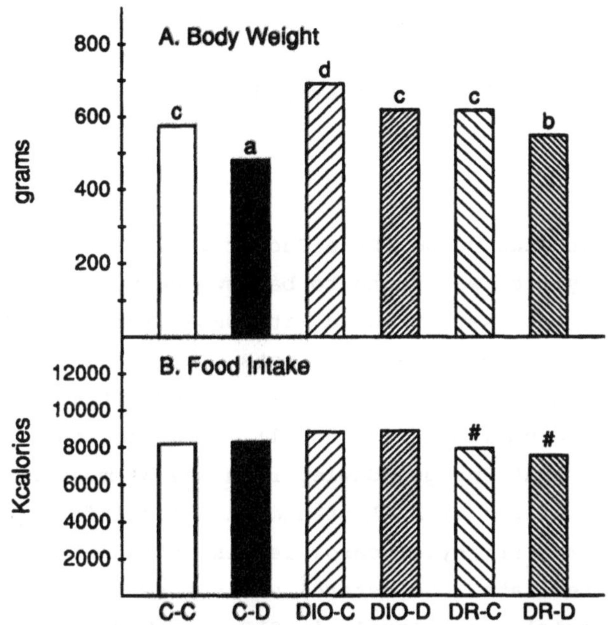

Figure 3. Effects of DHEA treatment in diet-induced obese and diet-resistant Sprague-Dawley rats. A, body weight and B, food intake. CC=chow-control, C-D=chow DHEA, DIO-C=diet-induced obese control, DIO-D=diet-induced obese DHEA, DR-C=diet-resistant control, DR-D=diet-resistant DHEA. Columns with different superscripts are significantly different from each other. # indicates DR rats' values significantly different from chow and DIO rats.

The combined results from these two experiments indicate that rats with diet-induced obesity have higher fat cell numbers than either chow-fed or DR rats. Whether this is due to filling of preexisting fat cells or proliferation of new cells or a combination of these two factors is not known. DHEA treatment appears to interfere with whatever this process is. Experiments using 3T3 preadipocytes have shown that addition of DHEA prevents differentiation of confluent cells to lipid-filled adipocytes (45,46). Other studies have shown that DHEA can lower proliferation rates of cells in culture and in skin and breast cells in intact mice (79,80).

Serum insulin levels were elevated in the DIO-control rats compared to both the chow-control and DR-control groups. DHEA treatment lowered insulin in all diet groups to values that were similar to each other and were lower than those of all the control rats. The DIO-control rats had higher serum cholesterol levels than all other groups. Thus, DIO-DHEA rats were the only group to show a response to treatment. Triacylglycerol levels responded in a manner similar to the insulin levels being elevated in DIO-control rats compared to chow-control rats while all DHEA-treated rats had values lower than did all control rats.

In previous studies using obese Zucker rats, serum cholesterol but not triacylglycerol was found to be lowered by DHEA treatment (48). In that case, rats were treated with DHEA from 12 to 18 months of age. Work with other strains of rats undergoing different nutritional manipulations have shown varied effects of DHEA treatment on serum lipids (27,55,81). The one consistent finding appears to be that Sprague-Dawley rats may be most responsive to a hypolipidemic effect of DHEA treatment. Tagliaferro et al. (27) reported lowered triacylglycerol levels in DHEA-treated Sprague-Dawley chow-fed rats (cholesterol was not reported). We have shown lowered triacylglycerol levels in DHEA-treated chow-fed as well as in condensed milk-fed Sprague-Dawley rats. The elevated serum cholesterol levels found in DIO Sprague-Dawley rats were lowered with DHEA treatment. These findings suggest that cholesterol levels may only be affected by DHEA when already elevated. DHEA treatment has been shown to lower serum cholesterol levels in both obese and lean dogs without effect on triacylglycerol (82) and has been reported to lower LDL levels in humans (53,83). In addition, an antiatherogenic effect of DHEA treatment in rabbits has been proposed (84,85). In general, these various findings suggest that there are genetic and nutritional interactions that need to be understood to establish whether DHEA may have a role as a hypolipidemic and/or antiatherogenic agent.

Conclusions

In the 30 years since the initial interest suggesting that DHEA may be involved in regulation of body weight, there have been few studies in humans undertaken to address this. Whether DHEA can be used to treat human obesity remains to be determined. A recent 28-day study using young male subjects showed no apparent negative effects of treatment and lowered body fat was found (53). DHEA treatment has also been shown to lower body weight in obese dogs (82). From a practical standpoint, application to wide scale human treatment would probably necessitate the development of analogues of DHEA that do not have estrogenic or androgenic potential.

Another aspect associated particularly with pharmacological treatment of obesity will be both the medical and social acceptance of at least some types of obesity as needing long-term therapy. This may involve nutritional, behavorial or pharmacological intervention or perhaps some combination of these. The risks of the metabolic consequences of obesity versus problems associated with drug therapy will need to be evaluated. The results to date of DHEA treatment in rodents indicate that genetics and nutrition and perhaps other metabolic factors interact to affect the degree to which DHEA treatment can alter body weight. Such findings may also occur in humans as their obese conditions can be and probably are due to a number of different factors, i.e., genetic, behavioral, environmental. Thus, we must be prepared for possible conflicting results from early human studies.

Regardless of whether DHEA has eventual application to treatment of human obesity, the recent findings of the effect of this compound in rodents demonstrates that body weight and fat changes can occur by altering an organism's metabolism. Thus, there are other means of altering the equation energy input = energy output than changing either food intake or level of physical activity.

Acknowledgements

This work was supported by NIH Grant HL08214 and the Hormel Foundation. I thank the many Drexel University students whose efforts contributed to the early DHEA studies. I also appreciate the scientific inputs of Drs. J. T. Billheimer, J. L. Sartin, T. Gansler, D. A. Diersen-Schade and A. G. Schwartz. Special thanks to Dr. P. F. Mohan for his excellent work over the past three years.

References

1. Šonka, J., Gregorová, I., Páv, J., and Škrha, F. 1964. Lancet ii, July 4, 44-45.

2. Šonka, J., and Gregorová, I. 1965. J. Physiol. (Paris) 56, 650-651.

3. Fehér, T., and Halmy, L. 1975. Can. J. Biochem. 53, 215-222.

4. Halmy, L. and Fehér, T. 1976. Endokrinologie 67, 184-191.

5. Hendrikx, A., Heyns, W., Steeno, O., and De Moor, P. 1965. Excerpta Med. (Amsterdam), 101, 63-70.

6. Marks, P.A., and Banks, J. 1960. Proc. Natl. Acad. Sci. USA 46, 447-452.

7. Oertel, G.W., and Benes, P. 1972. J. Steroid Biochem. 3, 493-496.

8. Beitner R., and Naor, Z.W. 1972. Biochim. Biophys. Acta 286, 437-440.

9. Ziboh, V.A., Dreize, M.A., and Hsai, S.L. 1970. J. Lipid Res. 11, 346-354.

10. Oertel, G.W., Menzel, P., and Bauke, D. 1970. Steroidologia 1, 2-7.

11. Levy, H.R. 1963. J. Biol. Chem. 238, 775-784.

12. McKerns, K.W., and Kaleita, E. 1960. Biochem. Biophys. Res. Commun. 2, 344-348.

13. Yen, T.T., Allan, J.A., Pearson, D.V., Acton, J.M., and Greenberg, M.M. 1977. Lipids 12, 409-413.

14. Schwartz, A.G. 1979. Cancer Res. 39, 1129-1132.

15. Cleary, M.P., Seidenstat, R., Tanner, R.T., and Schwartz, A.G. 1982. Proc. Soc. Exp. Biol. Med. 171, 276-284.

16. Weindruch, R., McFeetus, G., and Walford, R.L. 1984. Exp. Gerontol. 19, 297-304.

17. Pashko, L.L., Fairman, D.K., and Schwartz, A.G. 1986. J. Gerontol. 41, 433-438.

18. Granholm, N.H., Staber, L.D., and Wilkin, P.J. 1987. J. Exp. Zool. 242, 67-74.

19. Coleman, D.L. 1985. Endocrinology 117, 2179-2283.

20. Coleman, D.L., Schwizer, R.W., and Leiter, E.H. 1984. Diabetes 33, 26-32.

21. Coleman, D.L., Leiter, E.H., and Schwizer, R.W. 1982. Diabetes 31, 830-833.

22. Coleman, D.L., Leiter, E.H., and Applezweig, N. 1984. Endocrinology 115, 239-243.

23. Willmer, J.S., and Foster, T.S. 1965. Can. J. Biochem. 43, 1375-1377.

24. Tepperman, H.M., De La Garza, S.A., and Tepperman, J. 1968. Am. J. Physiol. 214, 1126-1132.

25. Hilgertová, J., and Šonka J. 1973. Horm. Metab. Res. 5, 286-289.

26. Cleary, M.P., Shepherd, A., Zisk, J., and Schwartz, A. 1983. Nutr. Behav. 1, 127-136.

27. Tagliaferro, A.R., Davis, J.R., Truchon, S., and Van Hamont, N. 1986. J. Nutr. 116, 1977-1983.

28. Mayer, D., Weber, E., Moore, M.A., Letsch, I., Filsinger, E., and Bannasch, P. 1988. Carcinogenesis 9, 2039-2043.

29. Zucker, L.M., and Zucker, T.F. 1961. J. Hered. 52, 275-278.

30. Johnson, P.R., Zucker, L.M., Cruce, J.A.F., and Hirsch, J. 1971. J. Lipid Res. 12, 706-714.

31. Zucker, T.F., and Zucker, L.M. 1962. Proc. Soc. Exp. Biol. Med. 110, 165-171.

32. Zucker, L.M. 1965. Ann. N.Y. Acad. Sci. 131, 447-458.

33. Zucker, L.M., and Antoniades, H.N. 1972. Endocrinology 90, 1320-1330.

34. Cushman, S.W., Zarnowski, M.J., Franzusoff, A.J., and Salans, L.B. 1978. Metabolism 27 (Suppl. 2), 1930-1940.

35. Zucker, L.M. 1967. J. Nutr. 91, 247-254.

36. Bray, G.A., York, D. A., and Severloff, R. S. 1973. Metabolism 22, 435-442.

37. Cleary, M.P., Muller, S., and Lanza-Jacoby, S. 1987. J. Nutr. 117, 353-360.

38. Johnson, P.R., Stern, J.S., Greenwood, M.R.C., Zucker, L.M., and Hirsch, J. 1973. J. Nutr. 103, 738-743.

39. Cleary, M.P., Vasselli, J.R., and Greenwood, M.R.C. 1980. Am. J. Physiol. 238, E284-E292.

40. Cleary, M.P., Shepherd, A., and Jenks, B. 1984. J. Nutr. 114, 1242-1251.

41. Cleary, M.P., Fox, N., Lazin, B., and Billheimer, J.T. 1985. Nutr. Res. 5, 1247-1257.

42. Cleary, M.P., and Zisk, J.F. 1986. Int. J. Obesity 10, 193-204.

43. Shepherd, A., and Cleary, M.P. 1984. Am. J. Physiol. 246, E123-E128.

44. Diersen-Schade, D.A., and Cleary, M.P. 1989. Horm. Metab. Res. 21, 356-358.

45. Gordon, G.B., Newitt, J.A., Shantz, L.M., Weng, D.E., and Talalay, P. 1986. Cancer Res. 46, 3389-3395.

46. Shantz, L.M., Talalay, P., and Gordon, G.B. 1989. Proc. Natl. Acad. Sci. USA 86, 3852-3856.

47. Cleary, M.P., Zabel, T., and Sartin, J.L. 1988. J. Nutr. 118, 382-387.

48. Cleary, M.P., Billheimer, J.T., Finan, A., Sartin, J.L., and Schwartz, A.G. 1984. Horm. Metab. Res. 16 (Suppl. 1), 43-46.

49. Muller, S., and Cleary, M.P. 1985. Metabolism 34, 278-284.

50. Gansler, T.S., Muller, S., and Cleary, M.P. 1985. Proc. Soc. Exp. Biol. Med. 180, 155-162.

51. Mohan, P.F., Inhen, J., Levin, B.E., and Cleary, M.P. (submitted manuscript).

52. Mohan, P.F., and Cleary, M.P. 1989. J. Nutr. 119, 496-501.

53. Nestler, J.E., Barlascini, C.O., Clone, J.N., and Blackard, W.G. 1988. J. Clin. Endocrinol. Metab. 66, 57-61.

54. Casazza, J.P., Schaffer, W.T., and Veech, R.L. 1986. J. Nutr. 116, 304-310.

55. Kritchevsky, D., Tepper, S.A., Klurfeld, D.M., and Schwartz, A.G. 1983. Pharmacol. Res. Commun. 15, 797-803.

56. McIntosh, M.K., and Berdanier, C.D. 1988. J. Nutr. 118, 1011-1017.

57. McIntosh, M.K., and Berdanier, C.D. 1988. Proc. Soc. Exp. Biol. Med. 187, 216-222.

58. Mohan, P.F., and Cleary, M.P. 1988. Am. J. Physiol. 255, E1-E8.

59. Lardy, H., Su, C-Y, Kneer, N., and Wielgus, S. 1989. In: Hormones, Thermogenesis and Obesity, (H. Lardy and F. Stratman, eds.), Elsevier, New York, pp. 415-426.

60. Song, M-K.H., Grieco, D., Rall, J.E., and Nikodem, V.M. 1989. J. Biol. Chem. 264, 18981-18985.

61. Garcea, R., Daino, L., Frassetto, S., Cozzolina, P., Ruggui, M. E., Vannini, M. G., Pascale, R., Lenzerini, L., Simile, M.M., Puddu, M., and Feo, F. 1988. Carcinogenesis 9, 931-938.

62. Finan, A.C., and Cleary, M.P. 1988. Nutr. Res. 8, 755-765.

63. Cleary, M.P. (1990) Int. J. Biochem. 22, in press.

64. Gordon, G.B., Shantz, L.M., and Talalay, P. 1987. Adv. Enzyme Regul. 26, 355-382.

65. Schwartz, A.G., Pashko, L., and Whitcomb, J.N. 1986. Toxicol. Pathol. 14, 357-362.

66. Mohan, P.F., and Cleary, M.P., submitted manuscript.

67. Kim, M-J. C., and Berdanier, C.D. 1989. Proc. Soc. Exp. Biol. 192, 172-176.

68. Mohan, P.F., Phillips, F., and Cleary, M.P., submitted manuscript.

69. Mohan, P.F., and Cleary, M.P. 1989. Int. J. Biochem. 21, 1103-1107.

70. Šonka, J., Hilgertova, J., Sevcik, J., and Kratochvil, O. 1978. Endrocrinol. Exp. 12, 165-168.

71. Cleary, M.P., Hood, S.S., Chando, C., Hansen, C.T., and Billheimer, J.T. 1984. Nutr. Res. 4, 485-494.

72. McIntosh, M.K., and Berdanier, C.D. 1987. Nutr. Rep. Int. 33, 181-188.

73. Berdanier, C.D. 1974. J. Nutr. 104, 1246-1256.

74. Gosnell, B.A. 1987. Physiol. Behav. 39, 687-691.

75. Levin, B.E., and Sullivan, A.C. 1988. Am. J. Physiol. 253, R475-R481.

76. Levin, B.E., Triscari, J., and Sullivan, A.C. 1983. Am. J. Physiol. 244, R347-R355.

77. Levin, B.E., Triscari, J., and Sullivan, A.C. 1986. Am. J. Physiol. 251, R433-R440.

78. Triscari, J., Nausse-Karol, C., Levin, B.E., and Sullivan, A.C. 1985. Metabolism 34, 580-587.

79. Pashko, L.L., Schwartz, A.G., Abou-Gharbia, M., and Sween, D. 1981. Carcinogenesis 2, 717-721.

80. Dworkin, C.R., Gorman, S.D., Pashko, L.L., Christafalo, V.J., and Schwartz, A.G. 1986. Life Sci. 38, 1451-1458.

81. Ben-David, M., Dikstein, S., Bismuth, G., and Sulman, F.G. 1967. Proc. Soc. Exp. Biol. Med. 125, 1136-1140.

82. MacEwen, E.G., Kurzman, I.D., and Haffa, A.L.M. 1989. In: Hormones, Thermogenesis and Obesity (H. Lardy and F. Stratman, eds.). Elsevier, New York, pp. 399-404.

83. Fassati, P., Fassati, M., Šonka, J., and Josifko, M. 1973. Agressologie 14, 259-268.

84. Gordon, G.B., Bush, D.F., and Weisman, H.F. 1988. J. Clin. Invest. 82, 712-720.

85. Arad, Y., Badimon, J.J., Badimon, L., Hembree, W.C., and Grensberg, H.N. 1989. Arteriosclerosis 9, 159-166.

DHEA AND MITOCHONDRIAL RESPIRATION

Carolyn D. Berdanier
Department of Foods and Nutrition, University of Georgia
Athens, Georgia 30602

Michael K. McIntosh
Department of Foods and Nutrition, University of North Carolina
Greensboro, North Carolina 27412-5001

Introduction

Mitochondria are thought to be the energy powerhouses of the cell. It is this organelle that produces the high energy nucleotide, ATP, as it uses oxygen to make water. Although some ATP is synthesized elsewhere as a result of phosphate group transfers to ADP (substrate phosphorylation), the majority of the ATP needed to run metabolic processes is synthesized in the mitochondria. Thus, if one is interested in the control of energy metabolism, or energy balance, one must also be interested in how mitochondrial metabolism is controlled. Hence, this chapter reviews the literature on mitochondrial metabolism as it relates to the effects of dehydroepiandrosterone (DHEA) on energy balance.

Mitochondrial Structure and Function

The number and size of mitochondria within the cell seem to be influenced by many factors: age, genotype, dietary factors, hormones, and nutritional state. However, within a given cell type, within the same animal, the number is relatively constant. For example, rat liver cells taken from a young, well nourished, non stressed, normal animal contain about 800 mitochondria/cell. These mitochondria are

Dehydroepiandrosterone (DHEA)

scattered throughout the cell and are in close proximity to other organelles that have a high ATP requirement. Thus, electronmicrographs of cell cross sections often show mitochondria in close proximity to the nucleus and ribosomes where protein synthesis is taking place. High rates of protein synthesis demand large amounts of ATP since every amino acid incorporated into the protein macromolecule must first be activated by ATP. Thus, if the protein being synthesized contains 300 amino acid residues, 300 molecules of ATP will be required. Similarly, one will find large numbers of mitochondria in muscle located in close proximity to the contracting muscle fibrils. The ATP molecules produced by the mitochondria need difuse only a very short distance to where they are used. Mitochondria are frequently located near fat droplets which serve as sources of metabolic fuel. Stored triacylglycerides are hydrolyzed to glycerol and fatty acids. The fatty acids are then transferred, via the acylcarnitine transferase mechanism, into the mitochondria and oxidized. There is a large body of literature on the importance of this process in regulating energy balance.

Specialized fat cells called brown fat are localized in pads under the sternum, in the subscapular area, and at the base of the neck. These fat pads have their characteristic dark color because they contain many more mitochondria and are more highly innervated and vascularized than white fat pads. They are also unique in that they produce a protein which uncouples respiration from ATP synthesis with the result of an increase in heat production by these tissues (1-5). In animals placed in cold environments this heat production allows the animal to maintain optimal body temperature. (1, 6-8). Brown fat thermogenesis as it is called, may also serve an important role in maintaining body weight. Some researchers (e.g. 6-8) have proposed that genetically obese animals are those animals that can not produce this uncoupling protein and therefore can not get rid of excess intake energy. They also propose that genetically lean individuals are lean because they can produce this uncoupling protein and thus can get rid of excess intake energy as heat. If the energy is lost as heat it is not available for use in muscle contraction, synthetic reactions, or for storage.

As mentioned, hepatocytes contain large numbers of mitochondria. Up to 20 percent of the cytoplasmic volume may be mitochondria. In the heart, with its high energy demand for muscle contraction, up to 50 percent of the cytoplasmic volume may be mitochondria.

Mitochondria are of many sizes and shapes. In the brown fat cell, they are spherical; in hepatocytes they are oval shaped; in the kidney they are cylindrical and in fibroblasts they are thread like. The organelle consists of two membranes. The outer membrane surrounds the inner one which is highly convoluted. Within the inner membrane is the matrix. This matrix can change in volume and organization as it changes its respiratory activity. The matrix contains, in addition to its enzymes, some DNA and ribosomes with which it synthesizes approximately one third of its complement of respiratory chain enzymes (9).

The membranes of the mitochondria differ not only in appearance but in function as well. The outer membrane is smooth and some what elastic whereas the inner membrane is, as mentioned, convoluted. These convolutions serve to increase the surface area of the inner membrane. The inner membrane is covered with regularly spaced spheres on stalks called inner membrane spheres. These spheres are not present on the outer membrane. The two membranes can be separated and their functional components studied. Table 1 lists some of the important components of these mitochondrial fractions.

Whereas the outer membrane is nearly 50% lipid, the inner membrane contains about 20% lipid. This lipid is phospholipid. The inner membrane is particularly rich in cardiolipin. These membranes are actually phospholipid bilayers in which proteins are embedded. Some of these proteins are dependent on the relative fluidity of this bilayer for their function whereas other proteins function equally well in sol and gel environments. The proteins of the electron

Table 1

Matrix[1]
Citric acid cycle enzymes
Fatty acid oxidizing enzymes
Carbamyl phosphate synthetase
Ornithine carbamoyl transferase
Pyruvate dehydrogenase complex
Inner membrane[2]
NADH dehydrogenases
Iron sulfur proteins
Cytochrome b, c, C_1, a_3
ATPase/ATP synthetase complex
Succinate dehydrogenase
β hydroxybutyrate dehydrogenase
Carnitine acyl transferase
creatine phosphate
Outer membrane[3]
Phospholipase A_2
Acyl Co A synthetase
NADH dehydrogenases
Monoamine oxidase

[1] More than 150 different proteins have been identified in the matrix fraction

[2] More than 60 different proteins have been identified in this membrane

[3] More than 100 different proteins have been found in this membrane

transport chain, or respiratory chain, seem to be examples of the latter whereas the proteins of the coupling complexes seem to be examples of the former.

Respiration

Respiration is the process by which aerobic cells obtain energy from the oxidation of fuel. Carbon dioxide and water plus the energy trapped in the ATP molecule or released as heat are the products of this process. Although the whole cell is involved in respiration, this review will only consider that portion of the process that occurs in the mitochondria and which is linked to the production of ATP.

Glycolysis as well as fatty acid and amino acid oxidation results in the production of an activated two carbon residue, acetyl CoA. This common metabolic intermediate is joined to oxalacetate to form citrate and is then processed by the citric acid cycle. This cyclic sequence of reactions is found in every cell type that possesses mitochondria. The cycle is catalyzed by a series of enzymes and yields reducing equivalents (H^+) and carbon dioxide. Reducing equivalents are produced when α ketoglutarate is produced from isocitrate, when α ketoglutarate is decarboxylated to produce succinyl CoA, when succinate is converted to fumarate, and when malate is converted to oxalacetate. These reducing equivalents, one pair for each step, are carried to the respiratory chain by way of either NAD or FAD. Reducing equivalents from succinate are carried by FAD while those produced by the oxidation of the other substrates are carried by NAD. Once oxalacetate is formed, it can then pick up another acetate group from acetyl CoA and begin the cycle once again by forming citrate. Thus, for every turn of the cycle, two carbon dioxide molecules and eight pair of reducing equivalents are produced. As long as there is sufficient oxalacetate to pick up the incoming acetate, the cycle will continue to turn and reducing equivalents will continue to be produced and these, in turn, will be joined with molecular oxygen to produce water, the end product (with carbon dioxide) of the catabolic process.

The above simplistic description of the citric acid cycle implies that it is free of controls and proceeds unhindered given adequate supplies of substrates, enzymes, and molecular oxygen. This is not true. There are numerous controls in place that regulate the cycle. However, it is outside the scope of this review to list and describe each one of them. Suffice it to say that the citric acid cycle provides reducing equivalents to the respiratory chain in a regulated manner so as to ensure the optimal utilization and conservation of the cell's fuel resources. It is important to note, however, that the citric acid cycle's activity produces the reducing equivalents needed by the respiratory chain and that the respiratory chain must transfer these hydrogen ions and their associated electrons to the oxygen ion to produce water. In doing so, it generates the electrochemical gradient necessary for the formation of the high energy bonds of ATP. Obviously, then, these processes, citric acid cycle, respiratory chain and ATP synthesis are regulated coordinately. Before this coordinate regulation is discussed, however, the respiratory chain and the mechanism for ATP synthesis must be described.

Respiratory Chain

The pairs of electrons produced at the four steps described above in the citric acid cycle as well as electrons transferred into the mitochondria via other processes are passed down the respiratory chain to the ultimate acceptor, molecular oxygen. The respiratory chain is shown in Figure 1.

The enzymes of the respiratory chain are particularly complex. They are embedded in the mitochondrial inner membrane and are difficult to extract and study. They catalyze the series of oxidation-reduction reactions which comprise the respiratory chain. Each reaction is characterized by a redox potential which can be calculated using the

Figure 1. Sequential Reactions of The Respiratory Chain (Abbreviated)

$NADH + H^+ + FMN \rightleftharpoons NAD^+ + FMNH_2$
$FMNH2 + 2Fe\ S(III) \leftarrow FMN + 2Fe\ S(II) + 2H^+$
$2Fe\ S(II) + 2H^+ + Q \rightleftharpoons 2Fe\ S(III) + QH_2$
QH_2 + 2 cytochrome b (III) $\rightleftharpoons$ Q + $2H^+$ + 2 cytochrome b (III)
2 cytochrome b (II) + 2 cytochrome c (III) $\rightleftharpoons$ 2 cytochrome b (III) + 2 cytochrome c (II)
2 cytochrome c (II) + 2 cytochrome a (III) $\rightleftharpoons$ 2 cytochrome c (III) + 2 cytochrome a (II)
2 cytochrome a (II) + 2 cytochrome a_3 (III) $\rightleftharpoons$ 2 cytochrome a (III) + 2 cytochrome a_3 (II)
2 cytochrome a_3 (II) + 1/2 O_2 + $2H^+$ $\rightleftharpoons$ 2 cytochrome a3 (III) + H_2O

Nernst equation:

$$E_h = E^1_0 + 2.30PRT \log n \frac{[\text{electron acceptor}]}{[\text{electron donor}]}$$

Where E_h=observed potential, E^1_0=standard redox potential (pH=7.0, T=25°, 1.0 M concentration), R=gas constant (8.31 J deg^{-1} mol^{-1}), T=Temperature (°K), N=Number of electrons being transferred, P=Faraday (23,062 cal, V^{-1}=96, 406 J V^{-1}).

The more positive the potential, the greater is the affinity of the negatively charged acceptor for the positively charged electrons. It is this potential that drives the respiratory chain reactions forward towards the formation of water. Each succeeding acceptor donor pair has a higher affinity for the electrons than the preceding pair until the point in the chain where water is formed. By comparison to the preceding pairs, water itself has little tendency to give up its electrons and thus, the chain reaction stops. Of course, water does not accumulate in the mitochondria; it leaves as quickly as it is formed and thus does not feed back to inhibit the chain. If it did accumulate it would change the concentration of the chain components diluting them. When water accumulates, the mitochondrion swells; if in excess, the mitochondrion bursts and dies. Obviously, then, water must exit the organelle as soon as it is formed so as to maintain optimal osmotic conditions.

Oxidative Phosphorylation

As electrons pass down the respiratory chain, ATP is synthesized. This process is called oxidative phosphorylation. It is here that scientists disagree as to how the respiratory chain is linked to the synthesis of ATP. Several mechanisms have been proposed and discussed at length (10). Perhaps the best accepted mechanism is that proposed by Mitchell (11) in 1961. Mitchell proposed that the reactions of the respiratory chain do not form (as was proposed by others) a common chemical intermediate. Rather, he proposed that the sequential reactions of the respiratory chain generate an electrochemical gradient of H^+ ions across the inner mitochondrial membrane. According to Mitchell, this gradient serves as the means for coupling the energy flow from electron transport to the formation of ATP. In this proposal, the inner mitochondrial membrane is an integral part of the mechanism which couples energy gradient of the electron flow in the respiratory chain to the synthesis of ATP. A priori, the membrane must be intact and in the form of a continuous closed vesicle for oxidative phosphorylation to take place. If the membrane is disrupted, coupling will not occur. Respiration may occur but ATP synthesis will not. Mitchell proposed that the enzymes of the respiratory chain are arranged so as to transport hydrogen ions from the matrix across the inner membrane such that the gradients will develop in close proximity to the Fo F_1 ATPase complex and provide a proton gradient sufficient to drive ATP synthesis by causing a dehydration of ADP and Pi.

Functionally, the various enzymes of oxidative phosphorylation can be divided into five complexes. Complexes I, II, III, and IV plus ubiquinone (Q) and cytochrome c make up the respiratory chain. The electron carriers are the quinoid structures (FMN, FAD, Q) and the transition metal complexes. At three stages in the chain, oxidative energy is conserved (Figure 2) via coupled vectorial proton translocations and creation of a membrane proton gradient ($\Delta\mu H^+$). This protonic energy is what drives the synthesis of ATP. Should this energy be dissipated as happens when agents such as dinitrophenol are present, ATP synthesis does not occur or occurs to a very limited extent. Uncoupling occurs because protons are transported through the membrane from the intramembrane space to the matrix. This "short circuits" the normal flow of protons through complex 5. In such a circumstance electrons continue down the respiratory chain but the protonic energy thus generated is not captured in the high energy phosphate bond. Instead, this energy is released as heat. There are variations or degrees of protonic energy capture in vivo which are related to the presence or graded absence of naturally occurring agents which serve to either inhibit the action of complex V (13) or uncouple the respiratory chain energy generation process from the energy capture process of ATP synthesis. As mentioned, some brown fat cell mitochondria have the ability to synthesize a protein which functions as an uncoupler of oxidative phosphorylation. It is reasonable to assume that other cell types such as hepatocytes or muscle cells may also have mechanisms that control the tightness or looseness of coupling. Naturally occurring substances such as long chain fatty acids have some role in this respect. Otto and Ontko (14) for example, showed that the influx of long chain fatty acids into isolated mitochondria resulted in a mild swelling of the organelle and a reduction in the amount of ATP produced. Williams et al as well as others (15-22) reported a change (decrease) in the ratio of ATP produced to oxygen consumed in rats fed hydrogenated coconut oil or fat free diets. These reports indicated that the fatty acids in the mitochondrial membrane (which reflected somewhat the fatty acids of the diet) changed the degree to which the protonic energy developed by the respiratory chain could be captured in the synthesis

Figure 2. Complexes in the oxidative phosphorylation scheme and sites at which sufficient energy is generated to synthesize ATP

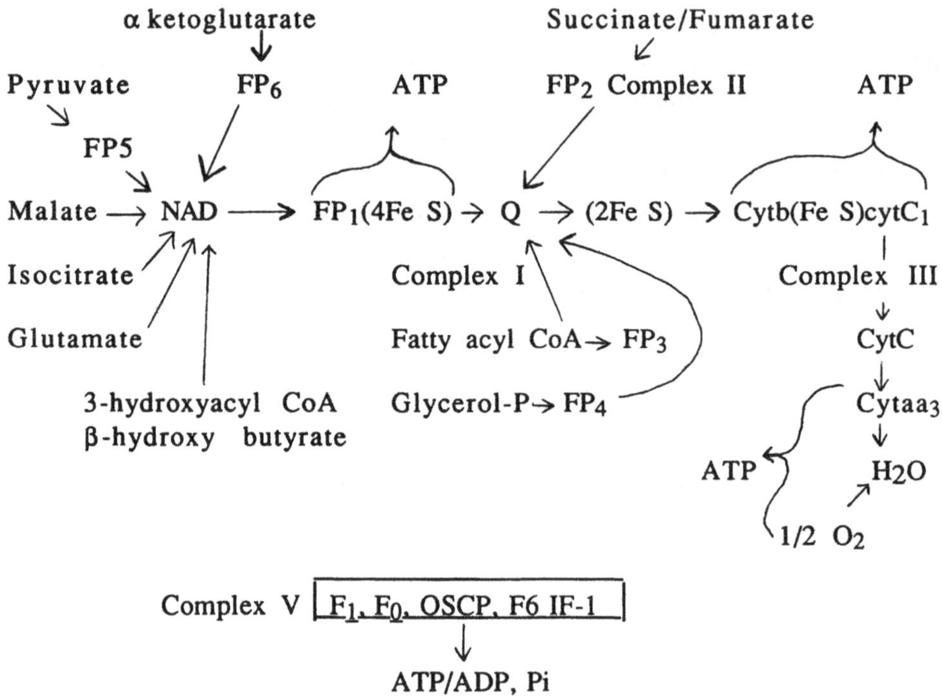

Complex I consists of all the NAD linked reactions which donate electrons to the first series of iron sulfur centers of the chain. Complex II consists of all the FAD linked reactions which donate electrons via ubiquinone (Q).
Complex III consists of cytochrome b_{562}, b_{566}, iron sulfur to cytochrome C_1 electron transfers.
Complex IV consists of the cytochrome a to a_3 electron transfers.
Complex V is the ATP synthase/ATPase complex. OSCP is the oligomycin sensitive carrier protein(s). The F_1 consists of 5 nonidentical protein subunits and is water soluble. The F_0 consists of 3-4 subunits and is hydrophobic. F_6 is one of these subunits whose activity depends on to phospholipid environment.

of ATP. That which was not so captured, was released as heat. Hence, animals fed diets devoid of essential fatty acids evidence a reduction in fuel efficiency (the gain in weight related to the amount of food consumed) and, although not reported, these rats should also evidence an increase in heat production.

Other naturally occuring compounds, e.g. hormones and nutrients aside from fatty acids (23), also influence mitochondrial respiration and ATP synthesis. The thyroid hormones, glucocorticoids, sex hormones, catecholamines, insulin, glucagon, parathyroid hormone, and growth hormone have all been studied. The literature is vast. Some of these hormones have direct effects on mitochondrial respiration and coupling through affecting the synthesis and activation of the various protein constituents of the five complexes. Other hormones have their effects on the exchange of divalent ions, notably calcium, and/or on the phospholipid fatty acid composition of the inner mitochondrial membrane. Some hormones, thyroid hormones for example, influence enzyme synthesis/activation, calcium flux, and membrane lipid composition. Clearly, Mother Nature intended to create a very carefully regulated system with checks and balances to ensure the continuity of life. Without this careful control, survival during times of energy stress could not be assured. Of interest in this review are the effects of the sex steroid hormone intermediate, DHEA, on mitochondrial metabolism.

Steroid Effects on Mitochondria

Knowledge about the effects of steroid hormones on mitochondrial function has been accumulating over the last thirty years. Litwack and Singer (24) reviewed the literature up to 1972 on glucocorticoids. They reported that injections of this hormone resulted in mitochondrial swelling and decreased respiratory activity by isolated hepatic mitochondria. Kimura and Rasmussen (25) reported that the administration of dexamethasone, a synthetic glucocorticoid, to rats markedly reduced calcium ion uptake and increased calcium release by subsequently isolated hepatic mitochondria. Mitochondria from glucocorticoid treated rats translocated ATP from inside to outside mitochondria at a faster rate than mitochondria from control rats. Kimura and Rasmussen suggested that the decrease in calcium transport and retention was not due to an alteration in the calcium carrier per se but was due to an effect of the hormone on the regulation of ATP synthesis and transport. It was not clear from this paper how this steroid hormone affected ATP synthesis. Subsequently, Allan et al (26) reported that dexamethasone treatment resulted in an increase in oxidative capacity with decreased ATP synthesis. This meant that dexamethasone treatment served to induce a degree of uncoupling of respiration from ATP synthesis. This was characterized by an increase in the uncoupler-dependent ATPase activity with no change in total ATPase activity. As mentioned earlier, the ATP synthetase/ATPase is a multiprotein complex having multiple internal and external controls. Dexamethasone may function to stimulate the dissipation of the proton gradient, may selectively stimulate the synthesis of certain mitochondrial proteins and/or influence the flux of divalent ions notably Ca^{++} which would activate or inactivate complex V.

The adrenal glands produce not only the glucocorticoids and aldosterone but also small amounts of DHEA, an intermediate in sex steroid synthesis. The gonads also release this metabolite to the circulation. Just as the glucocorticoids affect mitochondrial respiration and ATP synthesis, so too does DHEA. However, the mechanism by which DHEA has this effect is not known.

One of the first reports of an effect of DHEA on mitochondrial activity appeared in an extensive monograph by Sonka et al (27) in 1976. He reported that respiration by isolated heart or liver mitochondria was inhibited as the level of DHEA in the incubation media rose. Significant inhibition of respiration was also observed when progesterone, etiocholanolone or testosterone were added to the media. If DHEA-sulfate or cholesterol were added, no inhibition occurred. The inhibition was reported to occur whenever site I substrates (pyruvate, α ketoglutarate, or citrate) were used but not when a site II substrate (succinate) was used. DHEA has been reported to inhibit glucose 6 phosphate dehydrogenase in vivo so perhaps it is acting on the dehydrogenases which comprise site 1 or complex I of the respiratory chain. Unfortunately, the report of Sonka was one which covered many aspects of the metabolic effects of DHEA on metabolism and which omitted the details of these in vitro studies. Subsequently, Mohan and Cleary (28) reported that short term DHEA administration (3 or 7 days of daily injections) resulted in a stimulation of succinate supported and glutamate-malate supported respiration in hepatic tissue of genetically obese rats. In lean rats, DHEA treatment for three days was without effect; but after seven days of treatment, respiration was stimulated in these rats as well. These in vivo results are in contrast to those in vitro results reported by Sonka (27) and in contrast to those reported for BHE rats by Berdanier and McIntosh (29,30). In the latter papers, state 4 pyruvate, and α ketoglutarate supported respiration and state 3 α ketoglutarate and succinate supported respiration were less in DHEA treated rats than in control rats. The difference in the results reported by Mohan and Cleary and by Berdanier and McIntosh and Sonka might have been due to the strain of rat used, and to the amount, duration, and mode of administration of DHEA. Strain differences in response to a single DHEA dose level as well as differences due to dose level have been reported (31).

The differences in mitochondrial respiration might be due to inherent differences between in vivo and in vitro studies. Mohan and Cleary (32) examined the in vitro effects of DHEA on mitochondria isolated from the adrenals, heart, kidneys, brain, liver, and brown adipose tissue. Like Sonka (27), they found that, in vitro, DHEA inhibited site 1 substrate supported respiration but was without effect on succinate supported respiration. Also, like Sonka, we have found that in vitro DHEA, but not DHEA sulfate (DHEA-S), estrone, progesterone, or testosterone inhibited site 1 substrate supported respiration in a dose dependent manner while having little effect on succinate supported respiration (Figure 3).

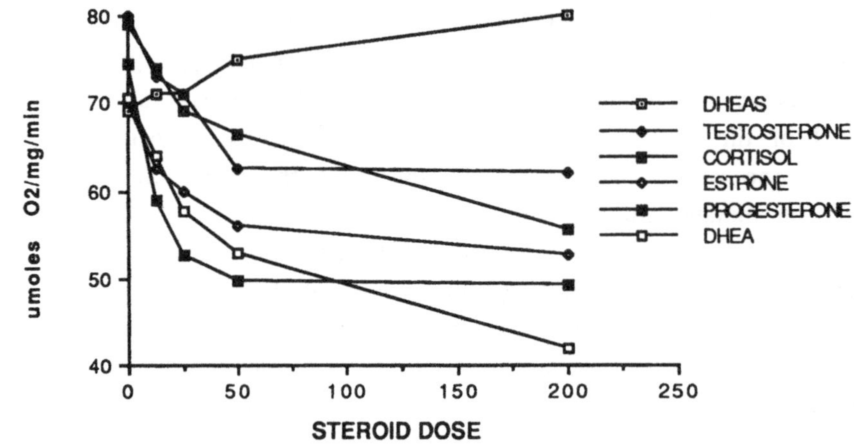

Figure 3. Effect of Steroid Dose on State 3 Malate + Pyruvate Supported Respiration

---

The observed differences in respiration due to DHEA treatment between in vivo and in vitro experiments may be attributed to the presence (in vivo) or absence (in vitro) of the sex steroid sulfotransferases (ST) and sulfohydrolases (STS). These cytosolic enzymes are responsible for either the deactivation (by sulfurylation) or activation (by sulfur hydrolysis) of androgens and estrogens in nongonadal and nonadrenal tissues such as the liver (35). According to Leiter (35), sulfurylation of androgens and estrogens by ST enzymes removes their receptor activity whereas this activity can be restored via hydrolysis of sulfur group by STS enzymes. Methods used to isolate mitochondria undoubtedly remove these important enzymes, removing the heptatocytes ability to regulate androgen and estrogen activity. This would explain why McIntosh and Berdanier, Sonka, and Mohan and Cleary were able to demonstrate that although DHEA inhibited mitochondrial respiration in vitro in a dose dependent manner when site I substrates were used, DHEA-S had no inhibitory effects on respiration. Without the cytosolic STS enzymes, DHEA-S added to isolated mitochondria would be inactive and unable to affect respiration. The in vitro environment would also be devoid of ST enzymes. Without the ability to deactivate androgen and estrogen via sulfurylation, metabolically active steroids that are added to isolated mitochondria escape potential deactivation and carry out their inhibitory effects on respiration without regulation by ST enzymes normally present in intact hepatocytes. Variations in the levels of hepatic ST and STS enzymes may also help explain differences in the in vivo metabolism of androgens and estrogens between males and females and also between genetically lean and obese and between genetically nondiabetic and diabetic subjects (35). The antiobesity effects of DHEA, including DHEA's effects on mitochondrial respiration, may therefore be dependent not only on the amount, duration, and mode of administration of the steroid, but also on the cytosolic levels of the enzymes that activate or deactivate this preandrogenic hormone.

What do these observations mean with respect to understanding how DHEA affects metabolism? First, recall that measurements of respiration are simply measurements of the amount of oxygen consumed by isolated mitochondria when stimulated by the addition of small amounts of ADP. The assumption is that respiration will be quite active until all the added ADP is phosphorylated to ATP. The mitochondria will then return to the prestimulated resting state. The measurement of oxygen consumption presumes that all of the ADP is phosphorylated but does not measure the ATP produced. If some uncoupling of respiration had occurred or if some internal ATP/ADP recycling had occurred, oxygen consumption might increase without a related increase to ATP production. If that were the case, then one might expect an increase in heat production or effects on metabolic pathways, such as protein synthesis or substrate cycling, which depend on ATP. Tagliaferro et al (33) have reported that DHEA treatment resulted in an increase in heat production and hepatic and peripheral protein synthesis. Leighton et al (34) hypothesized that DHEA increases heat production and energy wasting by increasing hepatic peroxisomal beta oxidation of fatty acids which results in the expenditure of chemical energy as heat. Berdanier and McIntosh (30) found an increase in whole body oxygen consumption in DHEA-treated rats even though hepatic malate+pyruvate-supported mitochondrial respiration was lower. These data would indicate that DHEA may reduce the efficiency of energy trapping by intact mitochondria, resulting in less ATP synthesized per molecule of oxygen consumed.

The in vitro inhibitory effects of DHEA on mitochondrial respiration when site 1 NADH-linked substrates (pyruvate, malate, isocitrate, glutamate) but not site 2 FADH-linked substrates (succinate, glycerol, fatty acylCoA) (see Figure 2) demonstrated by us and by Sonka and Mohan and Cleary would indicate that DHEA may be directly inhibiting one of the two NADH dehydrogenase enzymes which are vital constituents of Complex I of mitochondrial respiration. NADH ubiquinone reductase and NADH ferricyanide reductase are the two initial dehydrogenases in the electron transport chain that transfer electrons from reducing equivalents (NADH, NADPH) generated either locally in the mitochondria primarily from the TCA cycle or those that are shuttled in from cytosolic reactions via the malate-aspartate shuttle and the α glycerophosphate shuttle. Mohan and Cleary (32) demonstrated that 100 μM DHEA inhibited the malate-aspartate shuttle without having any effects on the alpha glycerol phosphate shuttle. NADH ubiquinone reductase can be inhibited by rotenoids, piericidin A, amytal, seconal, and demoral and therefore possibly by DHEA as well. The direct inhibition of these NADH dehydrogenases by DHEA requires further investigation.

Another possible indirect mechanism by which DHEA may lower mitochondrial respiration would be by reducing the availability of NADH and NADPH within the mitochondria, thereby limiting the supply of reducing equivalents for entry into the electron transport chain at site 1. Berdanier and McIntosh (29-31) and Cleary et al (36-38) have shown that DHEA-treated rats have lower glucose 6 phosphate dehydrogenase (G6PD) activities than control rats, thereby limiting pentose shunt activity and the availability of NADPH reducing equivalents not only for fatty acid synthesis, but also for entry into the respiratory chain via mitochondrial transhydrogenation at Complex I. This DHEA-mediated reduction in cytosolic reducing equivalents could contribute to the lower state 3 respiration rates observed in DHEA-treated rats.

Another possible explanation for a reduction in NADH and NADPH reducing equivalents entry into the electron transport chain at site 1 could be an increased microsomal activity associated with DHEA treatment. Detoxification by microsomal P450s requires a large supply of reducing equivalents for drug inactivation. This phenomena would not only reduce the amount of NADPH available for Complex I electron transfer, but could also reduce the flow of NADH reducing equivalents produced during glycolysis from the cytosol to the mitochondria via the malate-aspartate shuttle. Indeed, ATP-dependent transhydrogenases can reduce NADP+ with NADH to replenish microsomal NADPH during drug detoxification. This partitioning of reducing equivalents away from mitochondrial oxidative phosphorylation may partially account for DHEA's inhibition of the NADH-linked malate-aspartate shuttle but not the alpha glycerol phosphate shuttle reported by Mohan and Cleary (32). Substrate cycling has also been investigated as a mechanism by which DHEA reduces energy efficiency, promoting energy wasting in the cell. Mohan and Cleary (28) demonstrated an increase in the activities of fatty acid CoA hydrolase and synthetase activities in hepatocyte homogenates and mitochondria from DHEA-treated rats. They hypothesized that this is a futile cycle which promotes energy wasting and contributes to DHEA's antiobesity effects by partitioning energy away from storage and towards oxidation. A futile cycle of fatty acid synthesis and degradation would not only require the input of ATP, but it would also require a supply of reducing equivalents, thereby limiting the supply of NADH and NADPH for entry into Complex I of oxidative phosphorylation.

Protein synthesis, ATP synthesis, ATP/ADP cycling, substrate cycling, sex steroid ST and STA activities, Complex I NADH dehydrogenase activities, reducing equivalent shuttling, and other aspects of metabolic control need to be examined as we try to understand how DHEA can have such diverse effects on different genotypes, cell types, and cell functions. Surely there must be some underlying mechanism of action that will place all of these diverse reports in an understandable order concerning DHEA's effects on mitochondrial respiration and energy partitioning.

References

1. Nicholls, D.G., R.M. Locke. 1984. Thermogenic mechanisms in brown fat. Physiol.Rev. 64, 1-64.

2. Sundin, U., B. Cannon. 1980. GDP-binding in brown fat mitochondria of developing and cold adapted rat. Comp.Biochem.Physiol. 65B, 463-471.

3. Kopecky, J., F. Juerriri, P. Jezek, Z. Drahota, J. Houstek. 1984. Molecular mechanism of uncoupling in brown adipose tissue mitochondria. FEBS 170, 186-190.

4. Kozak, L.P., J.H. Britton, U.C. Kozak, J.M. Wells. 1988. The mitochondrial uncoupling protein gene. Correlation of exon structure to transmembrane domains. J.Biol.Chem. 263, 12274-12277.

5. Bouillard, F., S. Raimbault, D. Ricquier. 1988. The gene for rat uncoupling protein: Complete sequence, structure of primary transcript and evolutionary relationship between exons. B.B.R.C. 157, 783-792.

6. Himms-Hagen, J. 1985. Brown adipose tissue metabolism and thermogenesis. Ann.Rev.Nutr. 5, 69-94.

7. Krief, S., R. Bazin, F. Dupuy, M. Lavau. 1988. Role of brown adipose tissue in glucose utilization in conscious pre-obese Zucker rats. Biochem.J. 263, 305-308.

8. Trayhurn, P. 1986. Brown adipose tissue and energy balance. In: Brown Adipose Tissue. (P. Trayhurn and D.G. Nicholls, eds.) Edward Arnold, London, pp. 299-338.

9. Tzagoloff, A., A.M. Myers. 1986. Genetics of mitochondrial biogenesis. Ann.Rev.Biochem 55, 249-285.

10. Lehninger, A. 1975. Biochemistry, 2nd Edition, Worth Publishers Inc., Baltimore, pp. 443-542.

11. Mitchell, P. 1968. Chemiosmotic Coupling and Energy Transduction. Glynn Research; Bodmin, UK.

12. Hatefi, Y. 1985. The mitochondrial electron transport and oxidative phosphorylation system. Ann.Rev.Biochem. 54, 1015-1069.

13. Schwerzmann, K., P.L. Pedersen. 1986. Regulation of the mitochondrial ATP synthase/ATPase complex. Arch.Biochem.Biophys. 250, 1-18.

14. Otto, D.A., J.A. Ontko. 1982. Structure-function relations between fatty acid oxidation and mitochondrial inner membrane-matrix regions. Eur.J.Biochem. 129, 479-485.

15. Williams, M.A., R.C. Stancliff, L. Packer, A.D. Keith. 1972. Relation of unsaturated fatty acid composition of rat liver mitochondria to oscillation period, spin label motion permeability and oxidative phosphorylation. Biochem.Biophys.Acta. 267, 444-456.

16. Stancliff, R.C., M.A. Williams, K. Utsumi, L. Packer. 1969. Essential fatty acid deficiency and mitochondrial function. Arch.Biochem.Biophys. 131, 628-642.

17. Smith, J.A., H. De Luca. 1964. Structural changes in isolated liver mitochondria or rats during essential fatty acids deficiency. J.Cell.Biol. 21, 15-26.

18. Divakaran, P., A. Venkataraman. 1977. Effect of dietary fat on oxidative phosphorylation and fatty acid profile in rat liver mitochondria. J.Nutr. 107, 1621-1631.

19. Robblee, N.M., M.T. Clandinin. 1984. Effect of dietary fat level and polyunsaturated fatty acid content on the phospholipid composition of rat cardiac mitochondrial membranes and mitochondrial ATPase activity. J.Nutr. 114, 263-269.

20. Abuirmeileh, N.H., C.E. Elson. 1980. The influence of linoleic acid intake on membrane bound respiratory activities. Lipids 15, 918-924.

21. Rafael, J., J. Patzelt, H. Schafer, I. Elmadfa. 1984. The effect of essential fatty acid deficiency on basal respiration and function of liver mitochondria in rats. J.Nutr. 114, 255-262.

22. Deaver, O.E., Jr., R.C. Wander, R.H. McCusker, C.D. Berdanier. 1986. Diet effects on membrane phospholipid fatty acids and mitochondrial function in BHE rats. J.Nutr. 116, 1148-1155.

23. Aw, T.Y., and D.P. Jones. 1989. Nutrient supply and mitochondrial function. Ann.Rev.Nutr. 9, 229-251.

24. Litwack, J., S. Singer. 1972. Subcellular actions of glucocorticoids. In: Biochemical Actions of Hormones II (G. Litwack, ed.). Academic Press, New York, pp. 114-163.

25. Kimura, S., H. Rasmussen. 1977. Adrenal glucocorticoids, adenine nucleotide translocation and mitochondrial calcium accumulation. J.Biol.Chem. 252, 1217-1225.

26. Allan, E.H., A.B. Chesholm, M.A. Titheradge. 1983. The stimulation of hepatic oxidative phosphorylation following dexamethasone treatment of rats. Biochem.Biophys.Acta. 725, 71-76.

27. Sonka, J. 1976. Dehydroepiandrosterone, metabolic effects. Monograph LIII Acta Universitatis Carolinae Medica. Charles University Prague, pp. 1-171.

28. Mohan, P.F., M.P. Cleary. 1988. Effect of short term DHEA administration on liver metabolism of lean and obese rats. Am.J.Physiol. 255, E1-E8.

29. Berdanier, C.D., M.K. McIntosh. 1989. Genotypic differences in metabolic responses to DHEA. In: Hormones, Thermogenesis and Obesity (H. Lardy and F. Stratman, eds.). Elsevier Science Publishing, pp. 385-397.

30. Berdanier, C.D., M.K. McIntosh. 1989. Further studies on effects of DHEA on hepatic respiration in BHE rats. Proc.Soc.Exp.Biol.&Med. 192, 186-194.

31. Berdanier, C.D., M.K. McIntosh. 1988. Lessons learned from genotypic differences in response to dehydroepiandrosterone treatment. In: Frontiers in Diabetes Research Lessons from Animal Diabetes (E. Shafrir and A. Renold, eds.). John Libby, London, pp. 356-361.

32. Mohan, P.F., M. P. Cleary. Dehydroepiandrosterone and related steroids inhibit mitochondrial respiration in vitro. Int.J.Biochem. 21, 1103-1107.

33. Tagliaferro, A., J. Davis, S. Truchon, N. Van Hamont. 1986. Effects of DHEA acetate on metabolism, body weight, and composition of male and female rats. J.Nutr. 116,1977-1983.

34. Leighton, B., A.R. Tagliaferro, E.A. Newsholme. 1987. Effect of dehydroepiandrosterone acetate on liver peroxisomal enzyme activities of male and female rats. J.Nutr. 117,1287-1290.

35. Leiter, E. 1989. The genetics susceptibility in mice. FASEB J. 3, 2231-2241.

36. Cleary, M., J. Zisk. 1986. Anti-obesity of two different levels of DHEA in lean and obese middle-aged female Zucker rats. Int.J. Obesity 10,193-204.

37. Cleary, M., A. Shephard, B. Jenks. 1984. Effect of DHEA on growth in lean and obese Zucker rats. J.Nutr. 114,1242-1251.

38. Cleary, M., J. Billheimer, A. Finan, J. Sartin, A. Schwartz. 1984. Metabolic consequences of DHEA in lean and obese adult Zucker rats. Horm.Metab.Res. 16(suppl), 43-46.

EFFECT OF DEHYDROEPIANDROSTERONE ON RODENT LIVER MICROSOMAL, MITOCHONDRIAL, AND PEROXISOMAL PROTEINS

R.A. Prough and H-Q. Wu

Department of Biochemistry, University of Louisville School of Medicine, Louisville, Kentucky 40292

L. Milewich

Department of Obstetrics & Gynecology, The University of Texas Southwestern Medical Center, 5323 Harry Hines Blvd., Dallas, Texas 75235

Introduction

Dehydroepiandrosterone (DHEA), administered *per os*, is effective in the prevention of a variety of genetic and induced disorders in mice and rats. For example, DHEA prevents or delays the onset of diabetes (1), obesity (2-4), lupus (5), hemolytic anemia (6), acute lethal viral infections (7), spontaneous viral-induced breast cancer (8), chemical-induced malignancies of skin (9), lung (10), and colon (11). At the present time, it has not been possible to ascertain whether the molecular mechanisms by which DHEA acts to exert its many beneficial therapeutic effects (1-11) are the same or even related. It seems reasonable to entertain the possibility that there are some common biochemical phenomena that underlie the chemopreventive effects of DHEA.

Various hypotheses have been proposed to explain the underlying biochemical mechanism(s) by which DHEA exerts its beneficial properties. The first hypothesis is based on the inhibitory effect of DHEA on mammalian glucose-6-phosphate dehydrogenase *in vitro* (12). Inhibition of glucose-6-phosphate dehydrogenase with possible reduction in cytosolic NADPH levels could result

Dehydroepiandrosterone (DHEA)

in decreased lipogenesis, which would have an impact on obesity and, possibly, other conditions where DHEA treatment is effective. Under basal physiologic conditions, however, DHEA levels in mice plasma are approximately 1-3 ng/ml, *i.e.* 4 to 12 nmol/l, and those of mice treated with DHEA *p.o.* are approximately 24-41 ng/ml, *i.e.* 80-140 nmol/l (13). Clearly, these levels are too low to produce significant inhibition of mammalian glucose-6-phosphate dehydrogenase since the K_i for DHEA is approximately 18 μmol/l (12). Thus, it would appear that the effects exerted by DHEA *in vivo* may be mediated by mechanisms other than those involving the direct inhibition of glucose-6-phosphate dehydrogenase. However, this possibility cannot be ruled out since tissue levels of this steroid have not been determined and may actually be considerably higher than those found in plasma.

The ability of DHEA to inhibit glucose 6-phosphate dehydrogenase in rodent tissues such as liver has been difficult to demonstrate. Recently, however, Garcea *et al.* (14) reported that while high concentrations of DHEA do directly inhibit glucose 6-phosphate dehydrogenase activity in isolated rat hepatocytes, the levels of cytosolic NADPH noted in these cells were not altered by treatment with DHEA. This may be explained by the fact that there are several methods by which cytosolic NADPH levels are maintained; *e.g.*, malic enzyme or $NADP^+$-specific isocitrate dehydrogenase. Inhibition of a single NADPH-producing pathway does not apparently alter the redox status in the liver cell cytosol which normally maintains a NADPH/$NADP^+$ ratio of approximately 8:1. Of interest, Garcea *et al.* also noted that DHEA administration blocks the progression of liver cells into cell types normally associated with preneoplastic lesions in animals treated with carcinogens (14). Further, the levels of ribulose 5-phosphate were decreased in livers of animals treated with DHEA suggesting that an inhibitory effect on glucose 6-phosphate dehydrogenase activity does occur *in vivo* leading to depressed levels of sugar phosphates, but not of reduced pyridine nucleotides. In addition, intraperitoneal injection of carcinogen-initiated animals with ribonucleosides and deoxyribonucleosides, required for nucleic acid synthesis, reversed the effect of DHEA on the progression of initiated cells; a greater number of preneoplastic lesions were observed in the nucleoside-treated, DHEA-fed animals. These results suggest that DHEA inhibition of glucose 6-phosphate dehydrogenase does occur *in vivo* and results in a lack of sufficient precursors necessary for nucleic acid synthesis and proliferative growth. The ability of DHEA to inhibit tumor cell growth in cell culture or in host animals transplanted with tumors correlates well with its ability to limit the amount of ribose phosphates necessary for rapid, proliferative growth. However, this single effect on chemical carcinogenesis is difficult to extrapolate to the other pathophysiological conditions affected by DHEA.

The second hypothesis involves endocrine effects that may result from *in vivo* conversion of DHEA to androgens and estrogens (13) and the subsequent action of these sex steroid hormones on target tissues. However, there appears to be contradictory evidence in support of this proposition. For example, in female (NZB x NZW)F_1 mice, glomerulonephritis is attenuated and survival is prolonged by treatment with either DHEA (5) or androgens (15,16), but the severity of lupus in these animals is worsened and life-span is shortened by estrogen treatment (15,16). DHEA or estrogen treatment, however, is effective in reducing blood glucose levels to normal levels in diabetic CBA/Lt J-*db/db* mice, while androgens worsen the condition (17). Nonetheless, it is possible that some of the therapeutic benefits that accrue with the administration of DHEA are brought about by way of sex steroid hormones or other metabolites produced by extraglandular metabolism of DHEA.

A third hypothesis involves the action of DHEA on cells of the immune system. Thus, DHEA administration to mice leads to a decrease in plasma immunoglobin levels of approximately 50% (unpublished observation) and to the inhibition of the ability of stem cells in sub-lethally irradiated mice to repopulate NK and T lymphocytes, which may offer an explanation, at least in part, as to why DHEA inhibits autoimmune disease (18).

A fourth hypothesis is based on the finding of induction of peroxisomes and peroxisome-associated enzymes in liver of mice and rats treated with DHEA, which would lead to an increase in fatty acid β-oxidation (19-22). Considerable attention has been focused on several unrelated set of drugs and toxic chemicals which possess a common property; namely, that they all cause an increase in the number and size of peroxisomes, also accompanied by marked changes in mitochondrial and microsomal proteins. Hypolipidemic drugs (22), including ciprofibrate, agents like aspirin and nicotinic acid, and phenoxyacid herbicides (23) also cause peroxisome proliferation. Certain nutritional states, such as high-fat diets (24) and diets deficient in vitamin E (25) have been shown to cause peroxisome proliferation in rodent liver. A major part of the induction of proteins associated with peroxisome proliferation is the peroxisomal β-oxidation enzyme system. Properties of this induction phenomenon include hepatomegaly, proliferation of smooth endoplasmic reticulum (and the cytochrome P450IVA subfamily), proliferation of peroxisomes and changes in enzyme composition of the peroxisomes (enzymes of β-oxidation and catalase), and alterations in mitochondria (number, structure, and changes in the composition of certain enzymes). In the microsomal compartment, specific forms of microsomal cytochrome P450 are induced; one form has been purified and shown to exhibit a unique specificity for the 12-hydroxylation of lauric acid (26). Induction of this enzyme subfamily has also been associated with peroxisome proliferation in rodent liver (27).

Based on the results of our enzymatic studies (19) and morphologic data (20,28), we propose that the effects of DHEA administration are similar to those produced by the action of clofibrate, di(2-ethylhexyl)phthalate (DEHP), and other peroxisomal proliferators. Similar enzymic/morphological results have been observed when rats were fed DHEA acetate (29). The beneficial effects of DHEA must be carefully assessed, since it is known that agents which serve as peroxisome proliferators also cause hepatic cancer in rodents (22,28). However, these deleterious effects have not been observed in mice fed DHEA for 22 months (data not shown).

Results

Effects of DHEA feeding on mouse and rat liver

The liver is exposed to relatively large amounts of DHEA when administered *per os* in the diet (0.45% DHEA in AIN 76A diet) and, thus, the biochemical changes that occur in this organ as a consequence of DHEA action may play a role in the beneficial effects exerted by this steroid. An equivalent dosage *i.p.* would be over 225 mg/kg body weight; this requirement for high doses of DHEA accounts for why feeding is a more effective route of administration. DHEA treatment of mice and rats leads to the development of hepatomegaly in these animals (2,3,20,28) and to changes in liver color from pink to deep mahogany (23); this color change also is evident in liver mitochondria of DHEA-treated mice. The color is not due to an apparent increase in the concentration of hemoproteins and the cause(s) for the color change is unknown at the present time. The increase in liver size and change in color persisted for as long as the mice were treated with DHEA (20).

To establish whether DHEA action was correlated with alterations in the relative levels of hepatic proteins, we conducted

studies with liver tissue obtained from mice and rats treated with either a special diet (0.45% DHEA in AIN 76A chow) or AIN 76A diet alone. The proteins were separated by polyacrylamide gel electrophoresis in the presence of sodium dodecyl sufate (SDS-PAGE) either on 10% polyacrylamide gels or on gel gradients (3% to 10% polyacrylamide) and were detected by Coomassie blue protein staining. A protein of relative molecular mass of approximately 72,000 (M_r 72K) was induced markedly by DHEA action in liver of rats and mice of various strains (20,28,30 as illustrated in Fig. 1). This protein was detected in liver homogenates, 200 - 20,000xg particulate fractions, cytosolic fractions, but not in microsomal fractions (20,28,30).

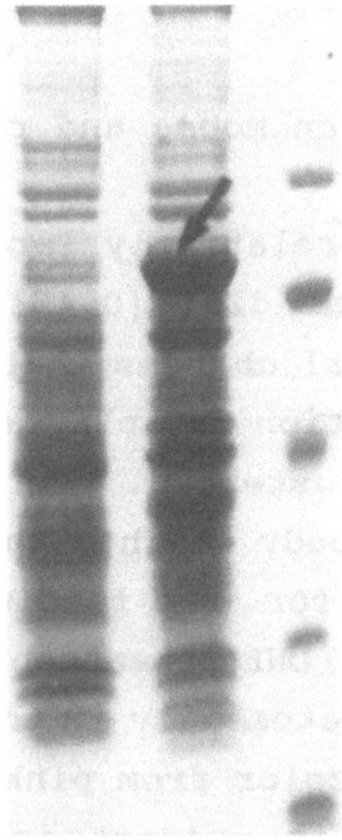

Figure 1. Induction of a M_r Approximately 72K Protein (Peroxisomal Enoyl-CoA Hydratase) in Rat Liver by DHEA Treatment. Protein patterns obtained by SDS-PAGE (10% polyacrylamide) and Coomassie blue staining of rat hepatic cytosols. Cytosols were prepared from liver of male Sprague-Dawley rats (100 g body weight) that were fed either a control diet (AIN 76A) or a DHEA-containing diet (0.45%, w/w) for 7 days. Control diet, *left lane*; DHEA-containing diet, *middle lane*; molecular weight markers, *right lane*. Notice the intensity of the 72 kDa protein induced by DHEA action (indicated by the arrow).

With the methodology used in this study, the 72 kD protein was detected in mice within 2 days after initiation of DHEA feeding and was induced maximally by day 7 of treatment. The high content of this protein was maintained throughout the entire period of DHEA treatment, up to seven months. When DHEA treatment of mice was discontinued after 2 weeks and the food substituted with a DHEA-free diet for an additional 2 weeks, the induced 72 kD protein was no longer detectable by Coomassie blue staining. Thus, the induction of the 72 kD protein can be reversed by removal of DHEA from the diet (30).

Additional differences in the patterns of liver proteins of control and DHEA-treated mice were detected by 2-D gel electrophoresis. In initial studies, conducted under equilibrium conditions in the first dimension by use of broad-range pH ampholites (pH 3 to 10), we were unable to detect the 72 kD protein in liver of either control or DHEA-treated mice even by use of the very sensitive silver-staining technique. When the first dimension was run under nonequilibrium conditions, the 72 kD protein was detected by silver staining both in control liver and in liver of DHEA-treated mice (30); higher apparent content of the protein was observed in livers from DHEA-fed animals. These results indicated that the isoelectric point of the 72 kD protein was that of a basic protein.

Cytosolic proteins in mouse and rat liver

The effect of DHEA feeding on cytosolic proteins involved in foreign compound metabolism were studied (19). A major enzyme system noted to be induced by many chemicals, but decreased by peroxisome proliferators, is the enzyme family of the glutathione *S*-transferases (27). Cytosolic glutathione *S*-transferase activity was decreased to 68% of the control value when DHEA was administered in the diet for 7 days. However, no change in cytosolic or microsomal NAD(P)H-quinone oxidoreductase (DT-diaphorase) activity was observed in these studies (Table 1). These observations suggest that the cytosolic enzymes may be under different control mechanisms than the microsomal or peroxisomal proteins. In addition, the specific activity of NADP-linked malic enzyme was increased 2- to 3-fold in liver cytosol, but those of glucose 6-phosphate dehydrogenase, NADP-linked isocitrate dehydrogenase, and ATP-citrate lyase were not affected significantly by DHEA treatment of mice (28,30,31).

Table 1. Effect of DHEA Feeding on Selected Cytosolic and Peroxisomal Enzyme Activities in Rodent Liver

Activity	Fold Change
cytosolic enzymes	
Glutathione *S*-transferase (rat)	0.68
NAD(P)H-quinone oxidoreductase (rat)	1.0
Malic (NADP-linked) enzyme	
(female C57BL/6 mouse)	1.9
(male C57BL/6 mouse)	1.2
Isocitrate dehydrogenase	
(female & male C57BL/6 mouse)	1.1
ATP-citrate lyase (female & male C57BL/6 mouse)	1.0
Glucose 6-phosphate dehydrogenase	
(female & male C57BL/6 mouse)	1.3
peroxisomal enzymes	
Catalase (mouse & rat)	1.9
Palmitoyl CoA oxidase (rat)	9.2
Enoyl-CoA hydratase (mouse & rat)	≤38.
Carnitine acetyl-CoA transferase (mouse & rat)	≤12.

The activities were measured as described previously (19-21,31).

Microsomal proteins in mouse and rat liver

Upon feeding DHEA in AIN 76A diet (0.45% w/w) to mice or rats, NADPH-cytochrome c(P450) reductase activity and cytochrome b_5 content were increased 1.8- and 1.4-fold, respectively (19). Maximal activity for the flavoprotein reductase (1.8-fold increase) was seen after 4 days of feeding. The level of cytochrome P450, determined spectrophotometrically as the ferrous-carbon monoxide complex, was not significantly changed by feeding DHEA. The levels of cytochrome b_5 and NADH-cytochrome *c* (b_5) reductase decreased slightly between day 0 and day 4 of treatment, but returned to their original levels at day 7.

DHEA feeding results in decreased weight gain and this property of the steroid has been suggested to be associated with caloric restriction. Studies in pair-fed animals demonstrated that the induction was not due to caloric restriction due to DHEA feeding, but to some other induction mechanism (19). Maximal induction (1.7-fold) of NADPH-cytochrome P450 reductase was observed after DHEA administration by *i.p.* injection at 100-160 mg/kg body weight, but only a slight increase of cytochrome b_5 content was observed when the steroid was administered by this route (19). Interestingly, NADH-cytochrome *c* (b_5) reductase activity was decreased after treatment with DHEA intraperitoneally, and there was a small, but significant, induction of cytochrome P450 (20%). These results (19) demonstrate that either route of administration leads to essentially the same level of induction of the flavoprotein reductase, but the specific content of total cytochrome P450 and b_5 or activity of NADH-cytochrome b_5 reductase were not markedly altered.

Western immunoblot analysis of the content of NADPH-cytochrome *c* (P450) reductase was conducted with liver microsomal fractions from rats that were administered DHEA either in the diet or by *i.p.* injection (19). Administration of DHEA by either route resulted in an approximate 1.8-fold increase in the content of the flavoprotein (measured by densitometric scanning), similar to the increase in enzyme activity seen with liver microsomes from rats treated with phenobarbital or dexamethasone (19). These results demonstrate that the increase in enzyme activity by action of DHEA results from the increased content of the flavoprotein, not just due to direct stimulation of enzyme activity by DHEA (or a metabolite).

Since DHEA feeding alters the course of chemical carcinogenesis in rodents, the changes in metabolism of various model substrates *in vitro* after DHEA feeding were studied to ascertain whether or not specific forms of cytochrome P450 were induced (Table 2). These substrates, which appear to be specific for certain forms of cytochrome P450 in rat liver, were chosen based on the work of Guengerich *et al.* (32) and Waxman (33). The substrates studied were: benzphetamine (P450IIB1), ethylmorphine (P450IIIA1), aniline (P450IIE1), ethoxyresorufin (P450IA1), lauric acid (P450IVA), and testosterone/androstenedione (various constitutive P450s). The nomenclature of Nebert *et al.* (34) was utilized to describe the various forms of P450 using a standardized gene designation. Feeding of DHEA did not affect most of the activities, except the ethylmorphine N-demethylase (50% decrease) and laurate 12-hydroxylase (1,680% increase) activities (Table 2).

Constitutive cytochromes P450 have been characterized by their ability to metabolize two steroid substrates *in vitro*, androstenedione and testosterone (35). Using liver microsomes from DHEA-fed rats, the pattern of hydroxylated metabolites formed were determined by HPLC analysis. There were major changes in the distribution of hydroxylated metabolites of androstenedione and testosterone when liver microsomes from DHEA-fed animals were compared to those of untreated animals; the overall rates of steroid hydroxylation were not apparently altered. The

16α-hydroxylase activities determined with either androstenedione or testosterone as substrates were significantly decreased in liver microsomes from DHEA fed rats. The rate of 16β-hydroxylation of androstenedione and testosterone by microsomes from DHEA treated rats was increased 600%. The data in Table 2 clearly suggests that forms of cytochrome P450 induced by various foreign chemicals, such as polycyclic aromatic hydrocarbons or barbiturates, were not induced by administration of DHEA. DHEA treatment did, however, cause the content of some constitutive cytochrome P450s to decrease, while the cytochrome P450IVA associated with laurate 12-hydroxylase (ω-hydroxylase) activity was increased. Microsomal steroid 16β-hydroxylase activity was increased, but is thought not to be due to induction of P450IVA (26).

TABLE 2. Effects of DHEA Feeding on Various Hepatic Microsomal Enzyme Activities in the Rat

Activity	P450 Isozyme	Fold Change
Ethoxyresorufin O-deethylase	IA1	0.98
Benzphetamine N-demethylase	IIB1	0.98
Ethylmorphine N-demethylase	IIIA1	0.48
Aniline hydroxylase	IIE2	0.87
Laurate 12-hydroxylase	IVA	16.8
Androstenedione		
6β-hydroxylase		1.3
16α-hydroxylase		0.21
16β-hydroxylase		6.5
Overall hydroxylation		0.97
Testosterone		
6β-hydroxylase		1.7
16α-hydroxylase		0.20
16β-hydroxylase		6.0
2α- or 2β-hydroxylase		0.29
Overall hydroxylation		0.98

The activities were measured as described previously (19).

Using Western immunoblot analysis to further characterize the induction of constitutive cytochromes P450, cytochrome P450IVA content was markedly increased by administration of DHEA either by feeding or *i.p.* injection (19). This increase in P450IVA is similar to the changes observed when rats are fed a diet containing 0.2% DEHP (36) or ciprofibrate by gavage (27). The time course of induction of cytochrome P450IVA displayed maximal induction of this form of the hemoprotein by day 4 of DHEA feeding at 120-160 mg/kg body weight (19).

Since DHEA is a sex-steroid hormone precursor; *i.e.*, it can be converted to testosterone or estradiol-17ß *in vivo* (13), it was of interest to determine which other precursors of these biologically potent steroids might lead to the induction of liver microsomal and peroxisomal enzymes. The synthesis of testosterone (T) from cholesterol *in vivo* may involve two different pathways depending on the animal species and tissue sites of metabolism; *viz*, the Δ^4 and Δ^5 pathways (37). DHEA is the direct product of 17α-hydroxypregnenolone (17-OHP) metabolism; DHEA can be converted to 5-androstene-3ß,17ß-diol (ADIOL) in a reaction catalyzed by 17ß-hydroxysteroid oxidoreductase. DHEA is also converted *in vivo* to 4-androstene-3,17-dione (ADIONE) by action of 3ß-hydroxysteroid oxidoreductase, but the reverse reaction does not occur readily *in vivo*. Estradiol-17ß (EDIOL) also is formed *in vivo* from orally-fed DHEA (13). We administered these steroids *i.p.* daily in corn oil for 4 days to rats at a dose of 120 mg/kg body weight and subsequently isolated various hepatic protein fractions for enzyme assay. The data presented in Figure 2 serves to illustrate the effect of the direct DHEA precursor, 17-OHP, and those of possible DHEA metabolites (ADIOL, ADIONE, T, and EDIOL) on induction of microsomal NADPH-cytochrome P450 reductase, P450IVA and peroxisomal palmitoyl-CoA oxidase activity. DHEA and ADIOL markedly induced palmitoyl CoA oxidase (29- and 22-fold, respectively); 17-OHP, the precursor of DHEA, caused a 4-fold increase in this activity (Figure 2). In a similar manner, catalase activity was increased approximately 2-fold by either DHEA or ADIOL, but not by the other steroids (data not shown). Although some induction of

NADPH-cytochrome P450 reductase was seen after treatment with 17-OHP, ADIONE, T, and EDIOL, DHEA and ADIOL were far better at inducing this flavoprotein reductase. Studies to evaluate the dose-dependency of induction of microsomal and peroxisomal enzymes by various steroids suggested that ADIOL was more effective (3- to 6-fold lower ED_{50}) in inducing NADPH-cytochrome P450 reductase than in inducing catalase or palmitoyl CoA oxidase (data not shown).

In other studies conducted by Western analysis of P450IVA, only DHEA and ADIOL were found to increase the protein content of P450IVA in liver microsomes. Initial studies to evaluate the dose dependencies of induction of P450IVA by either DHEA and ADIOL suggest that the content of P450IVA is nearly equal in animals treated with equimolar amounts of the two steroids. These results are similar to those seen for microsomal NADPH-cytochrome P450 reductase and peroxisomal palmitoyl CoA oxidase (data not shown).

Rat kidney and lung contain appreciable amounts of cytochrome P450IVA, and therefore, we also evaluated the effects of DHEA and ADIOL on this enzyme by Western analysis. DHEA induced markedly the content of this immunoreactive protein in kidney cortex microsomes, while ADIOL had little or no effect (data not shown). In contrast, ADIOL increased the content of immunoreactive cytochrome P450IVA in lung, while DHEA had no effect. These findings suggest that DHEA and ADIOL may display tissue specific induction of cytochrome P450IVA in these two extrahepatic tissues. Further studies are needed to distinguish tissue specific induction from differences in pharmacokinetic distribution of DHEA and ADIOL.

The possible role of the microsomal ω-hydroxylase in prevention of chemical carcinogenesis, diabetes, and obesity and other pathologies (1-11) will require additional studies. An understanding of the mechanisms of induction of these microsomal and peroxisomal proteins will be useful in unraveling these relationships.

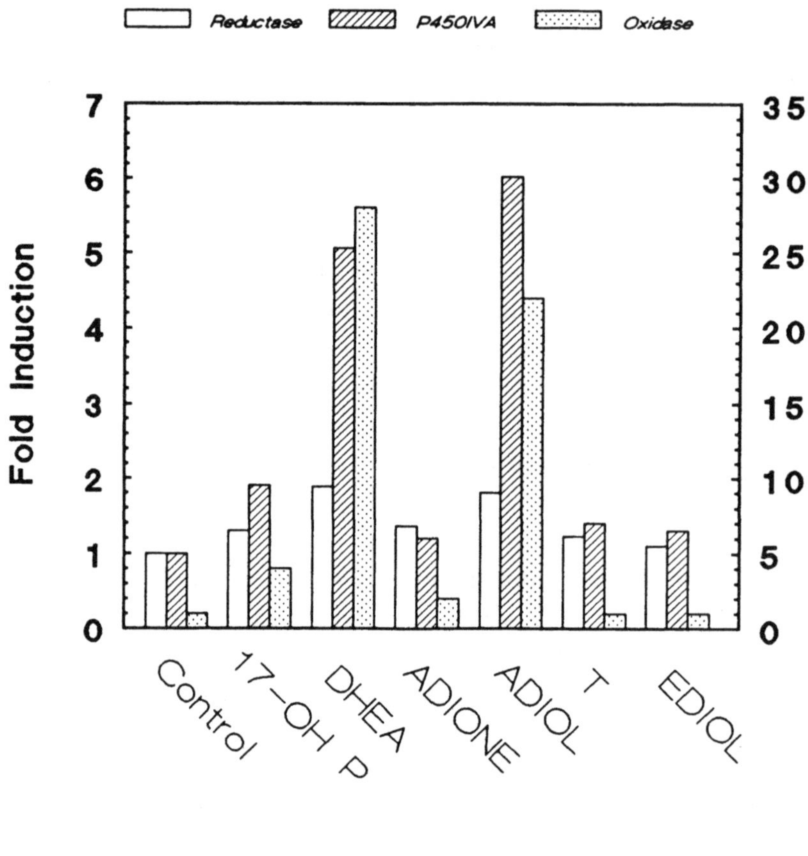

Figure 2. The Effect of Treatment with Various Steroids on the Induction of Microsomal and Peroxisomal Enzymes. The animals were injected *i.p.* with various steroids at a dose of 120 mg/kg body weight daily for 4 days. After peroxisomal and microsomal tissue fractions were obtained, the enzyme activities and hemoprotein content was determined as described previously (19). *Left axes*, NADPH-cytochrome P450 reductase (open bar) and cytochrome P450IVA (striped bar); *Right axes*, palmitoyl-CoA-oxidase (dotted bar). 17-OH P, 17-hydroxypregnenolone; DHEA, dehydroepiandrosterone; ADIONE, 4-androstene-3,17-dione; ADIOL, 5-androstene-3ß,17ß-diol; T, testosterone; and EDIOL, estradiol-17ß.

Peroxisomal enzymes in mouse and rat liver

Cytosolic fractions prepared from liver homogenates of rats that were fed either with the DHEA-containing diet (0.45%, w/w) or the control AIN 76A diet for 7 days were used to identify the induced protein of M_r 72K (20). Liver proteins were separated by SDS-PAGE, transferred electrophoretically to nitrocellulose paper, and stained with Ponceau-S. The portion of the nitrocellulose paper that contained the 72kD protein was excised and digested *in situ* with trypsin. The tryptic fragments were separated by reverse-phase high-pressure liquid chromatography, the individual peptides collected onto glass fiber discs, reduced and alkylated, and finally treated by amino terminal sequence analysis by gas phase sequencing. Comparison of the amino acid sequences obtained for five different peptide fragments with the sequences of known proteins enabled us to identify this protein as 2-enoyl-CoA hydratase/3-hydroxyacyl-CoA dehydrogenase/Δ^3,Δ^2-enoyl-CoA isomerase (20), a component of the fatty acid β-oxidation system (38,39) of liver peroxisomes (Table 1). The trifunctional protein has an actual M_r of 77K (40,41) rather than 72K as estimated by SDS-PAGE (30,31). In liver of rodents fed the control diet, 2-enoyl-CoA hydratase/3-hydroxyacyl-CoA dehydrogenase/Δ^3,Δ^2-enoyl-CoA isomerase could not be visualized by staining with Ponceau-S (20), as found in previous studies by use of Coomassie blue (30,31); there was no stained protein in the corresponding region of PAGE of cytosolic proteins from rodents not fed DHEA.

Enoyl-CoA hydratase activity was determined by use of homogenates and cytosols prepared from liver tissue of rats and mice treated either with the control diet or the same food that contained DHEA (0.45%, w/w). Assays also were conducted with cytosols prepared from liver tissue of rats injected *i.p.* with either DHEA (100 mg/kg body weight daily) in corn oil or the vehicle alone for 4 consecutive days. The specific activity of enoyl-CoA hydratase in rat liver cytosol was approximately 40-fold greater than that of controls (Table 1). The changes in enoyl-CoA hydratase activity appeared to be correlated directly with the amount of protein of M_r 72K (by SDS-PAGE). We found that approximately 7% of the total protein in the cytosol corresponded to this protein (Figure 1). In the mouse, the specific activity of liver enoyl-CoA hydratase was increased approximately 4-fold in cytosols and 7-fold in homogenates by DHEA action (20,21). The specific activity of 3-hydroxyacyl-CoA dehydrogenase in rat cytosol also was elevated 7-fold when compared with control liver, but not to the same extent as enoyl-CoA hydratase (data not shown).

On account of the finding of induction of peroxisomal 2-enoyl-CoA hydratase/3-hydroxyacyl-CoA dehydrogenase/Δ^3,Δ^2-enoyl-CoA isomerase by treatment with DHEA, we determined the specific activities of other enzymes known to be induced by clofibrate, a peroxisome inducer (27,42-46), *viz.* peroxisomal catalase, carnitine acetyl-CoA transferase, carnitine octanoyl-CoA transferase, urate oxidase and KCN-insensitive palmitoyl-CoA oxidase (19-21). Mitochondrial carnitine palmitoyl-CoA transferase (20) and microsomal ω-hydroxylase (19). The specific activity of catalase (Table 1) was increased 2-fold in liver homogenates and cytosols of DHEA-treated mice and rats (19-21). The specific activity of carnitine acetyl-CoA transferase in liver cytosols of DHEA-treated rats was increased 16-fold over control levels; similarly, approximately 3-fold and 8-fold increases were obtained with cytosols and homogenates prepared from liver of DHEA treated mice (20,21). The specific activity of carnitine octanoyl-CoA transferase was increased approximately 11-fold in rat liver cytosol, 3-fold in mouse liver cytosol, and 2-fold in mouse liver homogenates with DHEA treatment (20,21). The specific activity of KCN-insensitive palmitoyl-CoA oxidase was increased 9.3-fold in rat liver homogenates with DHEA treatment (19). The specific activity of urate oxidase was increased approximately 2-fold in liver homogenates of mice fed DHEA; however, this enzymatic activity was not detected in either mouse or rat liver cytosol (20,21). As expected, this enzyme is associated exclusively with the crystalline core of peroxisomes and not with the peroxisomal matrix. Enzymes associated with the peroxisomal matrix are easily released from the peroxisome during homogenization and are, therefore, found in the cytosolic fraction prepared by normal preparation procedures (46). Enzymes associated with the crystalline core of peroxisomes are concentrated in fractions containing mitochondrial organelles.

Similar increases in the specific activities of catalase and fatty acyl-CoA oxidase in rat liver, produced by DHEA acetate or DHEA treatment of rats, have been reported previously by Leighton *et al.* (29) and Wu *et al.* (19). We have found that the specific activity of mitochondrial carnitine palmitoyl-CoA transferase was also increased approximately 2-fold by DHEA treatment both in rat and mouse liver cytosols and 4-fold in mouse liver homogenates (20-21). In addition, both the specific activity and the relative concentration of mitochondrial carbamoyl phosphate synthetase I were decreased approximately 50% in livers of DHEA-treated animals (28).

Ultrastructural study of hepatocytes

Ultrastructural studies of mouse hepatocytes were conducted to establish whether the morphology and number of peroxisomes and mitochondria in these cells were altered by DHEA treatment. To detect peroxisomes by electron microscopy, liver tissue (1 mm^3) was fixed in phosphate buffer containing glutaraldehyde (2.5%), washed in buffer, and treated with 3,3'-diaminobenzidine to identify the catalase-containing peroxisomes. After embedding and sectioning, comparable areas of liver cells were photographed at the same magnification and analyzed morphometrically (Figure 3). Peroxisomes (*black arrows*) in the hepatocytes of the DHEA-treated animal had proliferated, were larger in cross-sectional area, and appeared in clusters (*panel 2*) when compared with the control (*panel 1*). Mitochondria (*white arrowheads*) were detected by use of liver tissue fixed as indicated above, but without treatment with 3,3'-diaminobenzidine. Notice the decrease in mitochondrial cross-sectional area in hepatocytes of the DHEA-treated animals (*panel 4*) compared with the control (*panel 3*) (23,33). These findings (Figure 3) are supportive of the view that DHEA administration to mice and rats causes proliferation of peroxisomes and mitochondria in hepatocytes; these changes are also reflected by increases in enzymatic activities observed in these organelles (20,21,31).

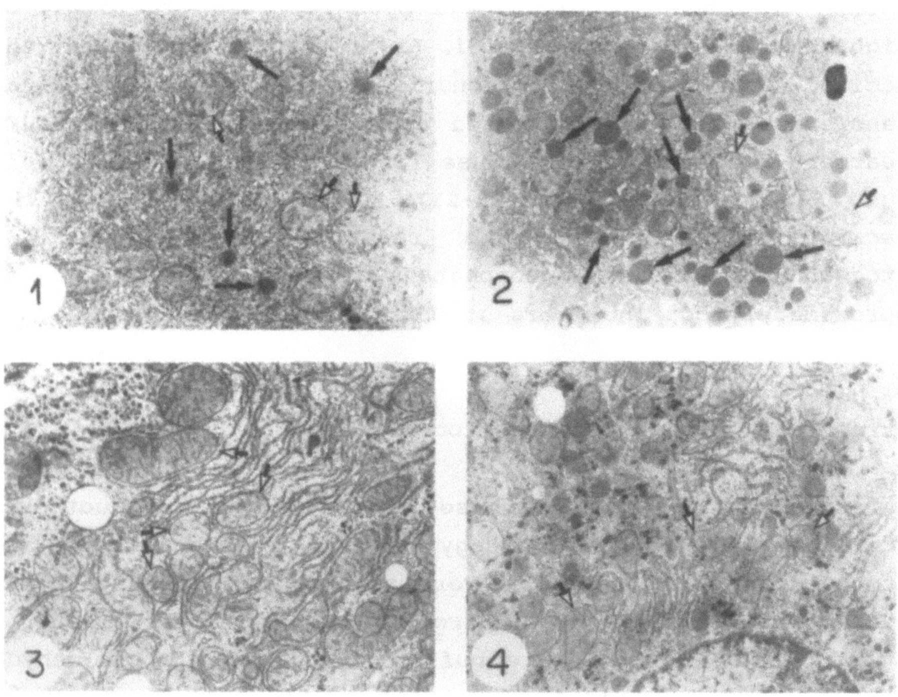

Figure 3. Peroxisomes and Mitochondria in Hepatocytes of Mice Treated with DHEA. BALB/c mice were fed either a control diet (AIN 76A) (*Panels 1 and 3*) or the same diet containing DHEA (0.45%, w/w) (*Panels 2 and 4*).

Plasma triglycerides and total cholesterol

The action of agents known to induce peroxisome proliferation in liver of mice and rats has been associated with hypolipidemia (47-49). To determine whether DHEA treatment of mice resulted in a similar effect, we determined triglycerides and total cholesterol levels in plasma of mice that were fed either the control AIN 76A diet or the diet containing DHEA (0.45% w/w). We found that, depending on the strain of mice, serum triglyceride levels were either significantly increased (C3H/HeJ and BALB/c mice, 47% and 92%, respectively), slightly decreased [(NZW x BXSB)F_1 and (C57BL/6 x DBA/2)F_1 mice] (23,52), or significantly decreased (C57BL/Ks *db/db*) (50).

Cholesterol levels in plasma of mice treated with DHEA were increased significantly over those of animals that received the control diet in the two strains of mice evaluated, *viz.*, BALB/c and (C57BL/6 x DBA/2)F_1; the increases in cholesterol levels in these mice by DHEA treatment were 87% and 57%, respectively (20,50).

Rates of hepatic fatty acid and cholesterol syntheses

The rate of fatty acid synthesis in liver of female BALB/c mice, determined by incorporation of tritium from intravenously administered ^3H_2O, was reduced approximately 70% in mice treated with DHEA when compared with control animals (31). This decrease may be due, at least in part, to increased β-oxidation of newly synthesized fatty acids by peroxisomal enzymes. Increased serum triglyceride levels were observed as a result of DHEA action on mice of the same strain and sex (BALB/c) [control, 93 ± 15 mg/dl (mean ± SD); DHEA, 179 ± 24 mg/dl] (20). The decrease in uptake and hydrolysis of triglycerides at tissue sites other than the liver may be explained by changes in lipoprotein lipase activity; this possibility, however, has not been investigated further.

In this same study (31), we also found that the rate of cholesterol synthesis in liver of DHEA-treated female BALB/c mice was increased approximately 44%. This finding correlates well with the increase in plasma cholesterol concentration in animals of the same strain and sex [control, 52 ± 8 mg/dl (mean ± SD); DHEA, 97 ± 11 mg/dl] (20).

Discussion

In studying some aspects of the involvement of the liver in the mechanism of action of DHEA, a steroid known to elicit therapeutic and chemopreventive effects in mice and rats of susceptible strains (1-11), we established that the major liver peroxisomal protein induced by action of this steroid in the mouse and rat was 2-enoyl-CoA hydratase/3-hydroxyacyl-CoA dehydrogenase/Δ^3,Δ^2-enoyl-CoA isomerase (20,21). The identification of this protein, a component of the peroxisomal fatty acid β-oxidation system, led us to investigate peroxisomes and peroxisome-associated enzymes and other enzymes in liver known to be induced by action of "peroxisome proliferators."

The findings presented in this review are an indication that there is close similarity between the hepatic effects produced by DHEA treatment of mice and rats and those produced by the action of unrelated compounds known by the generic name of peroxisome proliferators. Namely, the oral administration of peroxisomal proliferators, *e.g.* clofibrate, to rats and mice has been shown to result in (i) development of hepatomegaly (51-54), (ii) marked induction of a "80 kDa" protein (55), (iii) induction of peroxisomal 2-enoyl-CoA hydratase/3-hydroxy-acyl-CoA dehydrogenase/Δ^3,Δ^2-enoyl-CoA isomerase activity

(40,41,56), and (iv) induction of microsomal cytochrome P450IVA (19,26) as found with DHEA treatment. Moreover, we found that other effects produced in mouse and rat liver by DHEA treatment were similar to those produced by treatment with clofibrate and other peroxisome proliferators. Thus, the specific activities of various peroxisome-associated enzymes, in addition to enoyl-CoA hydratase (40,41,56), *viz.* catalase, carnitine acetyl-CoA transferase, carnitine octanoyl-CoA transferase, and urate oxidase (27,42-44,57-63), are increased by the action of peroxisome proliferators as found with DHEA treatment. In addition to the induction of peroxisomal enzymes, the specific activities of mitochondrial carnitine palmitoyl-CoA transferase (62), microsomal ω-hydroxylase (26,45,46), and cytosolic malic enzyme (63) also are increased significantly by clofibrate treatment, as demonstrated with DHEA (20,21,64). These findings of changes in liver mitochondrial morphology (20,28) and the changes in specific activity of the mitochondrial enzymes, carnitine palmitoyl-CoA transferase (20) and carbamoyl phosphate synthase I (30) are supported by those of Mohan and Cleary, who demonstrated that mitochondrial respiration and transport are impaired in livers of DHEA-treated rats (65).

Additional correlates between the effects of DHEA treatment and those obtained by administration of classical peroxisome proliferators such as clofibrate to mice and rats are reflected in morphologic changes of peroxisomes and mitochondria in hepatocytes. Thus, peroxisomes proliferate and cluster together as seen with DHEA treatment (20,21); the cross-sectional area, the number per unit of cytoplasmic area, and the volume density of peroxisomes are increased by clofibrate feeding (52-54,66-69). The changes on hepatocyte mitochondrial morphology induced by clofibrate treatment also are similar to those produced by DHEA (20,30), *viz.* these organelles have a decreased cross-sectional area and there is an increase in the number of mitochondria per hepatocyte (69-71).

There is a widely-held belief that peroxisomal proliferators produce hypotriglyceridemia; nonetheless, there is a great paucity of published data to support this presumption. For example, it was reported that clofibric acid treatment lowers plasma triglyceride levels in the rat by approximately 13% to 38% (48). In the same study, it was found that gemfibrozil, a peroxisomal proliferator widely used as an hypolipidemic agent in clinical medicine, increases plasma triglyceride levels in the rat by approximately 18% to 27% (48). Another widely accepted characteristic feature of peroxisomal proliferators is that these compounds exert a cholesterol-lowering effect. For example, cholesterol levels were reduced 15% to 50% in plasma of rats treated with clofibrate (49) and, in a study conducted with young men, it was found that DHEA administration (1.6 g/day) for 28 days resulted in a small reduction of total serum cholesterol (approx. 4%) and low-density lipoprotein levels (approx. 4%) (72). On the basis of these and similar reports, we determined triglyceride and cholesterol levels in plasma of mice that were treated either with a control diet or a diet containing DHEA (0.45% w/w) to establish whether the action of the steroid also resulted in hypolipidemia in these animals. In the limited number of mouse strains evaluated in this study, we found that plasma triglyceride levels in DHEA-fed mice were either decreased significantly in obese mice (unpublished results), and lowered slightly, but not significantly in lean mice; or, increased significantly. However, cholesterol levels were increased significantly by DHEA action compared with controls in all mice studies (20,50). Chronic treatment of female BALB/c mice with DHEA, however, resulted in a significant decrease of the rate of fatty acid synthesis and in an increase in the rate of cholesterol synthesis in liver, when compared with controls (31). It is possible that the apparent decrease in the rate of hepatic fatty acid synthesis is a reflection, at least in part, of increased fatty acid β-oxidation by peroxisome-associated enzymes.

The conclusions of this review are that DHEA treatment of mice and rats results in hepatomegaly that is temporally associated with changes in the number of peroxisomes and mitochondria and changes in microsomal, mitochondrial and peroxisomal enzymes. Whether the beneficial preventive and therapeutic effects that accrue with DHEA administration to susceptible rodents (1-11) are associated with these changes in liver remains to be established.

Acknowledgement

This research was supported in part by NIH grant U01 CA43311 (LM,RAP) and R01 ES 04244 (RAP).

References

1. Coleman, D.L., E.H. Leiter, and R.W. Schwizer. 1982. Diabetes 31, 830-833.

2. Yen, T.T., J.A. Allan, D.V. Pearson, J.M. Action, and M.M. Greenberg. 1977. Lipids 12, 409-413.

3. Shepherd, A., and M.P. Cleary. 1984. Am. J. Physiol. 24, E123-E128.

4. Granholm, N.H., L.D. Staber, and P.G. Wilkin. 1987. J. Exptl. Zoology 242, 67-74.

5. Lucas, J.A., S.A. Ahmed, M.L., Casey, and P.C. MacDonald. 1985. Clin. Invest. 75, 2091-2093.

6. Tannen, R.H., and A.G. Schwartz. 1982. Fed. Proc. 41, 463.

7. Loria, R.M., T.H. Inge, S.S. Cook, A.K. Szakal, and W. Regelson. 1988. J. Med. Virology 26, 301-314.

8. Schwartz, A.G. 1979. Cancer Res. 39, 1129-1132.

9. Pashko, L.L., G.C. Hard, R.J. Rovito, J.R. Williams, E.L. Sobel, and A.G. Schwartz. 1985. Cancer Res. 45, 164-166.

10. Schwartz, A.G., and R.H. Tannen. 1981. Carcinogenesis 2, 1335-1337.

11. Nyce, J.W., P.N. Magee, G.C. Hard, and A.G. Schwartz. 1984. Carcinogenesis 5, 57-62.

12. Ranieri, R., and H.R. Levy. 1970. Biochemistry 9, 2233-2243.

13. Leiter, E.H., W.G. Beamer, D.L. Coleman, and C. Longcope. 1987. Metabolism 36, 863-869.

14. Garcea, R., L. Daino, S. Frassetto, P. Cozzolino, M.E. Riggoi, M.G. Vannini, R. Pascale, L. Lenzerinni, M.M. Simile, M. Puddu and F. Feo. 1988. Carcinogenesis 9, 931-938.

15. Roubinian, J.R., N. Talal, J.S. Greenspan, J.R. Goodman, and P.K. Siiteri. 1979. J. Clin. Invest. 63, 902-911.

16. Roubinian, J.R., N. Talal, J.S. Greenspan, J.R. Goodman, and P.K. Siiteri. 1978. J. Exp. Med. 147, 1568-1583.

17. Leiter, E.H. 1981. Diabetes 30, 1035-1044.

18. Risdon, G.H., J. Cope, and M. Bennett. 1990. Am. J. Pathol. (in press).

19. Wu, H.-Q., J. Masset-Brown, D.J. Tweedie, L. Milewich, R.A. Frenkel, C. Martin-Wixtrom, R.W. Estabrook, and R.A. Prough. 1989. Cancer Res. 49, 2337-2343.

20. Frenkel, R.A., C.A. Slaughter, K. Orth, C.R. Moomaw, S.H. Hicks, J.M. Snyder, M. Bennett, R.A. Prough, R.S. Putnam, and L. Milewich. 1990. J. Steroid Biochem. 35 (in press).

21. Frenkel, R.A., R.A. Prough, S.H. Hicks, and L. Milewich. 1989. J. Cell Biol. 107, 122a.

22. Rao, M.D. and Reddy, J.K. 1987. Carcinogenesis 8, 631-636.

23. Vainio, H., Linnainmaa, K., Kahonen, M., Nickels, J., Hietanen, E., Marniemi, J., and Pelkonen, P. 1983. Biochem. Pharmacol. 32, 2775-2779.

24. Ishii, H., Fukumori, N., Horie, S, and Suga, T. 1980. Biochim. Biophys. Acta 617, 1-11.

25. Reddy, J.K, Lalwani, N.D., Dabholkar, A.S., Reddy, M.K., and Qureshi, S. A. 1981. Biochem. Internatl. 3, 41-49.

26. Tamburini, P.P., H.A. Masson, S.K. Bains, R.J. Makowski, B. Morris, and G.G. Gibson. 1984. Eur. J. Biochem. 139, 235-246.

27. Hawkins, J.M., Jones, W.E., Bonner, F.W., and Gibson, G.G. 1987. Drug Metab. Revs. 18, 441-515.

28. Leighton, B., A.R. Taglafierro, and E.A. Newsholme. 1987. J. Nutr. 117, 1287-1290.

29. Marrero, M., J.M. Snyder, R. A. Prough, P.C. MacDonald, and L. Milewich. 1987. The Endocrine Soc. Abstracts of Papers, 69th Annual Meeting, p. 208.

30. Marrero, M., R.A. Prough, R.S. Putnam, M. Bennett, and L. Milewich. 1990. Biochem. Pharmacol.

31. Marrero, M., R.A. Prough, R.A. Frenkel, and L. Milewich. 1990. Proc. Soc. Exptl. Biol. Med. 193, 110-117.

32. Guengerich, F.P., Dannan, G.A., Wright, S.T., Martin, M.V., and Kaminsky, L.S. 1982. Biochemistry 21, 6019-6030.

33. Waxman, D.J. 1987. In: Cytochrome P450. Structure, Mechanism, and Biochemistry. (P.R. Ortiz de Montellano, Ed.), pp. 525-539. New York: Plenum Press.

34. Nebert, D.W., Nelson, D. R., Adesnik, M., Coon, M.J., Estabrook, R.W., Gonzalez, F.J., Guengerich, F.P., Gunsalus, I.C., Johnson, E.F., Kemper, B., Levin, W., Phillips, I.R., Sato, R. and Waterman, M.R. 1989. DNA 8, 1-13.

35. Wood, A.W., Ryan, D. E., Thomas, P.E., and Levin, W. 1983. J. Biol. Chem. 258, 8839-8847.

36. Okita, R.T. and Chance, C. 1984. Biochem. Biophys. Res. Commun. 121, 304-309.

37. Smith E.L., R.L. Hill, I.R. Lehman, R.J. Lefkowitz, P. Handler and A. White. 1983. Principles of Biochemistry: Mammaliam Biochemistry, 7th ed., McGraw-Hill, New York, pp. 501-505.

38. Lazarow, P.B., and C. de Duve. 1976. Proc. Natl. Acad. Sci. U.S.A. 73, 2043-2046.

39. Osumi, T., N. Ishii, M. Hijikata, K. Kamijo, H. Ozasa, S. Furuta, S. Miyasawa, K. Kondo, K. Inoue, H. Kagamiyama, and T. Hashimoto. 1985. J. Biol. Chem. 260, 8905-8910.

40. Ishii, N., M. Hijikata, T. Osumi, and T. Hashimoto. 1987. J. Biol. Chem. 262, 8144-8150.

41. Palosaari, P.M., and J.K. Hittunen. 1990. J. Biol. Chem. 265, 2446-2449.

42. Markwell, M.A.K., E.J. McGroarty, L.L. Bieber, and N.E. Tolbert. 1973. J. Biol. Chem. 248, 3426-3432.

43. Moody, D.E., and J.K. Reddy. 1974. Res. Commun. Chem. Pathol. Pharmacol. 9, 501-510.

44. Markwell, M.A.K., L.L. Bieber, and E. Tolbert. 1977. Biochem. Pharmacol. 26, 1697-1703.

45. Orton, T.C., and G.L. Parker. 1982. Drug Metab. Dispos. 10, 110-115.

46. Bains, S.K., S.M. Gardiner, K. Mannweiler, D. Gillet, and G.G. Gibson. 1985. Biochem. Pharmacol. 34, 3221-3229.

47. Baudhuin, P., H. Beaufay, and C. deDuve. 1982. Ann. New York Acad. Sci. 386, 449-452.

48. Stegmeier,K., F.H. Schmidt, A. Reinicke, and H.D. Fahimi. 1982. Ann New York Acad. Sci. 386, 449-452.

49. Thorp, J.M., and W.S. Waring. 1962. Nature 194, 948-949.

50. Milewich, L., M. Bennett, R.S. Putnam, and R.A. Frenkel. 1989. J. Cell Biol 107, 645a.

51. Best, M., and C. Duncan. 1964. J. Lab. Clin. Med. 64, 634-642.

52. Hess, R., W. Staubli, and W. Reiss. 1965. Nature 208, 856-858.

53. Platt, D.S., and J.M. Thorp. 1966. Biochem Pharmacol. 15, 915-925.

54. Svoboda, D., and D.L. Azarnoff. 1971. Fed. Proc. 30, 841-847.

55. Reddy, J.K., and N.S. Kumar. 1977. Biochem. Biophys. Res. Commun. 77, 824-829.

56. Lazarow, P.B. 1978. J. Biol. Chem. 253, 1522-1528.

57. Lazarow, P.B. 1977. Science 197, 580-581.

58. Reddy, J.K., and N.S. Kumar. 1979. J. Biochem 85, 847-856.

59. Moody, D.E., and J.K. Reddy. 1978. Am. J. Pathol. 90, 435-448.

60. Kahonen, M.T. 1976. Biochem. Biophys. Acta 28, 690-701.

61. Leighton, F., L. Coloma, and C. Koenig. 1975. J. Cell Biol. 67, 281-309.

62. Halperin, M.L., and S.V. Pande. 1979. Meth. Enzymol 56, 368-379.

63. Westerfeld, W.W., D.A. Richert, and W.R. Ruegamer. 1968. Biochem. Pharmocol. 17, 1003-1017.

64. Casazza, J.P., W.T. Schaffer, and R.L. Veech. 1985. J. Nutr. 116, 304-310.

65. Mohan, P.F., and M.P. Cleary. 1989. Int. J. Biochem. 21, 1103-1107.

66. Svoboda, D., and D.L. Azarnoff.1966. J. Cell Biol. 30, 442-450.

67. Goldenberg, H., P. Kampfer, R. Kramar, and M. Pavelka. 1976. Histochemistry 46, 189-196.

68. Hruban, Z. M. Gothoh, A. Slesers, and S.F. Chow. 1974. Lab. Invest. 30, 64-75.

69. Paget, G. 1966. J. Atheroscler Res. 3, 729-733.

70. Anthony, L.E., Schmucker D.L., J.S. Mooney, and A.L. Jones. 1978. J. Lipid Res. 19, 154-166.

71. Krup, C.K.R., H.N. Aithal, and T. Kamasarma. 1970. Biochem. J. 116, 773-779.

72. Nestler, J.E., C.O. Barlascini, J.N. Clore, and W.G. Blackard. 1988. J. Clin. Endocrinol. Metab. 66, 57-61.

THE EPIDEMIOLOGY OF DHEAS WITH PARTICULAR REFERENCE TO CARDIOVASCULAR DISEASE: THE RANCHO BERNARDO STUDY

E. Barrett-Connor[1], K-T Khaw[2]

[1]Department of Community and Family Medicine, School of Medicine, University of California, San Diego, California 92093-0607

[2]Clinical Gerontology Unit, University of Cambridge, Addenbrooke's Hospital, Cambridge, England

Introduction

The function of dehydroepiandrosterone (DHEA) and its sulfate ester, dehydroepiandrosterone sulfate (DHEAS) is unknown, although this weak androgen is the major secretory product of the human adrenal gland and the steroid hormone produced in largest quantity. It has been postulated that DHEA or DHEAS is a discriminator of life expectancy and aging, inversely associated with obesity, cancer, and cardiovascular disease. We report here the distribution and determinants of DHEAS levels and selected covariates relevant to cardiovascular disease in 1135 men and 567 women from the Rancho Bernardo study of upper middle class Caucasian adults. Data are presented for men aged 30-79 and for women aged 50-79. To our knowledge this is the first population-based study of DHEAS.

Study Population and Methods

Between 1972 and 1974, 82 percent of a community of upper middle class older white adults in Rancho Bernardo, California, a geographically defined community, participated in a survey of risk factors for heart disease. All participants were healthy enough to come to the clinic; those with an acute illness were rescheduled.

Dehydroepiandrosterone (DHEA)

There were no other exclusionary criteria. All participants had a standardized interview that included medical history, current use of medication, including estrogen in women, and cigarette smoking habit. Blood pressure was measured with a standard mercury sphygmomanometer after the participant had been seated for at least five minutes. The measurement was repeated in those in whom the initial reading exceeded 160 mm Hg systolic or 90 mm Hg diastolic, and the lower of the two readings was recorded. Weight and height were measured with the subjects in light clothing and no shoes; obesity was estimated as the body-mass index (weight in kilograms/height in meters^2 x 100). All subjects were seen between 7 and 11 a.m.; 95 percent had fasted, as requested, for 12 hours before the venipuncture. Total plasma cholesterol was measured in a standardized Lipid Research Clinic laboratory. Fasting plasma glucose was measured in a hospital diagnostic laboratory using a hexokinase method. At a second visit, usually within 100 days of the first visit, approximately one-third of the cohort was interviewed and examined in more detail. At that time the current use of alcohol was assessed by a dietician as part of a 24 hour diet recall, and, separately, by a trained interviewer who asked about past week consumption. Lipoproteins were measured.

Followup for vital status was determined annually in 100%. Death certificates were obtained for all decedents and coded for underlying cause of death by a certified nosologist. In one-third, validation of death-certificate data whenever cardiovascular disease was mentioned anywhere on the death certificate was attempted by interviews with the next of kin, the physician, or both, and by reviews of hospital records. Packets containing historical data, electrocardiographs, and information on cardiac enzyme levels were sent to two of five members of a panel of cardiologists for classification with use of defined diagnostic algorithms. Cases in which the diagnoses established by the two primary reviewers differed were reviewed by the full panel until consensus was reached. The panel agreed with the diagnosis on the death certificate in approximately 85 percent of the cases.

DHEAS Assays

The plasma used for the DHEAS measurements reported here was obtained in 1972-1974 and frozen at -70°C until first thawed in 1985-1986 for determination of DHEAS levels. Specimens were chosen on the basis of availability without knowledge of baseline or subsequent vital or disease status. DHEAS (and other sex hormones) were measured by radioimmunoassay in the endocrinology research laboratory of S.S.C Yen. The sensitivity of DHEAS assay in this laboratory is 0.02 μg per milliliter, the intraassay coefficient of variation is 5.2 percent, and the interassay coefficient of variation is 10.0 percent. Previous work in this laboratory had shown that steroid hormone levels remain constant in frozen serum samples for up to 15 years. DHEAS, unlike DHEA, has relatively little circadian or individual variation.

Statistical Methods

Mean DHEAS levels were calculated according to ten-year age groups. Because of the strong inverse association of DHEAS with age, all other analyses were age-adjusted by five year age groups, using the direct method. Analysis of variance was used to test for statistically significant differences in mean DHEAS levels between high- and low-risk categories, after adjustment for age as a continuous variable. All analyses were also calculated with use of log values for DHEAS to account for the slight skew in distribution. Since essentially the same results were obtained with the two tests, untransformed means are shown.

To examine the relation of DHEAS to mortality, the mean DHEAS levels (in five-year age groups) were calculated according to whether death from cardiovascular disease and from ischemic heart disease occurred in the 12-year follow-up period. The 12-year mortality from cardiovascular disease and from ischemic heart disease was also calculated for the subjects with DHEAS levels below and above the approximate 50th percentile, with adjustments for age (in five-year age groups), using the Mantel-Haenszel direct-age adjustment and chi-square test for significance. Mortality rates were calculated after the exclusion of those with a personal history of heart disease at base line. The Cox proportional hazards model was used to assess the independent relation of DHEAS to subsequent death from cardiovascular and from ischemic heart disease, after adjustment for age, obesity, systolic blood pressure, plasma cholesterol level, fasting plasma glucose level, and current cigarette smoking status.

Age, Sex and Obesity

Table 1 shows the striking wide range of DHEAS levels within each decade in Rancho Bernardo men and women.

Table 1. Range of DHEAS (µg/dl) in Rancho Bernardo Men and Women

	Men	Women	
		without estrogen	with estrogen
30-39	45.2 - 810.4		
40-49	72.9 - 682.0		
50-59	31.1 - 738.7	18.3 - 472.1	6.8 - 229.0
60-69	24.3 - 572.8	18.3 - 384.3	4.7 - 243.1
70-70	11.5- 570.7	12.8 - 406.4	23.0 - 139.0

Figure 1 shows the age and sex specific distribution of average DHEAS levels. In both sexes, mean DHEAS levels decreased with age in a steady stepwise fashion. Levels were lower in women than in men in every age group, such that on average women in their 50s had levels similar to those seen in men in their 70s. Although DHEAS levels were only measured in women aged 50 and older, i.e. presumed to be postmenopausal, the lower levels in women did not appear to be due to a loss of ovarian production of estrogen. As shown, women taking replacement estrogen had lower DHEAS levels than women not taking estrogen in each age group. Because of the strong association of exogenous estrogen with the major heart disease risk factors, all cross-sectional analyses reported are restricted to women not using estrogen replacement therapy.

Figure 1. Rancho Bernardo mean DHEAS level by age, sex and estrogen use (in women)

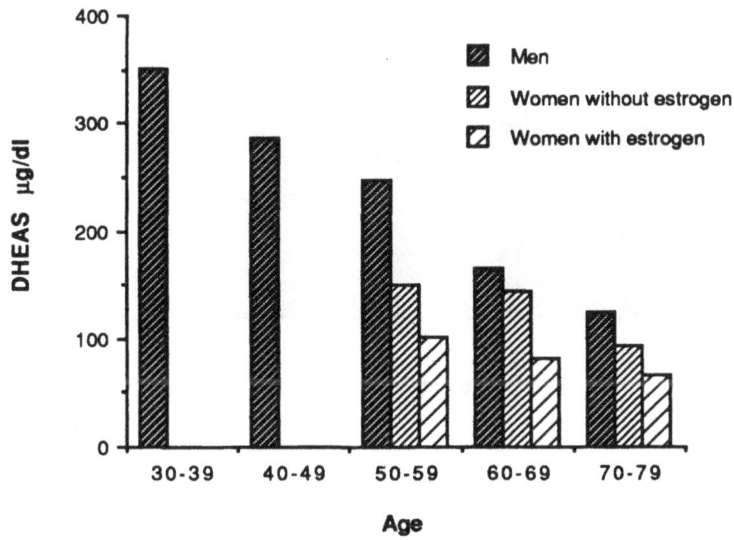

Figure 2 shows the age-adjusted relationship of DHEAS to obesity as estimated by quartiles of body mass index. Although the differences were small, in men DHEAS levels were significantly lower in those in the heaviest vs leanest quartile, and DHEAS was significantly correlated with body mass index ($r = -0.10$; $p < .001$). In women, the level of DHEAS was not related to body mass index by quartile or partial correlation analysis ($r = -0.01$; $p > 0.05$).

Figure 2. Age-adjusted DHEAS level by BMI in Rancho Bernardo men and women

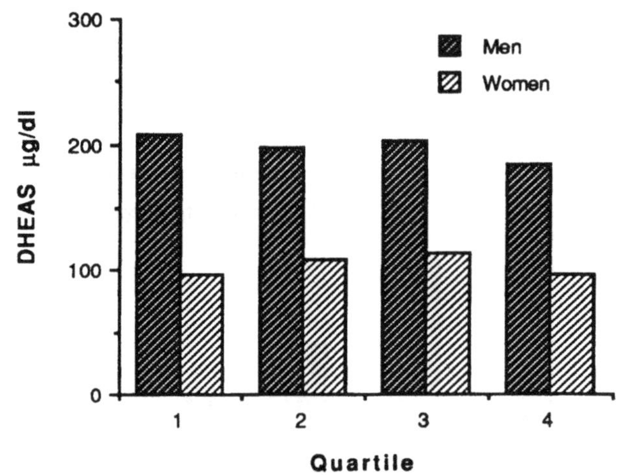

DHEAS and Heart Disease Risk Factors

The age-adjusted and age and body mass-adjusted correlation of DHEAS levels with four continuous risk factors for heart disease (total cholesterol, fasting plasma glucose, systolic and diastolic blood pressure) are shown in Table 2. In men, DHEAS correlated weakly but significantly with systolic blood pressure, and fasting plasma glucose; the former statistically significant association persisted after adjustment for obesity. In women, DHEAS levels were significantly correlated with total cholesterol and blood pressure, but not glucose; these associations were also independent of obesity. The association of DHEAS with cholesterol, plasma glucose or blood pressure was not particularly strong or stepwise, however, as shown in Figure 3.

Table 2. Age-adjusted (and BMI-adjusted) partial correlation coefficients for DHEAS with risk factors in Rancho Bernardo men and women

	Men		Women	
Cholesterol	0.08	(0.08)	0.10***	(0.10)
Fasting plasma glucose	0.10*	(0.11)	0.03	(0.05)
Systolic blood pressure	0.12**	(0.12**)	0.06*	(0.08**)
Diastolic blood pressure	0.01	(0.01)	0.06*	(0.08***)

*p<0.05 **p<0.01 ***p<0.001

Figure 3. Age-adjusted DHEAS level by systolic blood pressure, diastolic blood pressure, cholesterol and fasting plasma glucose in Rancho Bernardo men and women

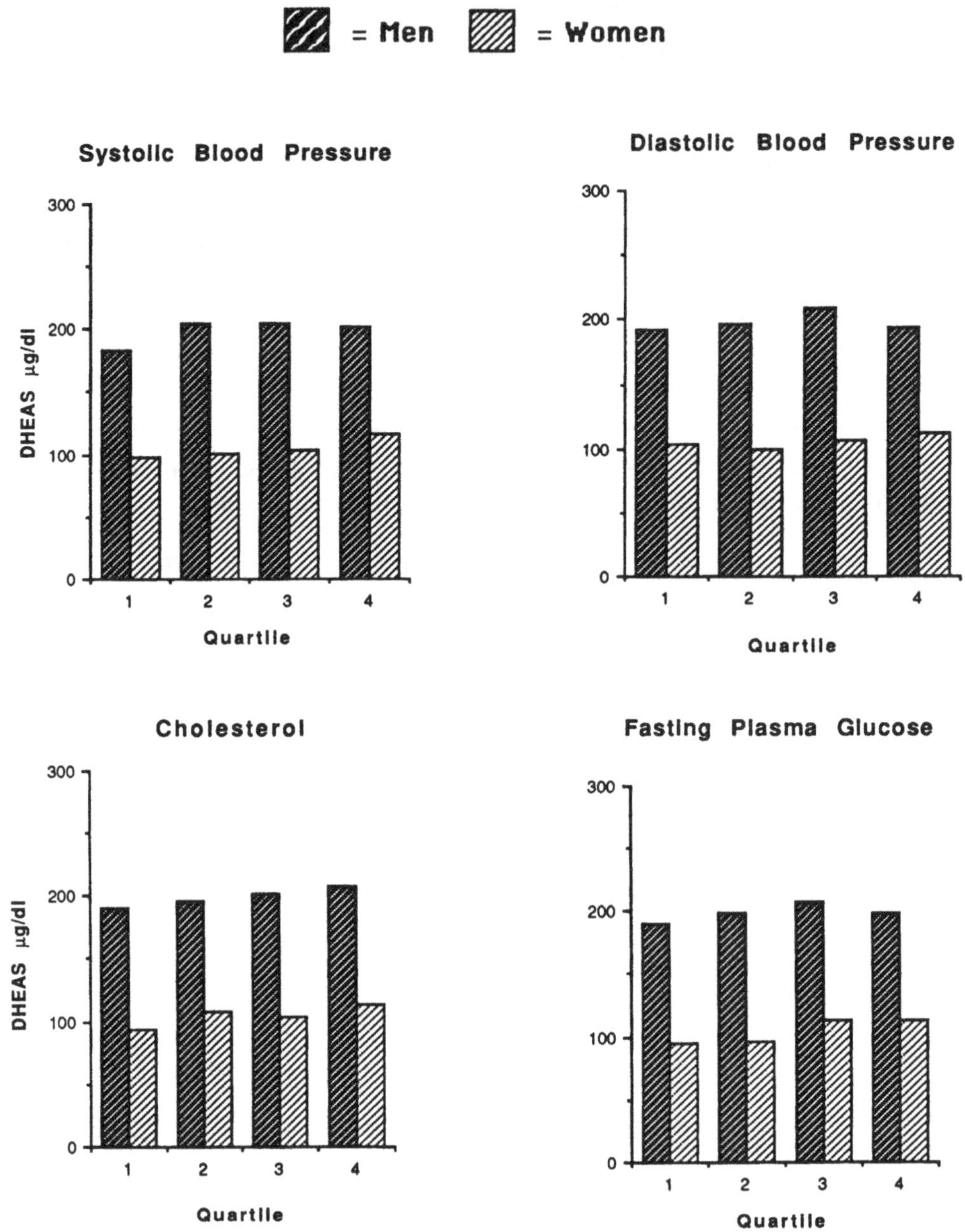

Table 3 shows the lack of any association of DHEAS with HDL, LDL or VLDL cholesterol in the 391 men who had lipoproteins measured (within 100 days of obtaining the blood for DHEAS). There were insufficient numbers of women not taking estrogen who had measurements of both lipoproteins and DHEAS for analyses.

Table 3. Association of age and BMI-adjusted lipoproteins by quartile of DHEAS and partial correlation coefficients, in 391 Rancho Bernardo men

	Quartile of DHEAS				
	1	2	3	4	r
HDL mg/dl	48	51	50	53	0.07
LDL mg/dl	157	156	158	162	0.07
VLDL mg/dl	24	19	21	20	-0.05

The association of DHEAS levels with cigarette smoking and alcohol consumption is shown in Figure 4. In both men and women, DHEAS levels were higher in current smokers than nonsmokers, and in current drinkers than nondrinkers. The association of DHEAS with smoking was independent of alcohol. DHEAS levels were highest in those who both drank and smoked, and higher in drinkers than nondrinkers among the nonsmokers. A graded positive dose-response of increasing DHEAS with increasing alcohol intake, shown in Table 4, was found whether alcohol intake was reported for the past 24 hours or the past week.

Figure 4. Age-adjusted DHEAS levels by smoking and alcohol intake in Rancho Bernardo men

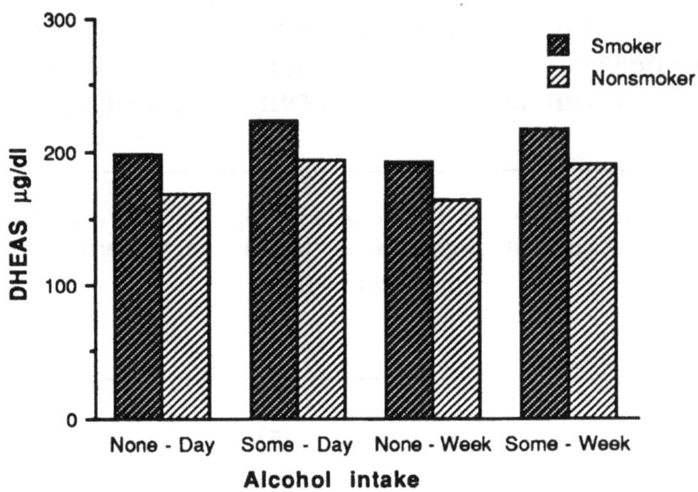

Table 4. Average (sd) dehydroepiandrosterone sulfate (µg/ml) by alcohol intake category in Rancho Bernardo men

	O ml/24h n=131	1-19ml/24h n=55	20-39ml/24h n=68	40ml/24h n=56
Alcohol intake estimated using 24 hour recall				
Crude	176 (99)	186 (119)	193 (130)	215 (131)
Age Adjusted	174	193	201	203
Age & BMI adjusted	174	194	200	205
Alcohol intake estimated using one week history				
Crude	169 (87)	191 (118)	193 (110)	210 (138)
Age adjusted	170	187	193	218
Age & BMI adjusted	170	187	194	218

Correlation with Other Sex Steroids

Table 5 shows the age-adjusted partial correlations of DHEAS with other sex steroids. In men, DHEAS levels were significantly and positively correlated with androstenedione and estrone, and weakly and negatively correlated with sex hormone binding globulin. In women, DHEAS levels were similarly associated with androstenedione and estrone, and were also significantly correlated with testosterone, an association not seen in men.

Table 5. Age-Adjusted partial correlations of DHEAS with sex hormones in Rancho Bernardo men and women

	Men n=1135	Women n=455
Androstendione	0.34***	0.43***
Testosterone	0.00	0.15
Estrone	0.17***	0.16***
Estradiol	0.04	-0.01
Sex Hormone Binding Globulin	-0.05*	-0.04

* $p < 0.05$ ** $p < 0.01$ *** $p < 0.001$

Social Class, Secular Trends and Seasonal Variation

DHEAS levels were unrelated to social class in this rather homogenous upper middle class population. There was a nonsignificant trend to have higher DHEAS levels in women with the least educated husbands, while husbands' own educational levels were associated positively with

DHEAS level. They were not significantly related to current marital status. There was also no significant seasonal variation in DHEAS, based on season of venipuncture, as shown in Figure 5. There also was no secular trend, according to the year the blood was obtained (data not shown).

Figure 5.

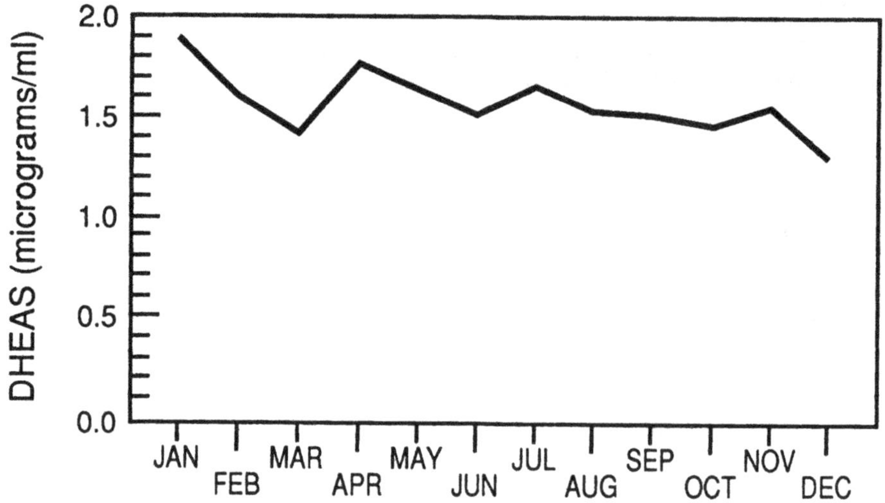

DHEAS and Cardiovascular Disease

Table 6 shows the difference in mean DHEAS levels in men and women who reported a personal history of cardiovascular disease at the time the blood was obtained for the DHEAS assay. Both men and women with cardiovascular disease (CVD) at baseline had significantly lower DHEAS levels than those without such a history.

Table 6. Mean age-adjusted DHEAS level (μg/dl) by personal history of CVD in Rancho Bernardo men and women aged 50-79 years

	No history		Positive history		P
	N	Mean	N	Mean	
Men	208	166	34	124	0.01
Women	248	99	21	61	0.01

The results in the prospective study were quite different for men and women, however. Table 7 shows the 12-year mortality from cardiovascular or ischemic heart disease (IHD) in a subset of 208 men and 248 women aged 50-79 who had no history of heart disease at baseline. As previously reported (1), in men the DHEAS level was significantly and inversely related to the risk of death from cardiovascular disease and death from ischemic heart disease. That is, men with higher DHEAS levels were significantly less likely to have a first cardiovascular event during the next 12 years. In contrast, as reported (2), women with the highest levels of DHEAS had the highest risk of cardiovascular disease mortality. Table 8 shows similar results, when the DHEAS is dichotomized to the approximate 50th percentile for each sex. When adjusted for other risk factors (systolic blood pressure, total cholesterol, fasting plasma glucose, body mass index, cigarette smoking and estrogen use in women) in a Cox model, the relationship of DHEAS to the risk of subsequent ischemic heart disease in men and women was essentially unchanged (relative risk for ischemic heart disease 0.39 in men and 2.03 in women).

Table 7. Mean age-adjusted DHEAS level (μg/dl) by 12-year CVD and IHD death in persons aged 50-79 years with no history of CVD at baseline

		No CVD death		CVD DEATH		P
		N	Mean	N	Mean	
Men						
	CVD	178	171	30	126	0.01
	IHD	184	169	24	131	0.04
Women						
	CVD	226	95	22	133	0.01
	IHD	236	95	12	158	0.01

Table 8. Mean age-adjusted 12-year rates of CVD and IHD death by baseline DHEAS category (μg/dl) in Rancho Bernardo men and women aged 50-79 years

	DHEAS	CVD Rate/100	RR	IHD Rate/100	RR
MEN	< 140	20.9	0.3*	16.8	0.3*
	≥ 140	6.4		5.3	
WOMEN		CVD Rate/100	RR	IHD Rate/100	RR
	< 84	5.1	2.4*	2.5	2.8*
	≥ 84	12.2		7.1	

*$p < 0.05$

Discussion

The decrease in DHEAS with each decade is unique: In this cohort no other sex hormone measured (estrone, estradiol, testosterone or androstenedione) showed a further decrease with age after age 50. The mechanism for the age-related decrease is unknown. It is not seen with other adrenal hormones. The similar decrement in DHEAS in both sexes, does not appear to be explained by obesity, since DHEAS was lower in overweight men but not in overweight women. Aging is associated decreasing weight, with an increasing proportion of body weight that is fat and increased centrality of obesity in both sexes. Percent body fat, which was not measured, could be related to the age related decline in DHEAS. Increased central obesity seems an unlikely explanation in that cigarette smokers and drinkers of alcohol had higher DHEAS levels and more central obesity in both sexes in this cohort (3).

In the followup study, DHEAS was inversely associated with CVD in men and positively associated with CVD in women. The data in women are based on few deaths and should be considered preliminary. Further, the analysis does not test whether the relationship is linear in women. Nevertheless, these data strongly suggest that a high level DHEAS is not predictive of a reduced risk of heart disease in women, as it is in men. A similar sex difference has also been reported in an angiographic study, where DHEAS was related to coronary atherosclerosis in men but not women (4). Experimental studies have shown reduced atherosclerosis in male animals treated with DHEA or DHEAS; female animals have not been studied or reported separately (5,6).

The explanation for this apparent sex difference in the DHEAS-associated risk of cardiovascular disease is unknown. The difference is unlikely to be due to gender differences in other risk factors, because DHEAS was at best only weakly associated with intrinsic heart disease risk factors. In addition, the association persisted after adjusting for a variety of biological and behavioral heart disease risk factors. If these sex differences are confirmed by others, they suggest that the DHEAS effect in men is not mediated by a common metabolic pathway, such as inhibition of the pentose cycle, but a sex-specific hormonal effect.

Women have less heart disease than men. The conventional view is that the female advantage is due to womens' estrogen levels. The majority of studies suggest that estrogen therapy reduces the risk of heart disease in postmenopausal women and increases the risk of vascular disease in men (7). These studies used unphysiological doses of estrogen; they nevertheless suggest that estrogens are good for women and not good for men.

If female hormones are good for women, as they appear to be, it is equally attractive t %%Trailer cd end %%Pages: 8 0 %%EOF od for men. Perhaps the loss of the female advantage after the menopause may be due to the now unopposed androgens produced by the adrenals, as much as estrogen deficiency. Studies in progress which are designed to determine the action of DHEAS may also help to elucidate the thus far unexplained greater risk of premature heart disease in men compared to women.

Summary

We report DHEAS levels and selected covariates in 1135 men and 567 women from the Rancho Bernardo study population of upper middle class Caucasian adults. Data are presented for men aged 30-79 and for women aged 50-79. In this cohort, DHEAS levels were significantly lower in women than in men. In both sexes DHEAS levels decreased steadily and significantly with age. This was true in women who were taking or not taking postmenopausal estrogen therapy; the former had lower DHEAS levels than untreated women.

DHEAS levels were significantly lower in overweight than leaner men but were not related to body mass index in women. In men age adjusted systolic blood pressure and fasting plasma glucose were positively and significantly correlated with DHEAS; total cholesterol and diastolic blood pressure were unrelated to DHEAS. In women, systolic and diastolic blood pressure and total cholesterol were correlated significantly with DHEAS, but fasting plasma glucose was unrelated to DHEAS. DHEAS levels were higher in cigarette smokers than non smokers in both sexes, and there was a dose response relationship. There was a similar dose-response relationship of DHEAS with alcohol intake that was independent of cigarette smoking.

Marital status and educational level attained were unrelated to DHEAS level. There was no evidence of seasonal variation or secular trend in DHEAS.

DHEAS levels were examined for their correlation with other sex hormones measured. In both sexes DHEAS levels were highly correlated with androstenedione and estrone levels. DHEAS was also strongly correlated with testosterone levels in women but not in men.

In cross-sectional analysis, men and women with a history of heart disease had lower DHEAS levels than those without such a history. In the prospective study, there was an inverse association between DHEAS level and risk of CVD death in men. In women, higher DHEAS levels predicted an increased, not decreased, risk of death from CVD.

Acknowledgement

This research was supported by grants (HV-1-2160-L and HL 34591) from the National Heart, Lung, and Blood Institute, and by a grant (AM 31801-02) from the National Institute of Diabetes, and Digestive and Kidney Diseases.

References

1. Barrett-Connor E, Khaw K-T, Yen SSC. A prospective study of dehydroepiandrosterone sulfate, mortality, and cardiovascular disease. N Eng J Med 1986; 315:1519-24.

2. Barrett-Connor E, Khaw K-T. Absence of an inverse relation of dehydroepiandrosterone sulfate with cardiovascular mortality in postmenopausal women. N Eng J Med 1987; 317:711.

3. Laws A, Terry RB, Barrett-Connor E. Behavioral covariates of waist-to-hip ratio in Rancho Bernardo. Amer J Pub Health 1990; Abstract. 2nd International Conference on Preventive Cardiology and the 29th Annual Meeting of the AHA Council on Epidemiology, Washington, D.C., June 18-22, 1989.

4. Herrington DM, Gordon GB, Achuff SC, et al. Dehydroepiandrosterone (DHEA), DHEA Sulfate (DHEAS), and angiographically defined coronary disease (CD). Abstract. 2nd International Conference on Preventive Cardiology and the 29th Annual Meeting of the AHA Council on Epidemiology, Washington, D.C., June 18-22, 1989.

5. Gordon GB, Bush DE, Weisman HF. Reduction of atherosclerosis by administration of dehydroepiandrosterone. A study in the hypercholesterolemic New Zealand white rabbit with aortic intimal injury. J Clin Invest 1988;82:712-20.

6. Arad Y, Badimon JJ, Badimon L, et al. Dehydroepiandrosterone feeding prevents aortic fatty streak formation and cholesterol accumulation in cholesterol-fed rabbit. Atherosclerosis 1989;9:159-66.

7. Bush TL, Barrett-Connor E. Noncontraceptive estrogen use and cardiovascular disease. Epidemiol Rev 1985;7:80-104.

DHEA EFFECTS ON CHOLESTEROL AND LIPOPROTEINS

E.G. MacEwen, A.L. Maki-Haffa, I.D. Kurzman
Department of Medical Sciences, School of Veterinary Medicine
University of Wisconsin, Madison, Wisconsin 53706

Introduction

Cardiovascular disease remains the chief cause of death in the United States and Western Europe, and atherosclerosis, the principal cause of myocardial and cerebral infarction, accounts for the majority of these deaths. High levels of total serum cholesterol and, particularly, of low density lipoproteins (LDL) are identified as high risk factors for the development of atherosclerosis in man (1). The development of atherosclerosis is the result of complex interactions among nutritional, environmental, hormonal, and genetic factors.

Atherosclerosis can be viewed, as proposed by Ross and Glomset and others (2-10), as a "response to injury" hypothesis associated with damaged endothelium, proliferating smooth muscle cells, and LDL-laden macrophages (foam cells) in the intima of the vessel. This progressive proliferative process is influenced by elevated levels of cholesterol (particularly LDL), adherence of platelets and macrophages, the release of chemoattractants and growth factors, and the proliferation of cells. DHEA may modify many of the factors associated with the atherogenic process.

Literature Review

Dehydroepiandrosterone (DHEA) is the major secretory product of the adrenal gland, and serum concentrations of its sulfate ester, DHEA sulfate (DHEA-S), are the most abundant of any steroid next to cholesterol. Although DHEA is the main precursor of placental estrogen and may be

Dehydroepiandrosterone (DHEA)

converted to active androgens and estrogens in peripheral tissues, there is no obvious biological role of either DHEA or DHEA-S in normal animals (11).

The sulfated form represents over 99% of the circulating steroid with less than 1% as free DHEA (12). Plasma concentrations of DHEA-S in humans peak at around 50 to 100 ug/dl at 20 to 30 years of age and decline thereafter. On the average, it is estimated that the decrease in DHEA production is approximately 20% for every decade after age 25 and that by 85-90, production remains constant at 5% (13). DHEA-S concentration exhibits circadian fluctuations. Recently, in a comparison of young (age 27 $\pm$ 1.5 years) and elderly (age 70.3 $\pm$ 7.4 years) men, it was shown that the circadian rhythm is lost in elderly men (14).

DHEA is also a gonadal steroid pre-hormone which can be converted to both androgens and estrogens. DHEA is derived from metabolism of pregnenolone following its hydroxylation to 17-beta-hydroxypregnenolone and cleavage of the C-17 side chain. Although androstenedione, the immediate precursor of testosterone and also a precursor of 17-beta-estradiol, is generally considered to be the immediate product of DHEA metabolism, a variety of other metabolites of DHEA may be produced depending upon species and tissue type.

The striking linear decrease of DHEA and DHEA-S with age as observed in cross-sectional studies of adult humans has led to the postulation that DHEA and DHEA-S are discriminators of life expectancy and aging. Decreased concentrations of DHEA or DHEA-S have been associated with atherosclerosis, cancer, mortality (in general), and death due to cardiovascular disease (15-18). For example, it has been reported that in men, an increase in plasma DHEA-S concentration was associated with a 36% reduction in mortality from any cause and a 48% reduction in mortality from cardiovascular disease (18). Additionally, exogenous administration of DHEA or certain analogues produces an array of beneficial effects on obesity, aging, diabetes, autoimmunity, and cancer initiation and proliferation in experimental animals.

The mechanism of action of the anti-atherogenic and anti-obesity effects of DHEA has not been elucidated. It has been postulated that a decrease in DHEA and DHEA-S concentrations will result in an increase in lipogenesis because DHEA is a potent non-competitive inhibitor of mammalian glucose-6- phosphate dehydrogenase (G6PD), the rate-controlling enzyme of the phosphate Penrose shunt and a major source of extra-mitochondrial NADPH (19,20). NADPH is required for the reductive synthesis of fatty acids and cholesterol (21,22). The rate-limiting step in the biosynthesis of cholesterol is the reaction in which hydroxymethylglutaryl-CoA is converted into mevalonate. Hydroxymethyl-glutaryl-CoA reductase (HMGCoA reductase), the enzyme which catalyzes this reaction, requires NADPH to function (23). NADPH is also necessary for the formation of D-ribose-5-phosphate, which is utilized in nucleotide biosynthesis via the intermediate 5-phosphoribosyl-1-pyrophosphate. Purine ribonucleotide and thymidylic acid biosynthesis are dependent upon tetrahydrofolic acid, which requires NADPH. Also, the enzymatic formation of deoxyribonucleotide diphosphates by ribonucleotide reductase is NADPH dependent. The atherogenic process is considered a proliferative process (8) and has been compared to a neoplastic process which requires DNA synthesis, which is inhibited by DHEA (10,24). DHEA has also been shown to inhibit the differentiation of 3T3 fibroblasts into adipocytes, a process requiring RNA and DNA synthesis (25).

The liver is the primary organ of lipid metabolism. Changes in liver composition and liver metabolism have recently been studied in the NZB/W F_1 female mouse model fed a diet containing 0.4% DHEA (26). This model is characterized by death due to disease similar to human systemic lupus erythematosis (SLE), usually before one year of age. The activity of key regulatory enzymes was determined to assess whether DHEA changed liver metabolism. During the first few days of feeding, there was a rapid decrease in pyruvate kinase activity and an increase in malic enzyme activity. These changes reflect a decreased rate of glycolysis and an increased rate of gluconeogenesis - changes seen during fasting. The animals initially ate less than normal; however, mice adapt to the diet. Most metabolic changes returned to normal, but pyruvate kinase was decreased 50% throughout the study. The low pyruvate kinase activity may indicate a shift from fat metabolism to carbohydrate metabolism in the liver. Phosphoenolpyruvate carboxykinase (PEPCK) activity was low after 2 months on the diet, but not different from control values at 4 and 6 months. (Low PEPCK equals decreased gluconeogenesis.) Fatty acid synthetase remained unchanged in DHEA fed mice. Liver mass was increased due primarily to parenchymal cell hypertrophy which was not edema, but rather was probably due to an increase in intracellular membranous organelles. No significant changes were seen in terms of liver protein, phospholipid, or cholesterol content, but there was a decrease in liver triacylglycerol content. The authors postulate that this indicates an increased rate of utilization of fatty acids by either beta-oxidation or peroxisomal oxidation (26).

Another mechanism by which DHEA could interfere with the atherogenic process is by affecting adherence of platelets and macrophages, the release of chemoattractants and growth factors, and the production of superoxide free radicals (9). DHEA has been shown to block superoxide anion production (27) and DHEA may reduce the pool of arachidonic acid which is metabolized to different eicosanoids through both the 5- and 12-lipoxygenase pathways, which generate various hydroxy fatty acids and leukotrienes, and the cyclooxygenase pathway, which is the source of specific prostaglandins, prostacyclins, and thromboxanes (27).

Cleary et al (28) hypothesize that a futile/substrate cycle of fatty acid deacylation/reacylation might be the mechanism of DHEA's anti-obesity action. Young obese Zucker rats treated with DHEA had near normal body weights, but body fat was not totally normalized. Adult obese Zucker rats lost weight after DHEA treatment. There was no reduction of food intake, therefore, the anti-obesity action is due to lowered caloric efficiency. Insulin levels were found to be lowered in both serum and pancreatic tissue following long-term DHEA treatment. The decrease is first manifested in serum suggesting effects on insulin secretion. Livers were enlarged. Both long-chain fatty acyl-CoA hydrolase and synthetase activities were increased which is the basis of Cleary's hypothesis that a futile cycle of fatty acid deacylation/reacylation accounts for DHEA's anti-obesity action (28).

Lardy et al have offered a mechanism for DHEA's anti-obesity effects. Because DHEA can cause weight loss without decreased food intake it may be called a thermogenic agent. Thermogenesis relates to the production of heat via a decreased efficiency of the tri-carboxylic acid (TCA) cycle. This study found that DHEA induced the synthesis of mitochondrial sn-glycerol 3-phosphate two to five fold in the liver of normal rats. This enzyme decreased metabolic efficiency by one third (29).

Anti-cholesterolemic effects of DHEA have been demonstrated in rats (30) and in another study decreased total body cholesterol in rats (31). Significant hypocholesterolemic activity was shown in non-obese euthyroid dogs with normal cholesterol levels, and in obese euthyroid hypercholesterolemic dogs (32). Gonadal steroids have been shown to affect blood and tissue lipids as well (33-35). It has been demonstrated that estrogen administration results in an increase in the concentration of high density lipoproteins (HDL) and a decrease in the concentration of LDL (36,37). The mechanism for this may involve the stimulation of LDL receptors in the liver, thus, increasing the clearance of LDL (38). In women, Aitken (39) reported an increase in serum cholesterol following oophorectomy, and Mann and Thorogood (40) have suggested that ovarian hormones may protect premenopausal women from Type IV hyperlipidemia. In one study, Hill et (41) reported that estradiol lowered plasma cholesterol and phospholipid levels in the rhesus monkey.

The anti-atherogenic effects of DHEA have been shown in the hypercholesterolemic New Zealand white rabbit model which is characterized by severe aortic plaquing (42). In this study, aortic injury was first introduced using an intra-aortic balloon. Exogenous DHEA, when administered to New Zealand white rabbits on a 2% cholesterol diet, reduced the plaque size in the aorta by 50%. The degree of fatty streak formation, when measured by cross-sectional plaque thickness, was reduced by 39% in rabbits fed both DHEA and cholesterol compared to controls fed only cholesterol. Fatty infiltration of the heart and liver were also markedly reduced. No significant changes were seen in body weight gain, food intake, and total plasma cholesterol or distribution among lipoprotein fractions (42).

A more recent study has reported that DHEA feeding prevents aortic fatty streaking and cholesterol accumulation in the cholesterol-fed New Zealand White rabbit without prior injury. Depending upon the method of analysis used, DHEA feeding resulted in 30-40% inhibition of fatty streak formation. This study may be closer to the human situation as overt injury does not need to be present to cause atherosclerosis (43).

In humans exogenous administration of adrenal androgens have resulted in changes in cholesterol and lipoprotein levels. Hellman et. al. (44) observed that androsterone administered to hypothyroid patients for 3-34 days resulted in an average decrease in serum cholesterol from 307.6 to 238.4 mg/dl. In a preliminary study, it was observed that a patient with a serum cholesterol of 416 mg/dl given 20 mg daily DHEA orally for 6 weeks had a reduction of cholesterol to 264 mg/dl (45). Another study using 200 mg DHEA acetate for 8 days to seven hypercholesterolemia patients resulted in an average decrease in serum cholesterol from 280 to 228 mg/dl (46). Other studies using DHEA and other androgens have shown variable results on cholesterol levels (47-49).

The administration of DHEA to normal men was shown to reduce serum low density lipoprotein (LDL) levels and body fat without altering insulin sensitivity (50). DHEA given orally for 28 days at a dose of 1600 mg produced a 7.5% fall in mean serum LDL cholesterol which is an estimated 14% reduction in risk for the development of cardiovascular disease. The decrease in body fat was without a loss of weight. The hyperinsulinemic-euglycemic clamp technique was used to assess tissue sensitivity to insulin. The results showed no change for either the placebo or DHEA groups. Serum DHEA-S and androstenedione levels rose in the treated group, however, no change was seen in serum concentrations of total testosterone, free testosterone, sex hormone-binding globulin, estradiol, or estrone. The change in the lipid profile is in sharp contrast to that seen with other androgens which typically lower high density lipoprotein (HDL) levels and increase LDL levels. It is more in line with an estrogen effect (51).

We have reported significant hypocholesterolemic activity in both normal euthyroid and obese hypercholesterolemic euthyroid dogs. Six normal and nineteen spontaneously obese dogs received DHEA for three months at an escalating dose of 30 to 75 mg/kg PO daily. A mean 9-12% reduction in total body weight in 68% (13 of 19) of the obese dogs was found, without restriction in food intake. The normal dogs did not lose weight or reduce food intake. Serum cholesterol in obese dogs went from 226 to 173 mg/dl post-treatment ($p < 0.02$). Analysis of lipoproteins in four normal dogs revealed that the marked reduction in cholesterol most significantly affected the LDL-HDL_1 fraction (32).

One of the earliest studies of DHEA's effect on lipid metabolism showed it to inhibit cholesterol biosynthesis in the lactating mammary gland (52). DHEA (3×10^{-5} to 3×10^{-4}M) added to minced guinea pig mammary gland incubated with acetate-1-^{14}C resulted in a marked decrease in the labeling of squalene, lanosterol, and cholesterol, and to a lesser extent to the glycerides. This study does not support the inhibition by DHEA of G6PD activity as the mechanism behind the inhibition of sterol synthesis, because the addition of 6-phosphogluconate did not reverse the inhibition of cholesterol synthesis. It did, however, reverse the inhibition of glyceride synthesis. There are two possible sites of inhibition during sterol synthesis, between acetate and mevalonate, and between mevalonate and squalene, but NADPH is only required as the specific hydrogen donor in the former. No alternative mechanism was postulated by the authors (52).

Lipid Metabolism and Transport

The liver uses cholesterol for the synthesis of lipoproteins which are then exported from the liver. Dietary cholesterol is used when available or the liver is capable of synthesizing its own by increasing the activity of HMGCoA reductase if dietary cholesterol is insufficient.

Plasma lipoproteins are circulating macromolecules consisting of varying amounts of lipid (triglycerides, cholesterol, and phospholipid), protein, and carbohydrate. The specific proteins of the lipoprotein, the apolipoprotein constituents, are responsible for regulating and controlling plasma lipoprotein metabolism. Lipoproteins are separated into various classes according to their lipid content which determines which density they will float at under ultracentrifugation. A greater lipid content yields a lower density and a greater tendency to move upward under ultracentrifugation. Human plasma lipoproteins are commonly divided into six major classes - chylomicrons, chylomicron remnants, very low density lipoproteins (VLDL), intermediate density lipoproteins (IDL), intermediate density lipoproteins (IDL), low density lipoproteins (LDL), and high density lipoproteins (HDL).

Chylomicrons, the largest class of lipoproteins, function to transport dietary triglyceride and cholesterol from the intestinal endothelium to various cells of the body. Lipoprotein lipase (LPL), which is attached to the endothelial surfaces, hydrolyzes the triglycerides and, the fatty acids which are liberated in the process are either used as an energy source by various cells or taken up and stored as triglycerides in adipose tissue. Chylomicron remnants are the resultant lipoprotein particle generated by the action of LPL. The liver rapidly clears these cholesterol-enriched particles under normal physiologic conditions.

VLDL function to transport triglycerides and cholesterol from the liver to various tissues. Its triglycerides are also hydrolyzed by LPL, and additionally by hepatic LPL. The resultant lipoprotein particles in this case are IDL and LDL. LDL function as the major cholesterol-transporting lipoprotein in the blood. Accumulations of LDL, the end product of VLDL metabolism, occurs in patients with familial hypercholesterolemia due to defective receptors for the apolipoproteins on the liver. The level of the accumulation correlates with the existence of accelerated coronary artery disease (53).

HDL are the smallest of the lipoprotein classes. They seem to function in reverse cholesterol transport whereby they acquire cholesterol from peripheral tissues and return it to the liver for redistribution or excretion. There exists a negative correlation with HDL concentrations and accelerated vascular disease (for review see 54).

The distribution of the lipoproteins shows a considerable amount of variation both among and between species, and the distribution is not static in an individual. Diets high in fat and cholesterol are capable of increasing LDL concentrations (55). After ingesting a meal containing fats, there is an increase in chylomicrons and VLDL and possibly a decrease in HDL (56). Adult human females typically have a greater percentage of total lipoprotein present as HDL and less as LDL than males (57,58). Increasing age causes an absolute and percent increase in the fractions of VLDL and LDL (59). With these qualifications in mind, the "normal" human distribution of lipoproteins is approximately 9% VLDL, 64% LDL, and 27% HDL.

Studies in the Rhesus Monkey

DHEA and DHEA-S are also abundant in the sera of non-human primates, including the rhesus monkey, and undergo similar age-related reductions. In one study, DHEA and DHEA-S peaked at 2 to 4 months of age (2600 ng/dl and 50 ug/dl between ages 6 to 36 months (60).

The rhesus monkey is a useful model to study endocrinological factors. A menstrual cycle similar to that seen in humans is observed. Also, there is evidence that the female rhesus monkey undergoes menopause with hormonal and menstrual characteristics that resemble those of women undergoing menopause (61).

The lipoprotein distribution of the rhesus monkey has been studied by a number of groups. As in the human population, there are the typically noted intraspecies differences due to age and sex, and one is able to change the lipid profile by altering the diet. The distribution seen in two normal adult males with an average serum cholesterol value of 165 mg/dl was found to be 1% VLDL, 1% IDL, 37% LDL, and 60% HDL (62). Unpublished data from our laboratory from 20 male and female adult animals with an average plasma cholesterol concentration of 144.6 $\pm$ 13 mg/dl was found to be 4% VLDL, 34% LDL, and 61% HDL. Although not consistent, most studies indicate that total serum cholesterol concentrations will increase with age in the rhesus monkey (63,64). Although arterial lesions associated with atherosclerosis among free-ranging rhesus monkeys have been shown to be rare (65,66) these animals are quite susceptible to the development of advanced atherosclerosis and its complications when fed high fat/high cholesterol diets (67-73).

Based on the above given observations in rhesus monkeys and other species, it was felt that DHEA is an important agent to evaluate in terms of its hypocholesterolemic effects and that the rhesus monkey is an appropriate animal model for such a study.

The hypothesis to be tested was that DHEA would lower serum cholesterol in the rhesus monkey. In particular, we hypothesized that the LDL fraction of the plasma lipoproteins would show the most significant decline in concentration. The aim of the present study was to evaluate the hypocholesterolemic effect(s) of exogenously administered DHEA in the rhesus monkey with hopes that DHEA might be a beneficial therapeutic agent for the treatment of high cholesterol and a potential preventative agent for atherosclerosis in humans.

Recent, yet unpublished, studies (in our laboratory and in conjunction with the Wisconsin Regional Primate Center, University of Wisconsin-Madison) have demonstrated that orally administered DHEA will lower plasma cholesterol and the LDL component.

Twelve non-obese healthy rhesus monkeys, six male and six female, ages 15-21 years were used for this study. The monkeys were on a standardized diet of Purina monkey chow biscuits and fruit. Body weights were measured and daily food intake was monitored.

Monkeys were randomized to receive either DHEA or placebo. There were six monkeys in each group, three males and three females. Monkeys received DHEA or placebo daily, for two months. DHEA was administered at a dose of 60 mg/kg/day divided b.i.d. for the first month and at a dose of 75 mg/kg/day divided b.i.d. for the second month. Due to the insolubility and unpalatability of DHEA, the vehicle of administration was a marshmallow. They are easy for the monkeys to handle without loss and are considered by the monkeys to be a desirable treat. The marshmallows with DHEA or without (placebo) were prepared by the investigators.

The lipoproteins were fractionated using the method of Rudel et al (74). Briefly, 20 ml of blood collected in the presence of EDTA was centrifuged at 2400 rpm for 15 minutes and plasma was collected. Fractionation of the plasma lipoproteins into various density classes (given throughout as g/ml) was accomplished by adding sodium bromide to the plasma and centrifuging in a Beckman LB-70 ultracentrifuge with a 70.1 Titanium rotor. VLDL were isolated at a plasma density of 1.006 by centrifuging at 2.34×10^6 ghr. Other lipoprotein fractions were isolated by sequentially raising the plasma density to 1.019 for IDL, 1.063 for LDL, and 1.225 for HDL. The IDL and LDL fractions were centrifuged for 2.34×10^6 ghr and the HDL fraction for 5×10^6 ghr. The concentration of cholesterol was determined for each fraction using an enzymatic, colorimetric assay.

Cholesterol levels showed a marked decrease in DHEA treated animals over the eight week treatment period. Cholesterol data for animals treated with either DHEA or placebo are shown in Figure 1. Mean total cholesterol in monkeys receiving DHEA decreased from 181 $\pm$ 19.2 mg/dl pre-treatment to 124 $\pm$ 15.4 mg/dl ($p = 0.02$) after four weeks of treatment and to 128 $\pm$ 9.9 mg/dl at eight weeks, a 30% reduction; $p = 0.011$. Monkeys on placebo also had a borderline significant change in cholesterol, but only after eight weeks. They went from 184 $\pm$ 16.1 mg/dl to 169 $\pm$ 10.5 mg/dl at four weeks ($p = 0.053$) to 159 $\pm$ 9.8 mg/dl at eight weeks, a 14% reduction; $p < 0.05$ (See Figure 1). In the treated group the reduction in plasma cholesterol most significantly affected the fraction containing the LDL cholesterol, from a pre-treatment mean of 62 $\pm$ 11.2 mg/dl to 27 $\pm$ 4 mg/dl at four weeks ($p = 0.009$) to 38 $\pm$ 3.3 mg/dl post-treatment, a 39% reduction; $p < 0.05$. Monkeys on placebo had a non-significant 1% increase in LDL cholesterol. The values were 70 $\pm$ 13 mg/dl, 47 $\pm$ 4 mg/dl, and 71 $\pm$ 7 mg/dl for pre-treatment, four weeks, and eight weeks post-treatment, respectively (See Figure 1). There was no significant change in cholesterol concentration in the fraction containing the HDL for the DHEA-treated monkeys. Those on placebo had a 22% reduction (104 ± 7.1 mg/dl pretreatment, to 104 ± 7 mg/dl at four weeks, to 82 ± 3.9 mg/dl post); p = 0.015.

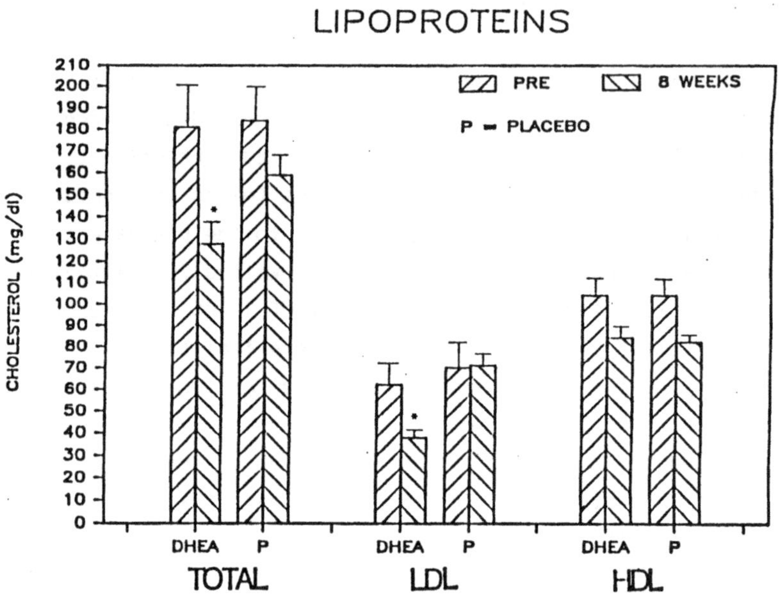

Figure 1: Monkeys receiving DHEA showed a significant decrease in total cholesterol from a pretreatment mean of 181 mg/dl to 128 mg/dl at eight weeks of treatment, a 29% reduction; $p < 0.02$. Monkeys on placebo had an unexplained 14% decrease in total cholesterol, however, when plasma lipoproteins were separated it was found that this reduction affected the HDL cholesterol and not the LDL cholesterol. Monkeys on DHEA had a significant decrease in LDL cholesterol, from a pretreatment mean of 62 mg/dl to 38 mg/dl post-treatment, a 39% reduction; $p < 0.05$. Results are shown in Figure 1.

Conclusion

DHEA is a hormone that remains an enigma. It possesses many characteristics suggesting that it is biologically active and yet its precise physiological function is unknown. There are numerous claims as to the benefits derived from DHEA including anti-obesity, hyperlipidemic, anti-diabetic, cancer preventative, and a host of other effects. This study in the rhesus monkey was stimulated by the finding that serum cholesterol levels in the canine decreased after oral administration of DHEA in an attempt to reduce body weight (32).

DHEA administration for eight weeks resulted in a marked decrease in serum cholesterol levels despite a lack of change in food intake and body weight. The lipoprotein fraction most affected was that containing LDL. These findings suggest that DHEA exerts its effects by altering some complex physiological mechanism(s).

Further studies of DHEA in the primate or other animal model should include how DHEA may possibly alter the changes typically observed on a high-fat, high-cholesterol diet. From the present study one might expect that the administration of DHEA would prevent many of the detrimental changes typically seen. In studies using the New Zealand rabbit (42), plaque size in the aorta was reduced following aortic injury coupled with cholesterol feeding when animals were given DHEA, however, no changes were seen in body weight, food intake, and total plasma cholesterol or distribution among lipoprotein fractions. It is apparent that closer analysis of the underlying biochemical processes needs to be undertaken.

Acknowledgement

This research was supported, in part, by American Heart Association, Wisconsin Affiliate, Grant No. 88-GA-51.

References

1. Brown, M.S., J.L. Goldstein. 1985. Sci. Am. Vol. 251, 58.
2. Ross, R., J.A. Glomset. 1976. N. Eng. J. Med. 295, 396, 420.
3. Ross, R., J.A. Glomset. 1973. Science 180, 1322.
4. Ross, R., L. Harker. 1976. Science 193, 1094.
5. Ross, R. 1981. Atherosclerosis 1, 293.
6. Ross, R. Harvey Lect. 1981-1982, 77, 161.
7. Ross, R. 1986. N. Eng. J. Med. 314, 488.
8. Benditt, E.P., Benditt, J.M. 1973. Proc. Natl. Acad. Sci. 70, 1753.

9. Rogers, K.A., A.R.C. Hoover, J.J. Castellot, et al. 1986. Am. J. Path. 125, 284.

10. Moss, N.S., E.P. Benditt. 1975. Am. J. Pathol. 78, 175.

11. Vande-Weile, P., S. Lieberman. 1960. In: Biological Activities of Steroids in Relation to Cancer. Pincus G. Vollmer E. (eds), New York: Academic Press, p. 93.

12. Zumoff, B., R.S. Rosenfeld, G.W. Strain, et al. 1980. J. Clin. Endocrin. Metab. 51, 330.

13. Orentrich, N., J.L. Brind, R.L. Rizer, et al. 1984. J. Clin. Endocrin. Metab. 59, 551.

14. Montanini, V., M. Simoni, G. Chiossi, et al. 1988. Hormone Res. 29, 1.

15. Bulbrook, R.D., J.L. Hayward. 1967. Lancet 1, 519.

16. Bulbrook, R.D., J.L. Hayward, C.C. Spicer. 1971. Lancet 2, 395.

17. Farewell, V.T., R.D. Bulbrook, J.L. Hayward. 1978. In: Early Diagnosis of Breast Cancer: Methods & Results, Grundmann, E. and Beck, L. (eds), Gustaf-Fischer Verlag, Stuttgart, pp. 43.

18. Barrett-Connor, E., K.T. Khaw, S.S.C. Yen. 1986. New Engl. J. Med. 315, 1519.

19. Marks, P.A., J. Banks. 1969. 1960. Proc. Natl. Acad. Sci. U.S.A. 46, 477-452.

20. Beitner, R., Z. Naor. 1972. Biochem. Biophysics Acta. 285, 437.

21. Lopez, S.A., W.A. Krehl. 1967. Proc. Sc. Exp. Biol. Med. 126, 776.

22. Oertel, G.W., P. Benes. 1972. J. Steroid biochem. 3, 493.

23. Lehninger, A.L. 1982. Principles of Biochemistry, New York: Worth Publishers Inc., p. 261.

24. Gordon, G.B., D.E. Bush, H. Weisman. 1988. J. Clin. Invest. 82, 712.

25. Gordon, G.B., J.A. Hewitt, I.M. Shantz, et al. 1986. Cancer Res. 46, 3389.

26. Miller, B.C., H.W. Lau, N.E. Tyler, and G.L. Cottam. 1986. Biochem. Biophysics Acta 962, 25.

27. Schwartz, A.G., J.M. Whitcomb, D.K.Fairman, et al. 1989. In: Hormones, Thermogenesis, and Obesity, H. Lardy, F. Stratman (eds), Elsevier Science Publ, NY, pp. 441-452.

28. Cleary, M.P. 1989. In: Hormones, Thermogenesis, and Obesity, H. Lardy, F. Stratman (eds), Elsevier Science Publ, NY, 365.

29. Lardy, H., C-Y Su, N. Kneer, S. Wielgus. 1989. In: Hormones, Thermogenesis, and Obesity, H. Lardy, F. Stratman (eds), Elsevier Science Publ, NY, p. 415.

30. Ben-David, M., S. Dikstein, G. Bismuth, et al. 1967. Proc. Soc. Exp. Biol. Med. 125, 1136.

31. Kritchevsky, D., S. Tepper, D. Klurfeld, et al. 1983. Pharmacol. Res. Commun. 15, 797.

32. Kurzman, I.D., E.G. MacEwen, A.L.M. Haffa. 1989. Intl. J. Obesity (in press) vol. 13.

33. McGill, H.C. Jr., M.P. Stern. 1979. In: Atherosclerosis Reviews, R. Paoletti, A.M. Gotto, Jr. (eds), Raven Press, NY, pp. 157.

34. Furman, R.H., P. Alaupovic, R.P. Howard. 1967. Prog. Biochem. Pharmacol. 2, 215.

35. Solyom, A. 1972. Lipids 2, 100.

36. Wallentin, R., U. Larson-Cohn. 1977. Acta Endocrinol. (Copenh) 86, 597-607.

37. Krauss, R.M., F.T. Lindgren, J. Wingred,et al. 1979. Lipids 14, 113.

38. Ma, P.T.S., T. Yamamoto, J.L. Goldstein, et al. 1986. Proc. Natl. Acad. Sci. 83, 792.

39. Aitken, J.M., A.R. Lorimar, D. McKayhart, et al. 1971. Clin. Sci. 41, 597.

40. Mann, J.I., M.Thorogood. 1975. Brit. Heart. J. 37, 790.

41. Hill, P., S.A. Tepper, D. Kritchevsky. 1978. Proc. Soc. Exp. Biol. Med. 157, 231.

42. Gordon, G.B., D.E. Bush, H. Weisman. 1988. J. Clin. Invest. 82, 712.

43. Arad, Y., J.J. Badimon, L. Badimon, et al. 1989. Arteriosclerosis 9, 159.

44. Hellman, L., H. Bradlow, B. Zumoff, et al. 1959. J. Clin. Endocrinol. Metab. 19, 936.

45. Kask, E. 1959. Angrology 10:358.

46. Felt, V. and L. Starka. 1966. Cor VASA 8, 40.

47. Mattson, L., G. Cullberg, P. Tang Keo, et al. 1980. Maturitas 2, 301.

48. Adlercreutz, H., J. Kersteli, K. Schaumann, et al. 1972 Eur. J. Clin. Invest. 2, 91.

49. Ehnholm, C., J. Huttunen, P. Kinnunen, et al. 1975. N. Engl. J. Med. 292, 1314.

50. Nestler, J.E., C.O. Barlascini, J.N. Clore, and W.J. Blackard. 1988. J. Clin. Endocrin. Metab. 66, 57.

51. Mooradian, A.D., J.E. Morley, S.G. Korenman. 1987. Endocrin. Rev. 8, 1.

52. Gianelly, A.A., C. Terner. 1968. Endocrin. 83, 1311.

53. Goldstein, J.L., M.S. Brown. 1982. Med. Clin. North Am. 66, 335.

54. Heiss, G., N.J. Johnson, S. Reiland, et al. 1980. Circulation 62 (Suppl. IV):116.

55. Nichols, A.V., V. Dobbin, J.W. Gofman. 1957. Geriatrics 12, 7.

56. Havel, R.J. 1957. J. Clin. Invest. 37, 848.

57. Barclay, M., R.K. Barclay, V.P. Skipski. 1963. Nature, Lond. 200, 362.

58. DeLalla, O.F., H.A. Elliot, J.W. Gofman. 1954. Am. J. Physiol. 170, 333.

59. Lewis, L.A., F. Olmstad, I.H. Page, et al. 1957. Circulation 16, 227.

60. Cutler, G.B. Jr., M. Glenn, M. Bush, et al. 1978. Endocrinol. 103, 2112.

61. Hogden, G.D., A.L. Goodman, A. O'Conner, D.K. Johnson. 1977. Am. J. Obstet. Gynecol. 127, 581.

62. Scanu, A.M., C. Edelstein, L. Vitello, et al. 1973. J. Biol. Chem. 248, 7648.

63. Lacko, A.G., J.L. Davis. 1979. J. Am. Geriatrics Society 17, 212.

64. Kessler, M.J., R.G. Rawlins. 1983. Experimental Gerontology 18:1, 1983.

65. Chawla, K.K., C.D.S. Murthy, R.N. Chakravarti, et al. 1967. Am. Heart J. 73, 85.

66. Lapin, B.A., L.A. Yakovleva. 1963. Comparative Pathology in Monkeys, Springfield, Illinois, Charles C. Thomas, 1963 p. 144.

67. Cox, G.E., L.G. Nelson, C.B. Taylor. 1954. Circulation 10, 603 (38).

68. Wissler, R.W., R.H. Hughes, L.E. Frazier, et al. 1965. Circulation, 32 (suppl. II): II-220.

69. Armstrong, M.L. 1976. Primates Med. 9, 16.

70. Fraser, R., L. Dubien, F. Musil, et al. 1972. Atherosclerosis 16, 203.

71. Portman, O.W., L.J. Sinisterra. 1957. J. Exp. Med. 106, 727.

72. Malinow, M.R., C.A. Maruffo, A.M. Perley. 1966. J. Pathol. Bacteriol. 92, 491.

73. Armstrong, M.L., M.B. Megan, E.D. Warner. 1974. Circ. Res. 34, 447-454.

74. Rudel, L.L., D.G. Greene, R. Shah. 1977. J.Lipid Res. 18, 734.

DIGITALIS-LIKE MATERIALS AND DHEA-SULFATE

F.I. Chasalow, S.L. Blethen
Schneider Children's Hospital, Long Island Jewish Medical Center, New Hyde Park, New York 11042

Introduction

The cardiac glycosides are an important group of natural steroids that are derived from plants. They affect mammalian electrolyte transport by direct interaction with NaK-ATPase. Based on the demonstration of a high affinity specific receptor for cardiac glycosides, it seems probable that they function by substituting for endogenous mammalian compounds. An example of this pattern is seen in the relationship between the opiates, which are derived from poppies, and the endogenous mammalian peptides, the endorphins, enkephalins, and dynorphins. By analogy, the endogenous (mammalian) cardiac glycosides could be identified by their ability to displace either digoxin or ouabain from receptors. Because these receptors are unstable and hard to purify, many investigators have studied the displacement of digoxin from its antibodies rather than from the digoxin receptor. Thus, the endogenous cardiac glycosides are usually identified and quantitated by their displacement of digoxin (or ouabain) from digoxin-specific antibodies. This chapter will consider the possible relationship between DHEA-S and endogenous cardiac glycosides.

Cardiac glycosides

Cardiac glycosides are a group of naturally occurring steroid conjugates with cardiotonic activity. They were first isolated

Dehydroepiandrosterone (DHEA)

from foxglove or *Digitalis*, a common English garden plant. In 1776, Withering described the use of extracts of digitalis to treat dropsy (edema) and other cardiac problems (1). The active principles from digitalis have since been isolated. Digoxin is currently the most widely used cardiac glycoside (2). Structurally, digoxin resembles DHEA-S in that it is a hydroxylated androgen conjugated to a polar moiety via the 3 position of the A-ring. It differs from DHEA-S by the presence of an extra lactone ring attached to the D-ring and the polar moiety is a tri-digitoxose sugar rather than a sulfate group.

Development of RIAs for digoxin

Digoxin is a very difficult drug to use because of its a narrow therapeutic ratio. Serum levels of < 0.8 ng/ml are ineffective but levels greater than 2.7 ng/ml often are toxic (3). Originally, rather cumbersome chemical methods were used to determine serum levels of digoxin (4), but at the present time, most clinical laboratories are using antibody based methods. In 1957, Lieberman and co-workers (5) were the first to consider a steroid (testosterone) as a hapten, chemically couple it to a protein (albumin) and use the conjugate as an antigen to prepare polyclonal antibodies (to testosterone) in a rabbit. Within 10 years, their technique was widely copied and had been applied to most steroids. In 1969, Smith *et al* (6) applied the technique to digoxin and developed antibodies by coupling BSA to digoxin via one of the hydroxyl groups of the digitoxose sugar groups in a non-specific manner. In 1970, New England Nuclear was the first to market a ligand-based assay kit for digoxin measurements in serum samples. Digitoxin (a related compound lacking a hydroxyl group at position 14) showed 5% cross-reactivity and the neutral steroids tested showed < 0.02% cross-reactivity. This RIA was widely used to prevent both inadequate digitalization and digitalis toxicity. Over the years, many manufacturers have produced kits each with its own unique antibody. Each kit adequately quantitates authentic digoxin added to serum samples from untreated, normal individuals. However, they differ in cross-reactivity with the digoxin-like materials.

Definition of digoxin-like materials

A practical definition of a digoxin-like material is: An endogenous substance found in mammalian samples (known not to contain digoxin, other cardiac glycosides, or other materials extracted from digitalis) that is detected as cross-reacting material by antibodies specific for digoxin (7).

The definition can be extended by including: (a) ability to displace digoxin from digoxin receptors or binding proteins or (b) ability to inhibit NaK-ATPase activity (8). The additional requirements may be overlapping because the major digoxin-binding protein is NaK-ATPase (9) and when digoxin binds to the NaK-ATPase, it inhibits ATP hydrolysis (10). The addition of these functions extends the definition of digoxin-like materials from one involving only immunochemical similarity to digoxin to one involving similarity of biochemical function. However, to confuse the nomenclature, many investigators actually use ouabain (instead of digoxin) in binding studies because it is more soluble in aqueous buffers than is digoxin. Thus, some investigators designate the endogenous material as ouabain-like material, rather than digoxin-like material (11). In this chapter, we will refer to these compounds as digoxin-like materials regardless of the specific ligand used.

Possible biological function of digoxin-like materials

As long ago as 1953, Szent-Gyorgyi suggested that the reason that there were specific binding proteins or receptors for cardiac glycosides was that the receptors were actually specific for an endogenous material and that the cardiac glycosides just substituted for this substance(s) (12). At present, there are at least 40 laboratories investigating endogenous digoxin-like materials. Most are concerned with identifying substances that interfere in clinical digoxin assays and eliminating their effects. Several laboratories have isolated and characterized digoxin-like materials from specific sources. To date, no laboratory has reported a unique chemical structure that might function as the endogenous regulator of the mammalian NaK-ATPase pump.

Recognition of digoxin-like materials

A few patients were found to have measureable serum digoxin levels (determined by RIA) prior to the initiation of therapy (13). This was particularly common in newborn infants (14) and patients with hypertension (15) or kidney disorders (16). Since the immunoreactive material was not digoxin but cross-reacted with digoxin, the unknown material that bound to the antibody was designated digoxin-like material (or some variant of that name).

Clinical laboratories attempted to reduce the effect of these endogenous cross-reacting substances in different ways. Some laboratories extracted and purified the glycoside fraction by thin layer chromatography prior to RIA (17); others changed reagents. Although each kit adequately quantitates authentic digoxin in most serum samples, each detects different amounts of digoxin-like materials, even when the same samples are assayed. For example, Graves et al (18) tested serum from women in the third trimester of pregnancy with 4 different digoxin RIA kits. The results ranged from very low levels (undetectable to 0.07 ng/ml) to much higher levels (the extreme values were 0.05 ng/ml to 0.6 ng/ml), even though none of the women had been treated with any cardiac glycosides (18). Similarly, Valdes (19) examined apparent serum digoxin concentrations in subjects with renal disease using 6 different kits. Although samples from normal subjects had digoxin levels < 0.02 ng/ml when assayed with all 6 kits, with 4 of the kits digoxin levels were >0.15 ng/ml when samples from patients with renal disease were assayed. There are 2 possible explanations for this variability: (a) antibodies developed to digoxin might be specific for different regions of the digoxin molecule and therefore might not have equal cross-reactivity for the equivalent regions of an endogenous digoxin-like material or (b) there might be more than one digoxin-like material in serum. Considered individually, the 2 studies described cannot distinguish between the 2 alternatives. However, 3 kits were used in both studies. One of these detected little or no digoxin-like material in both groups of subjects. This kit might be the best of the 3 for determining true digoxin concentration in serum. Of the other 2 kits used in both studies, one detected more endogenous digoxin activity in pregnant women while the other found more digoxin-like material in patients with renal disease. This pattern of different apparent digoxin concentrations in different clinical states cannot be attributed simply to differences in antibody specificity. This result would occur only, if, in addition to differences in antibody specificity, there were different digoxin-like immunoreactive materials present in pregnancy as compared to renal disease.

Steroid sulfates as digoxin-like materials

Although digoxin is a steroid derivative, only a few investigators have considered the possibility that a steroid or steroid derivative might be the putative digoxin-like material (20). Vasdev et al (21) isolated a digoxin-like material from normal human serum that was indistinguishable from DHEA-S in several chromatographic systems. Authentic DHEA-S was tested and shown to be detected as a digoxin-like material. On a quantitative basis, DHEA-S accounted for over 62% of the total digoxin-like material present in adult serum. In a separate study, Vasdev et al (22) examined serum from patients with renal disease. In spite of the fact that levels of immunoassayable digoxin-like materials were increased, DHEA-S levels were not increased in these samples. Thus, DHEA-S could not account for the increased levels of digoxin-like materials found in these patients.

The same paradox of increased serum digoxin-like materials and equivalent serum DHEA-S levels is also seen in the comparison of serum from newborns and normal adults. To explore this question, we studied digoxin-like materials in cord serum (a mixed pool of umbilical arterial and venous blood) (23). Using Sephasorb HP™ chromatography (methanol:water - 1:1), over 70% of the digoxin-like materials co-eluted with the steroid sulfate fraction. Enzymatic hydrolysis of the pooled steroid sulfate fraction followed by a second chromatography step (using Sephadex LH-20) resolved 4 digoxin-like materials each of which was also detected with assays for DHEA. Of these, 3 were identified as DHEA, 16α-hydroxy-DHEA, and 15β-hydroxy-DHEA. The fourth steroid has not yet been identified, but it may be 15β,16α-dihydroxy-DHEA on the basis of its cross-reactivity and relative retention. When authentic samples of these steroids were assayed for digoxin-like immunoreactivity, the total amount of digoxin-like material which could be attributed to these compounds, given their known concentrations in pooled cord serum, was similar to the total amount of digoxin-like material originally detected (23). In summary, at least 70% of the immunoreactive digoxin-like materials present in normal cord serum can be attributed to the presence of large amounts of steroid sulfates which are known to be present in such samples.

Chasalow and Bradlow (24) detected large amounts of digoxin-like materials in fluids obtained from women with gross fibrocystic disease of the breast. Chromatography on Sephasorb HP revealed 3 separate peaks of digoxin-like materials each of which contained about one-third of total digoxin-like activity. The third peak eluted in the same region as the steroid sulfates which are also known to be present in high concentrations in breast cyst fluids (25). The steroid sulfate fraction was isolated and hydrolysed; when analysed by RIAs specific for DHEA or DHEA-like materials, the hydrolysate contained materials that co-eluted with DHEA, 16α-hydroxy-DHEA, 15β-hydroxy-DHEA, and an additional unknown material (again this may be 15β,16α-dihydroxy-DHEA). Thus, the steroid sulfates account for the bulk of the digoxin-like materials present in the third peak, but, in contrast to cord serum, this peak represents about one-third of the total amount of digoxin-like materials in human breast cyst fluid.

In addition to the steroid sulfate peak, 2 other peaks of digoxin-like materials were separated by Sephasorb HP chromatography from extracts of human breast cyst fluid (24). Both peaks cross-reacted with some, but not all, antibodies specific for DHEA-S (DHEA-S like material) (25). When hydrolysates from Peak 2 were chromatographed, the same pattern of DHEA-like steroids was detected as was detected after hydrolysis of the steroid sulfate peak. Thus, this fraction contains the same DHEA-like moieties that were present in the steroid sulfate fraction. However, they must be conjugated to a different group at the 3 position in order to account for their different chromatographic mobility (24). The nature of the digoxin-like materials in the first peak has not yet been determined.

In summary, DHEA-conjugates and their hydroxy derivatives account for 65% or more of the radioimmunoassayable digoxin-like material detected in normal adult serum, mixed cord serum and breast cyst fluids. In the fetus, DHEA-S and its hydroxy metabolites are produced by the fetal adrenal and function as precursors for estrogen synthesis by the placenta. They have only a minimal effect on the activity of the NaK-ATPase and no other known function. The hydroxy DHEA-S derivatives are also present in low concentrations in serum from adults where their function is unknown, except during pregnancy (26).

One intriguing possibility is that DHEA-S may serve as a precursor for the true ligand for the digoxin receptor. The true ligand might be one of the other conjugates detected in the cyst fluid. This would be analogous to the role of vitamin D which serves as a precursor (prohormone) for 1,25-dihydroxy vitamin D.

Evidence for digoxin-like materials in serum that are not DHEA-S

As mentioned previously, patients with renal failure have high levels of digoxin-like material but do not have corresponding increases in the amount of DHEA-S present in their serum (22). Thus, although DHEA-S can be detected as digoxin-like material in digoxin assays and is present in high concentrations in normal and newborn serum, its presence does not account for the increased amounts of digoxin-like materials found in serum from patients with renal disorders (22). In their study, Vasdev et al (22) used a single antibody with specificity for the sulfate group at position 3 of DHEA-S. No chromatographic or other separations steps were employed. Thus, their methods would not have detected other conjugates of DHEA such as were found in breast cyst fluid by Chasalow and Bradlow (24).

In patients with poorly controlled salt-losing congenital adrenal hyperplasia due to 21-hydroxylase deficiency (but not those with the simple virilizing form), there are large amounts of digoxin-like materials present. The levels of digoxin-like materials decrease when 17-hydroxyprogesterone levels (an index of adrenal suppression) normalize. Although DHEA-S levels are also elevated in these patients, the increased amount of digoxin-like materials cannot be attributed solely to that compound (27).

Isolation of non-steroidal digoxin-like materials from mammals

Investigators have monitored the purification of mammalian digoxin-like materials by measuring the immunocrossreactivity with digoxin RIAs (28,29,30). Each study carefully notes that the source had not been pre-treated with cardiac glycosides and, therefore, that these were not plant materials. However, treatment with cardiac glycosides is not the only source of plant digoxin-like immunoreactivity in serum. Diet can also be involved. Kelly isolated digoxin-like materials from rat urine and showed that the amount present decreased when the rats' diet was changed. He also found similar digoxin-like materials in the rat chow itself. On the basis of these experiments, he concluded that the source of these digoxin-like materials was the rat chow (31). Thus, the digoxin-like material was really of plant rather than mammalian origin.

Tamura et al isolated a digoxin-like material from hog plasma and showed that it co-eluted with free fatty acids (32). They also tested authentic long chain fatty acids and showed that they cross-reacted in the assay for digoxin. Phosphatidyl choline has also been shown to qualify as a digoxin-like material (33). In fact, both free fatty acids and phosphatidyl choline have been shown to inhibit activity of NaK-ATPase. Thus, these lipids even fit the extended definition of digoxin-like material. However, only a small portion of total digoxin-like material in serum could be attributed to the presence of this class of lipids.

Graves attempted to isolate digoxin-like materials from human serum and tissues. He isolated a heat stable material with an apparent molecular weight of 150 to 250 daltons (34). Smith isolated four different materials from human serum, but did not completely characterize any of them (35). Tamura et al isolated a material with molecular weight of 336 daltons from bovine adrenal (36). The materials isolated by these investigators share several characteristics. Each was obtained by methanol extraction, which served to denature the specific binding proteins. Each could not be extracted from aqueous solutions at neutral pH. Each had characteristic chromatographic elution profiles on HPLC. But, because none of the columns used were truly equivalent, the question as to whether or not these different isolates are structurally related can not be answered. Thus, it is possible that these investigators are isolating the same material from different sources. In any case, their chemical identity remains unclear.

Summary

In this chapter we have defined the concept of endogenous mammalian digoxin-like materials and the chemical nature of of some of the materials that have been isolated. Although there are several isolates that are interesting, none has been completely identified and shown to have the properties expected of the true ligand for the digoxin receptor. DHEA-S is certainly a major contributor to digoxin-like activity in serum, especially in normal adults. DHEA-S and its hydroxy derivatives also appear to account for the bulk of digoxin-like material in newborn serum. But, other compounds are also present. Other DHEA conjugates, such as those found in breast cyst fluid, may actually be the active hormone. It remains to be seen if any material isolated as a digoxin-like material or a DHEA-S like material is the true endogenous regulator of NaK-ATPase or functions as a natriuretic hormone.

References

1. Withering, W.. 1785. An acount of the foxglove and some of its medical uses: With practical remarks on dropsy and other diseases. G. G. J. and J. Robinson, London.

2. Doherty, J.E. and J.J. Kane. 1975. Clinical pharmacology of digitalis glycosides. Ann.Rev.Med. 26, 159-171.

3. Ingelfinger, J.A. and P. Goldman. 1976. The serum digoxin concentration - Does it diagnose digoxin toxicity? N.Engl.J. Med. 294, 867-870.

4. Lowenstein, J.M.. 1965. A method for measuring plasma levels of digitalis glycosides. Circulation 31, 228-233.

5. Erlanger, B.F., F. Borek, S.M. Beiser and S. Lieberman. 1957. Steroid protein conjugates I. preparation and characterization of conjugates of bovine serum albumin with testosterone and with cortisone. J.Biol.Chem. 228, 713-727.

6. Smith, T.W., V.P. Butler and E. Haber. 1969. Determination of therapeutic and toxic digoxin concentrations by radioimmunoassay. N.Engl.J.Med. 281, 1212-1216.

7. Valdes, R.. 1985. Endogenous digoxin-like immunoreactive factors: impact on digoxin measurements and potential physiological implications. Clin. Chem. 31, 1525-1532.

8. Skou, J.C.. 1988. Overview: The NaK pump. In: Methods in Enzymology, Vol. 156 (S. Fleisher and B. Fleisher, eds.). Academic, San Diego, CA. pp. 1-25.

9. Wallick, E.T. and Schwartz. 1988. Interaction of cardiac glycosides with Na^+,K^+-ATPase. In: Methods in Enzymology, Vol 156 (S. Fleisher and B. Fleisher, eds.). Academic, San Diego,CA. pp. 201-213.

10. Forbush, B.. 1983. Cardiotonic steroid binding to NaK-ATPase. In: Current Topics in Membrane Transport, (J.F. Hoffman and B. Forbush, eds.), Academic, New York. pp. 167-201.

11. Lichtstein, D. and S. Samuelov. 1980. Endogenous 'ouabain-like' activity in rat brain. Bioch. Biophys. Res. Commun. 96, 1518-1523.

12. Szent-Gyorgyi, A.. 1953. Chemical physiology of contraction in body and heart muscle. Academic, New York, as cited in LaBella, F.S.. 1985. Fed. Proc. 44, 2780-2781.

13. Balzan, S., A. Cherizo, M. Grazia del Chieca, U. Montali and S. Quione. 1984. Digoxin-like immunoreactivity in normal human plasma and urine as detected by solid-phase radioimmunoassay. Clin. Chem. 30, 450-457.

14. Besch, H.R., S. Hufferd, M. Lake, R. Hurwita and A.M. Watanbe. 1976. False elevation of apparent digoxin levels in plasma of premature infants. Clin. Chem. 22, 1168.

15. Devynck, M.A., M.G. Pernoullet, J.B. Rosenfeld and P. Meyer. 1983. Measurement of digitalis-like compound in plasma: application in studies of essential hypertension. Br.Med.J. 287, 631-634.

16. Graves, S.W., B.A. Brown and R. Valdes. 1983. Endogenous digoxin-like substance in patients with renal impairment. Ann. Int. Med. 99, 604-8.

17. Skogen, W.F., M.R. Rea and R. Valdes. 1987. Endogenous digoxin-like immunoreactive factors eliminated from serum samples by hydrophobic silica-gel extractions and immunoassay. Clin. Chem. 33, 401-404.

18. Graves, S.W., R. Valdes, B.A. Brown, A.B. Knight and H.R. Craig. 1984. Endogenous digoxin-immunoreactive substance in human pregnancies. J.Clin.Endocrinol.Metab. 58, 748-751.

19. Valdes, R.. 1985. Endogenous digoxin-immunoreactive factor in human subjects. Fed.Proc. 44, 2800-2805.

20. Schreiber, V., J. Stepan and I. Atarka. 1981. Digoxin-like immunoreactivity of certain steroids and other hormones. Physl. Bohem. 30, 569-571.

21. Vasdev, S., L. Longerich, E. Johnson, D. Brent and M. H. Gault. 1985. Dehydroepiandrosterone sulfate as digitalis like factor in plasma of healthy adults. Res. Comm. Chem. Path. Pharm. 49, 387-399.

22. Vasdev, S., E. Johnson, L. Longerich, V. M. Prabhakaran and M. H. Gault. 1987. Plasma endogenous digitalis-like factors in healthy individuals and in dialysis dependent and kidney transplant patients. Clin. Nephrol. 27, 169-174.

23. Chasalow, F.I. and S.L. Blethen. 1990. Characterization of digoxin-like materials in human cord serum. In: Steroid formation, degradation and action in peripheral normal and neoplastic tissues, (L. Castagnetta, S. D'Aquino, H.L. Bradlow and F. Labrie, eds.). N.Y. Acad.Sci. (in press).

24. Chasalow, F.I. and H.L. Bradlow. 1990. Digoxin-like materials in human breast cyst fluids. In: Biochemistry of breast cyst fluids and cancer risk. (H.L. Bradlow, A. Angeli, F.I. Chasalow and L. Dogliotti, eds.). N.Y. Acad.Sci. (in press).

25. Chasalow, F.I., S.L. Blethen and H.L. Bradlow. 1988. Dehydroepiandrosterone sulfate (DHEA-S) and DHEA-S like compounds in fibrocystic disease of the breast. Steroids 52, 205-215.

26. Sekihara, H., J.A. Senett, G.W. Liddle, T.J. McKenna and L.R. Yarbro. 1976. Plasma 16β-hydroxydehydroepiandrosterone in normal and pathological conditions in man. J.Clin.Endocrinol. Metab. 43, 1078-1084.

27. Blethen, S.L., F.I. Chasalow and J.L. Kirkland. 1988. Digoxin-like material (DLM) in congenital adrenal hyperplasia. Pediatr.Res. 23, 273 (Abstract).

28. Gruber, K.A., J.M. Whitaker and V.M. Buckalew. 1980. Endogenous digitalis-like substance in plasma of volume expanded dogs. Nature 287, 743-745.

29. Schreiber, V., I. Gregorva, T. Pribyl and J. Stepan. 1981. Digitalis-like biological activity and immunoreactivity in chromatographic fractions of rabbit adrenal extract. Endocrinol.Exp. 15, 229-236.

30. Morgan, K., M.D. Lewis, G. Spurlock, P.A. Collins, S.M. Foord, K. Southgate, M.F. Scanlon and M.A. Mir. 1985. Characterization and partial purification of the sodium-potassium ATPase inhibitor from cultured rat hypothalamic cells. J.Biol.Chem. 260, 13595-13600.

31. Kelly, R.A.. 1986. Excretion of artifactual endogenous digitalis-like factors. Am.J.Physiol. (Heart Circ. Physiol. 20) 251, H205-H209.

32. Tamura, M., H. Kuwano, T. Kinoshita and T. Inagami. 1985. Identification of linoleic and oleic acids as endogenous Na^+/K^+-ATPase inhibitors from acute volume-expanded hog plasma. J.Biol.Chem. 260, 9672-9677.

33. Kelly, R.A., D.S. O'Hara, W.E. Mitch and T.W. Smith. 1986. Identification of NaK-ATPase inhibitors in human plasma as non-esterified fatty acids and lysophospholipids. J.Biol.Chem. 261, 11704-11711.

34. Graves, S.W.. 1988. Isolation and purification of digitalis-like factor. U.S. Patent 4,780,314.

35. Kelly, R.A., D.S. O'Hara, M.L. Canessa, W.E. Mitch and T.W. Smith. 1985. Characterization of digitalis-like factors in human plasma: Interactions with NaK-ATPase and cross reactivity with cardiac glycoside specific antibodies. J. Biol.Chem. 260, 11396-11405.

36. Tamura, M., T.T. Lam and T. Inagami. 1988. Isolation and characterization of a specific endogenous Na^+,K^+-ATPase inhibitor from bovine adrenal. Biochemistry 27, 4244-4253.

GLUCOSE-6-PHOSPHATE DEHYDROGENASE AND THE RELATION OF DEHYDROEPIANDROSTERONE TO CARCINOGENESIS.

F. Feo and R. Pascale

Istituto di Patologia Generale dell'Università. Sassari, Italy.

Introduction

Glucose utilization in mammalian cells mainly depends on its conversion to glucose-6-phosphate (G6P). This "mobilized" form of glucose enters all pathways of carbohydrate metabolism. Among them the hexose monophosphate pathway (HMP) plays a unique role in mammals: two of its first three steps are dehydrogenations which represent the major source of NADPH and the <u>only</u> source of ribulose-5-phosphate (Ru5P):

$$\text{D-glucose-6-P} + \text{NADP}^+ \longleftrightarrow \text{D-glucono-}\delta\text{-lattone} + \text{NADPH}$$

$$\text{D-glucono-}\delta\text{-lattone} + H_2O \longrightarrow \text{6-phospho-D-gluconate}$$

$$\text{6-Phospho-D-gluconate} + \text{NADP}^+ \longleftrightarrow \text{D-ribulose-5-phosphate} + \text{NADPH}$$

Since the discovery in 1931 of glucose-6-phosphate dehydrogenase (G6PD) by Otto Warburg, the pioneristic observations of F. Lipman, F. Dickens, S. S. Cohen, B. L. Horecker and E. Racker led to the description of the main steps of HMP. This pathway received much attention from researchers involved in researches on cancer biology and biochemistry. Transplanted hepatocellular carcinomas (HCC) exhibit high G6PD and HMP activities (1). High G6PD activites have also been found in preneoplastic and neoplastic lesions of the colon (2), skin (3), oral mucosa (4) and liver (5,6), as

Dehydroepiandrosterone (DHEA)

well as in regenerating liver and hyperplasia (7). It has been hypothesized that high G6PD and HMP activities are necessary for growth and inhibition of these activities could control tumor growth. Experimental evaluation of this hypothesis has mainly been performed by using dehydroepiandrosterone (DHEA), a non-competitive inhibitor of G6PD (8,9).

Dehydroepiandrosterone and Cancer

DHEA and DHEA sulfate are major secreted products of human adrenals (10). Their plasma concentrations undergo age-related decreases to 10-20% of their maximum level in the 6th-7th decade (11). DHEA sulfate may be converted into active androgens and it is the major precursor of placental estrogens (12). A clue to possible relationship between DHEA and cancer, was the observation of a negative correlation between plasma levels and urinary excretion of DHEA, DHEA sulfate and their metabolites (13-15) and breast cancer in women, or other types of cancer in women and men (16,17). In a prospective study, about 5000 women were scrutinized, for 9 years, for the development of breast cancer (18). A lower urinary excretion of androsterone and etiocholanolone (derived primarily from DHEA) was found in women who developed malignant breast cancer than in matched controls. Premenopausal women with benign breast lesions who seem to have a greater probability of subsequent development of breast cancer (19), exhibit lower plasma levels of DHEA sulfate than the controls (20).

In 1975 Schwartz and Perantoni (21) discovered that DHEA protects *in vitro* cultured rat liver epithelial-like cells and hamster embryonic fibroblasts against 7,12-dimethylbenzo(a)anthracene- (DMBA) and aflatoxin B1-induced cytotoxicity and transformation. DHEA and 16α-bromo-epiandrosterone (Epi-Br), an inhibitor of mammalian G6PD more potent than DHEA (8), inhibit

EBV-induced morphologic transformation and stimulation of DNA synthesis in human lymphocytes _in vitro_ (22). In keeping with these observations Feo et al. found that _in vitro_ cultured human skin fibroblasts (HSF) from individuals carrying the Mediterranean variant of G6PD (G6PD-Med) deficiency, are less sensitive than control HSF to the cytotoxic effect of dimethylnitrosamine (23,24) and benzo(a)pyrene (BaP, 25,26). BaP-treated G6PD-deficient cells were less prone than normal HSF to undergo anchorage-independent growth (26). Treatment of normal HSF with DHEA mimicked the effect of G6PD deficiency (23,24). Similar results were obtained with lymphocytes (HL) from G6PD-deficient individuals (27). These observations suggest a cancer chemopreventive effect of G6PD-deficiency.

Cancer Chemoprevention by Dehydroepiandrosterone.

Cancer prevention by DHEA was first documented for spontaneous breast cancer of female C3H (A^{vy}/e, obese) and C3H A/A mice, subjected to long-term treatment with DHEA (28,29). Topical application of DHEA and the synthetic steroid 3β-methylandrost-5-en-17-one to mouse skin blocks the formation of skin papillomas and carcinomas induced by multiple topical applications of DMBA (30) or by single application of DMBA followed by a promoting treatment with 12-O-tetradecanoylphorbol-13-acetate (TPA, 31). A marked reduction in incidence and number per animal of DMBA- and urethan-induced lung adenomas occurs in A/J mice subjected to a 14-week feeding a diet containing 0.6% DHEA (32). However, long-term feeding of the same diet does not influence the occurrence of nodular lung lesions, but inhibits the development of thyroid tumors and preneoplastic liver lesions in dihydroxy-di-N-propylamine (DHPN)-initiated F344 rats (33). It also inhibits the appearance of _in situ_ and invasive tumors of colon, and of microscopic precursor lesions, induced by 1,2-dimethylhydrazine in Balb/c mice (34). DHEA, however, was reported to not influence the development of azomethane-induced colon carcinoma (35). Recently, Sprague-Dawley rats fed for several weeks a diet containing 0.25% DHEA, during and subsequent to a 7-week treatment with N-nitrosomorpholine (NNM), underwent a reduction in the incidence of liver hemangiosarcoma (36). In the same rats DHEA caused phenotypic modulation of HCC which showed a higher degree of differentiation, lower mitotic rate, and reduced potential for metastases than the HCC developing in rats which did not receive DHEA. Rao (37) has recently reported chemoprevention of methylcholanthrene-induced carcinomas of uterine cervices of adult and virgin Swiss mice by a 18-week feeding a diet containing DHEA percentages as low as 0.05% and 0.1%.

Prevention of cancer development by DHEA, is preceded by inhibition of development of preneoplastic lesions (6,33,38-41). The availability of several experimental models of multistep carcinogenesis has permitted the recognition of multiple developmental stages, sensitive to inhibition by DHEA.

Experimental models for chemoprevention studies

Multistage carcinogenesis experiments start with initiation. This is achieved by topic or systemic treatment with a carcinogenic agent. If a subthreshold dose of carcinogen is given to animals tumor would not develop. However, initiated cells remain latent and preneoplastic lesions and tumors are generated during the stage called promotion. Multistep carcinogenesis models, more frequently used for chemoprevention studies with DHEA or analogues steroids, are the skin model, the "resistant hepatocyte" (RH) and triphasic models and the stop model. In the classical two-stage skin carcinogenesis model (42, Fig. 1A), repeated applications after initiation,

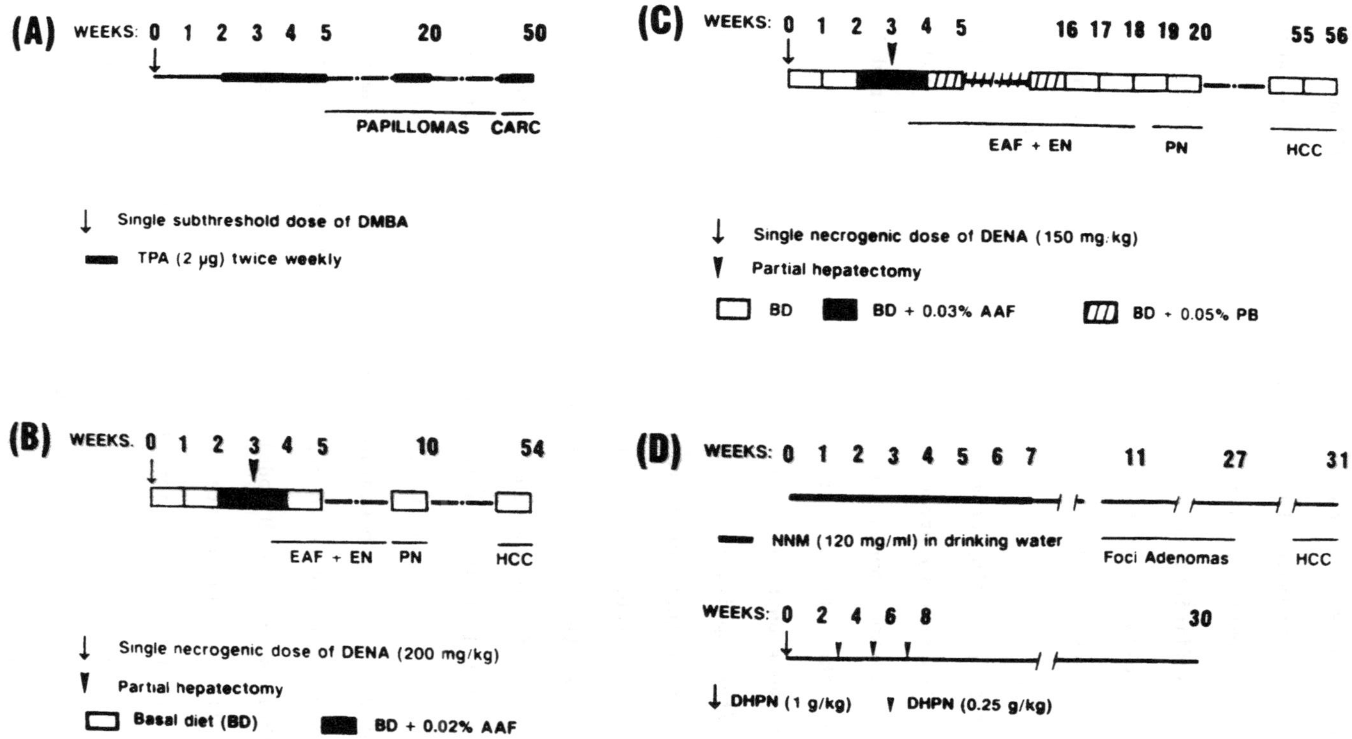

Figure 1. Experimental models more frequently used for the study of tumor chemoprevention by DHEA. (A) Skin multistage model. (B) "Resistant hepatocyte" (RH) model. (C) Triphasic model. (D) Stop model. Abbreviations: AAF, N-acetylaminofluorene; DENA, diethylnitrosamine; DHPN, dihydroxy-di-N-propylamine; DMBA, 7,12-dimethylbenzo(a)anthracene; EAF, enzyme altered foci; EN, early nodules; HCC, hepatocellular carcinoma; PB, phenobarbital; PN, persistent noduyles.

of a skin promoter, such as TPA, lead to the appearance of papillomas, largely dependent on the promoter for growth, and then of carcinomas which grow independently of promoter application (43). Limited TPA treatment of initiated skin, is insufficient to support papilloma development. However short TPA treatment, followed by skin irritants such as 12-O-retinoyl-phorbol-13-acetate, which do not cause tumor development in the absence of TPA, is highly tumorigenic (44). On this basis it has been recognized (43, 44) a conversion step, preceding promotion, which probably induces the promotability of initiated skin by promoters lacking convertogenic activity. A widely used model of experimental hepatocarcinogenesis is the RH model (45). In this model, the rats are initiated by a single, necrogenic dose of carcinogen; reparative growth is thought to fix DNA damage induced by the carcinogen (46). After repair, the rats are fed for 15 days a diet containing 0.02-0.03% N-acetylaminofluorene (AAF), with a partial hepatectomy at the midpoint of this treatment (selection, Fig. 1B). Initiated cells do not metabolize actively AAF (47), thus being relatively resistant to mitoinhibition by AAF (46). Since surrounding cells are sensitive to AAF, it has been suggested that initiated cells undergo rapid, selective growth, under a powerful "promoting" stimulus (reparative growth, 46). There is no complete agreement on this interpretation of events occurring in the RH model; AAF is a carcinogen and could represent a second hit which follows initiation. Nonetheless, in the RH model many centers of focal growth develop a few days after partial hepatectomy. They are recognizable histologically or histochemically by various biochemical alterations, such as the presence of γ-glutamyltranspeptidase (GGT) and glutathione-S-transferase, placental (GST-P), activities, the rise in G6PD activity, the decrease in G6Pase activity, etc. (enzyme altered foci, EAF). They increase in number and diameter, during a few weeks, becoming early nodules (EN), larger than a liver lobule and compressing surrounding parenchyma. Thereafter, most EAF and EN undergo a decrease in DNA synthesis and remodel to normally appearing liver, while relatively few maintain high DNA synthesis (46,48) and progress to non-remodeling, persistent nodules visible to the naked eye (PN, hyperplastic nodules, neoplastic nodules, adenomas), which progress to cancer (Fig. 1B).

A variant of the RH model, the triphasic model (49,50), consists of the administration of a diet containing 0.05% phenobarbital (PB), starting at the end of the selection treatment (Fig. 1C). Treatment with PB enhances the development of EAF, PN and HCC and inhibits remodeling. A certain resistance of EAF and EN to PB occurs after prolonged administration (50).

The stop model (33,36, Fig. 1D) is based on the administration for various weeks of several doses of a carcinogen to rats. Then the animals are tested, during several weeks, for the development of preneoplastic and neoplastic lesions. Histological evaluation of these lesions has led to the recognition of several cell types (51,52): clear/glycogen storage cells, acidophilic cells, rich in smooth endoplasmic reticulum, amphophilic cells, showing dense granular acidophilia with randomly scattered basophilia, basophilic cells, rich in rough endoplasmic reticulum and free of glycogen. Cell types with intermediate patterns may also be found. Recently, rat liver lesions with increased basophilia, in which the basophilic material forms characteristic bodies arranged in bands with a stripped pattern, have been called tigroid cell foci. These foci are thought to be precursors of nodules and HCC (52). Many other models, developed for the study of multistage carcinogenesis (cf. ref. 46), have not yet been used for chemoprevention studies with DHEA. It should be noted that the experimental approach to multistage carcinogenesis is largely artificial and does not reproduce human conditions. The multistep approach is only a model situation which permits the study of the mechanisms of carcinogenesis and tumor chemoprevention. In the skin carcinogenesis, RH and triphasic models, the stages of initiation, promotion (or selection/promotion) and progression are recognizable. The use of RH model for chemoprevention studies could be problematic due to the early and extensive spontaneous remodeling of EAF and EN. Preneoplastic nature of spontaneously remodeling lesions could be questioned. However, when Wistar rats are used instead of Fisher rats, the development of preneoplastic tissue proceeds relatively slowly and spontaneous remodeling is almost absent, for a few weeks after selection, especially in PB-treated rats (triphasic model, 50). Moreover, all hepatocarcinogenesis models are characterized by the appearance of many early lesions and, even if they do not remodel spontaneously, only a few of them progress to cancer. On the other hand, RH and triphasic models produce large PN which may be easily collected and used for biochemical and molecular analyses.

Effect of dehydroepiandrosterone on tumor initiation

The first hint of possible existence of an anti-initiating effect of DHEA is the observation that epithelial-like cells, cultured *in vitro* in the presence of DMBA, produce lower amounts of total water-soluble DMBA metabolites when incubated with DHEA (21). Similarly, primary cultures of HSF and HL carrying G6PD-Med, incubated with BaP, produce lower amounts of water-soluble BaP metabolites than control cells (26,27). The synthesis of BaP metabolites, mutagenic for his- *S. typhimurium*, is lower in normal HL, cultured *in vitro* in the presence of DHEA, and in HL carrying G6PD-Med, with respect to control HL (27). DHEA further reduces synthesis of mutagenic metabolites in the deficient cells. Finally, G6PD-deficient HL and HSF, cultured *in vitro* in the presence of BaP, show a great fall in the production of single water- and organic-soluble BaP metabolites (53-55). Deficent HL are less able, than control HL, to form BaP-7,8-diol-9,10-epoxide and BaP-DNA adducts (53,54). Accordingly, low DNA methylation occurs in the deficient HSF, incubated with dimethylnitrosamine (23), while dietary (56) or topical treatment (30) with DHEA inhibits DMBA binding to mouse skin.

These observations indicate a decreased ability of G6PD-deficient cells to metabolize carcinogens. Genetically transmitted (26,27,53-55) and DHEA-induced (26,27) G6PD deficiency do not affect water-soluble/organic-soluble metabolite ratio, in HSF and HL, indicating that this deficiency does not enhance carcinogen detoxification in these cells. In contrast with these findings, a 14-day feeding a diet containing 0.8% DHEA was recently reported to cause a fall in $[^3H]$DMBA-derived radioactivity, linked to rat liver DNA, and a rise in that linked to total protein (57). This was coupled with a rise in glutathione-S-transferase activity and in urinary excretion of $[^3H]$DMBA-derived radioactivity, in DHEA-treated rats. These data suggest an increase in DMBA detoxification in DHEA-treated rats. However, no DMBA metabolites and DMBA-DNA adducts were determined so that the nature of $[^3H]$DMBA-derived radioactivity is unknown and adsorption of radioactive derivatives to DNA and protein cannot be excluded. Nonetheless, the results of *in vitro* and *in vivo* studies on the effect of DHEA on carcinogen metabolism suggest an anti-initiating effect of G6PD deficiency: a fall in DNA adduct formation could reduce the number of initiated cells and the probability of later tumor development.

However, *in vivo* studies on anti-initiating effect of DHEA gave contradictory results. A single topic application of DHEA, 1 h before the initiating dose of DMBA, significantly reduces the yield of skin papillomas (30). Garcea et al. (6,38) have studied the effect of DHEA on the development of EAF and EN in the liver of Wistar rats, initiated with diethylnitrosamine (DENA) and then treated according to the RH or triphasic models. DHEA (0.6%) was included in the diet according to the following scheme: (A) from initiation to 7th week; (B) from initiation to the end of AAF feeding (4th week); (c) from 4th to 7th week. Rats were killed at the 7th week. Treatments (A) and (C) were equally effective in preventing the development of EAF and EN in both models. In contrast, treatment (B) was ineffective, indicating the absence of an irreversible inhibition by DHEA of the initiation and/or selection steps. In agreement with these findings Thornton et al. (58) did not found an anti-initiating effect of DHEA, in azaserine-induced acinar pancreatic and liver preneoplastic lesions. This however contradicts the recent finding that DHEA enhances initiation of lung carcinogenesis by DHPN, while it reduces liver initiation by the same carcinogen (59). DHEA administration, during DHPN treatment of Syrian golden hamsters infested two weeks before initiation with metacercariae of Opistorchis viverrini, enhances the generation of GST-P-positive EAF and cysts, in the liver, and of local proliferative lesions in the pancreas (60). It should be taken into account, in considering the above data, that the ratio between metabolic activation and inactivation may vary according to the carcinogen and cell type. Furthermore, biochemical and molecular changes, induced by DHEA, could be different in various tissues and species.

Anti-promotion effect of dehydroepiandrosterone

An anti-promotion effect of DHEA was first observed by Pashko et al. (30):

DEHA administration together with TPA partially inhibited papilloma development in DMBA-initiated skin. The interaction between the hormone and the promoter cannot be excluded in these experiments.

DENA-initiated rats, subjected to the RH or triphasic treatments, and fed for 4 weeks, after the end of AAF feeding (Fig. 1B and 1C), a diet containing 0.6% DHEA, showed a great fall in the percentage of liver occupied by EAF and EN as well as in DNA synthesis in preneoplastic cells (6). This was clearly not due to the interaction between DHEA and PB, since DHEA showed similar effects in RH and triphasic models. In agreement with these findings, a fall in number and size of putative preneoplastic lesions has been observed at the end of the selection step, in DENA-initiated rats treated according to the RH model, when 1% DHEA was given to the animals during the selection treatment (39). In these conditions, however, the interference of DHEA with AAF metabolism cannot be excluded.

In a series of researches utilizing the stop model (Fig. 1D), DHEA has been given after the end of carcinogen administration. Long-term treatment of DHPN-initiated rats with 0.6% DHEA inhibited the development of thyroid adenomas and urinary bladder nodular (papillary) hyperplasia (33). In the liver, this treatment caused phenotypic modulation, consisting of a fall in clear cell/glycogen storing foci and in EAF expressing GGT and GST-P activities, coupled with an increase in basophilic foci. In NNM-initiated rats, subjected to long-term treatment with 0.25% DHEA, a fall in mixed cell/glycogen storing foci was coupled with an increase in amphophilic/tigroid cell foci. The yield of liver tumors was not reduced, as instead did the malignant potential of HCC induced (36). DHEA also caused a fall in preneoplastic foci and nodules in the adrenal cortex of the initiated animals (36). Finally, a 4 week administration of 0.25% DHEA to rats, after a

26-week feeding a diet containing 0.05% dimethylaminoazo-benzene, caused a fall in liver GST-P-positive EAF, mixed cell foci, and, again, an increase in amphophilic/tigroid foci, with no changes in basophilic foci (61). These data seem to indicate that in the liver of DHEA-treated rats there occurs a loss of phenotypic stability of preneoplastic lesions, reminiscent of remodeling. This agrees with the reported remodeling of GGT-positive EAF in DHEA-treated Wistar rats subjected to RH treatment (40). A DHEA anti-promotion effect was also found in the liver of aza-serine initiated rats (58). However, DHEA enhanced the development of acidophilic lesions in the pancreas of the same animals.

In conclusion, even if with some exceptions, anti-promotion effect of DHEA is substantiated by numerous observations with various tissues and experimental models. This effect is largely coupled with a reduction in cell production (DNA synthesis) and, in some instances, increase in remodeling and is probably one of the major factors leading to a reduced incidence of malignant tumors in DHEA-treated rats.

Anti-progression effect of DHEA

A well defined progression step in multistage carcinogenesis models is not always obvious. In RH model (Fig. 1B), PN grow and progress to HCC without promoting stimuli. Some intermediate lesions, represented by carcinomatous changes inside PN (nodule in the nodule, cf. ref. 46) may be seen during the progression of PN to HCC. Transplantation of PN, but not surrounding liver, into homologous spleen is followed by carcinomatous transformation (62). Recently, in our laboratory, male Fisher rats were subjected to RH treatments and then fed, for 8 or 15 weeks, a diet containing 0.6% DHEA, starting at the 15th week after initiation, when PN are present in the liver. It has been observed (unpublished results) a fall in nodular G6PD activity and DNA synthesis. Percentage of liver parenchyma occupied by PN decreased proportionally to the length of DHEA treatment. However, rapid growth of nodular cells occurred again a few weeks after arresting DHEA treatment, although HCC carcinomas developed exclusively in the rats not treated with DHEA. These results, even if preliminary, indicate that DHEA treatment during nodule progression, reversibly inhibits nodule growth and causes a delay in their progression to malignancy.

Mechanistic Approach.

The study of the mechanism of the DHEA anti-tumor effect has mainly pointed to the evaluation of metabolic effects of the hormone. The attention has been focused on G6PD inhibition by DHEA. Among the primary functions of HMP, in mammals, is the production of NADPH and pentose phosphates. Secondary functions include lipid biosynthesis, glutathione metabolism, drug detoxification, carcinogen activation, phagocytosis, and synthesis of proline, DNA and RNA. Other sources of NADPH are mainly represented, in mammals, by malic enzyme (ME) and cytoplasmic isocitric dehydrogenase (ICD).

Inhibition by DHEA of rat liver G6PD is followed by a compensatory induction of ME and, to a minor extent, of ICD (40). As a consequence, NADPH level and NADPH/$NADP^+$ ratio do not decrease (40). Surprisingly, ME and ICD induction has also been reported in the liver of rats treated with DHEA, in conditions that do not induce G6PD inhibition (63,64). However, a great fall in NADPH level and NADPH/$NADP^+$ ratio occurs in human HSF and HL from individuals carrying G6PD-Med deficiency (26,27,55). This excludes a compensatory rise in NADPH producing enzymes.

Various steps of DNA and RNA synthesis are NADPH-dependent reactions, and NADPH is also required for cholesterol synthesis. Mevalonate, a cholesterol precursor, and some of its metabolites are required for DNA synthesis in growing tissues (65). In these tissues G6PD activity plays a unique role since HMP is the only source of R5P, used for nucleotide biosynthesis. Pre-cancerous tissue exhibits a relatively high growth rate (46), which could be influenced by a decreased yield of NADPH and R5P. Many studies focused on the evaluation of this hypothesis.

Role of G6PD inhibition

Metabolic activation of most carcinogens, by the mixed function oxidases, is a NADPH-dependent process. HSF and HL, carrying G6PD-Med exhibit a very low HMP activity which seems to function near its maximum (26,27). This could be responsible for the limited ability of these cells to metabolize relatively high carcinogen amounts (53-55). According to recent results, a rise in BaP concentration in the culture medium causes a higher stimulation of BaP metabolism in _in vitro_ cultured normal HSF than G6PD-deficient HSF (55). $NADPH/NADP^+$ ratio is lower in the deficient cells, than in normal cells (26,27,55). It does not undergo great changes, in normal HSF, after incubation with 0.25 μM and 2.5 μM BaP, while a great decrease occurs in the deficient cells, at both BaP concentrations (55). This suggests that NADPH is a limiting factor for BaP metabolism in G6PD-deficient cells. Accordingly, rat liver NADPH oxidation, caused by ethanol infusion (66), strongly inhibits BaP hydroxylation (67). Moreover, evidence obtained from a permeabilized whole-cell system suggests that NADPH is rate-limiting in the mixed-function oxidation of BaP (68).

Treatment of _in vitro_ cultured cells from animals or normal individuals with DHEA, mimics many alterations of carcinogen metabolism induced by the genetically-transmitted G6PD deficiency (24,26,27). This suggests a preeminent role of reduced G6PD activity in the decrease in carcinogen metabolism at least in extrahepatic tissues of DHEA-treated rats. Some experimental evidence shows that inhibition of tumor promotion by DHEA may be correlated with inhibition of DNA synthesis. Pashko et al. (30) have reported that Epi-Br, a steroid 30-50 folds more potent than DHEA as an inhibitor of G6PD, is about 25 folds more active than DHEA as an inhibitor of TPA-stimulated epidermal DNA synthesis in ICR mice. No inhibition of this synthesis was observed by using steroids which were inactive as G6PD inhibitors, with the exception of corticosterone which inhibited DNA synthesis, even if it did not affect G6PD activity.

DHEA slows the growth of various cell lines in culture (69). Interestingly, the addition of a mixture of ribonucleosides of adenine, guanine, cytosine and uracil (RN) to the reaction medium resulted in a partial reversal of growth inhibition by DHEA, while a complete reversal occurred with a mixture of four deoxyribonucleosides of adenine, guanine, cytosine and thymine (DRN, 70). DHEA and some of its analogs inhibit in vitro differentiation of 3T3-LI fibroblasts to adipocytes, a phenomenon linked to cell growth, lipid synthesis and morphologic changes (70). Structural requirement and relative potencies of various DHEA analogs, were closely similar for inhibition of G6PD activity and 3T3-LI cell differentiation. The DHEA effect was overcome by DRN, but not by RN, or by fusing the cells with liposomes containing 6-phosphogluconate (70,71). These results suggest that DHEA represses DNA synthesis by reducing nucleoside supply, as a result of G6PD inhibition.

The administration of a diet containing 0.6% DHEA, during the development of EAF and EN in rats, subjected to the RH and triphasic treatments, causes a fall in DNA synthesis in the focal cells, coupled with inhibition of G6PD activity in both preneoplastic and surrounding tissues (6,40). This leads to a great fall in the percentage of liver occupied by GGT-positive foci and foci expressing a high G6PD activity. G6PD inhibition by DHEA, although associated, as reported above, with a compensatory rise in the activity of other NADPH producing enzymes causes about a 50% decrease in Ru5P liver content (40). Histochemical evaluation confirmed the existence of a high malic enzyme activity in preneoplastic lesions of DHEA-treated rats (40). DRN and RN administration, along with DHEA, was shown to completely overcome DHEA-induced fall in the percentage of GGT-positive foci and DNA synthesis, without modifying inhibition by DHEA of histochemically determined G6PD activity. Free bases could not substitute for RN and DRN (unpublished results), so that it may be suggested that a reduced nucleoside availability, secondary to Ru5P deficiency in hepatocytes exhibiting low G6PD and HMP activities (40), is a mechanism of DNA synthesis and growth inhibition by DHEA in rat liver. Against this conclusion, however, are some contradictory results on the effect of DHEA on rat liver G6PD activity: a decrease (6,40,72.73), no changes (63,64), or a lowered activity only when G6PD is elevated by fasting-refeeding (74-76), sucrose feeding (77) or genetic predisposition (78), have been reported. Probably, G6PD inhibition is influenced by isoenzymatic forms, exhibiting different sensitivities to DHEA (9,79), rat strain and feeding schedule. In some studies only 0.3-0.4% DHEA was introduced with food (63) or rats had access to food, containing 0.6% DHEA, only a 3 h period per day (64). It has been calculated (40) that during this period of time food intake and, consequently, DHEA intake are 30-40% lower than unrestricted food intake. On the other hand, a 39% G6PD inhibition occurs only 12 h after feeding 0.6% DHEA (Fig. 2), whereas

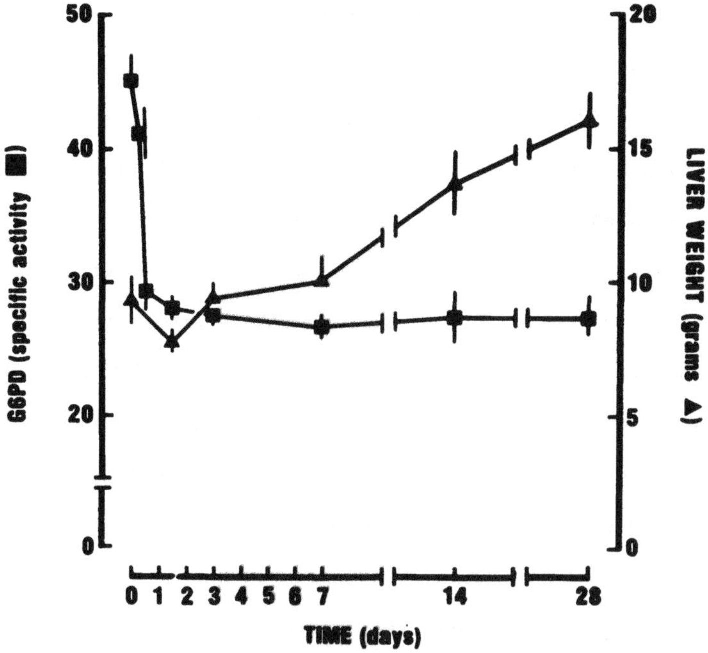

Figure 2. Variations of liver G6PD activity and liver weight during treatment with DHEA. Female Wistar rats were fed, for the periods of time indicated, a diet containing 0.6% DHEA. Specific activity: nmol/min, mg protein. Data are mean ± SD of 5 rats.

normal G6PD values are restored 6-8 h after stopping DHEA treatment (6). Thus, long time intervals between feeding periods and administration of relatively low DHEA amounts could cause G6PD inhibition by DHEA to be very low or absent. It should also be considered that, according to unpublished findings in our laboratory, freshly prepared diet, containing 0.6% DHEA, undergoes a 47% loss in DHEA content, at 22°C, 48 h after preparation. This emphasizes the need of adding DHEA to the diet shortly before use. Another point to consider is the fact that DHEA, included in the diet at 0.6-0.8% for several weeks, causes an increase in liver weight (6,59,63,80), coupled with a rise in total liver protein (80) and in cell size, without changes of nuclei number (6) and no change or low increase of liver DNA content (6,57,80). Various types of experiments provide strong evidence that liver hypertrophia does not interfere with inhibition of G6PD activity in DHEA-treated rats. In Wistar rats maximum G6PD inhibition occurs 3 days after starting DHEA administration; while liver weight increases significantly only a week after starting DHEA treatment (Fig. 2). Hepatocytes isolated from DHEA-treated animals show a lower HMP activity than normal hepatocytes, which is not stimulated by NADPH re-oxidation by methylene blue (40), a behavior in keeping with the existence of G6PD deficiency. G6PD decreases in Sprague-Dawley rats when DHEA is added to a high fat diet, but not to purina chow, even if the same increase in liver weight occurs under the two different regimens (63). Finally, G6PD inhibition by DHEA has recently been demonstrated, in our laboratory, to occur in liver, PN and HCC of F344 male rats in conditions that do not cause liver hypertrophia (unpublished results).

DHEA at the dose level of 0.25% causes only slight inhibition of liver G6PD activity which seems to mainly depend on non-parenchymal cells (73). As reported above, treatment with 0.25% DHEA, causes phenotypic modulation of preneoplastic lesions leading to a reduction in the malignant potential of HCC induced (36). It also causes various enzymatic modifications in the liver (73) and preneoplastic lesions (41,61) suggestive of a metabolic shift towards a pattern (reduction in glucose consumption and increase in glycogen and G6P hydrolysis) characteristic of normal cells (73). However, the possibility that abolition of metabolic patterns of preneoplastic tissue is a consequence rather than a cause of the phenotypic modulation induced by DHEA cannot be ruled out, at present.

Inhibition of superoxide radical formation

Free radical production and subsequent lipid peroxidation are thought to be one of the mechanisms by which some substances elicit tumor-promoting effects (81). Superoxide radical (O_2^-), and other forms of reactive oxygen may induce deletion mutations and chromosome aberrations (82,83). Interestingly, malignant transformation of in vitro cultured hamster embryo fibroblasts, induced by X-irradiation followed by TPA treatment, is prevented by treating with superoxide dismutase, along with TPA (84). Papilloma development, during skin carcinogenesis, is blocked by topical application, along with TPA, of copper(II)3,5-diisopropyl-salicylic acid, a compound which mimics the action of superoxide dismutase (85). TPA stimulates O_2^- formation by polymorphonuclear granulocytes (PMN) in vitro (86), a phenomenon which depends on NADPH reduction of oxygen (87). DHEA and Epi-Br inhibit this stimulation (88,89). TPA stimulation of both O_2^- formation and HMP activity is reduced in PMN carrying G6PD-Med (89). Inhibition of O_2^- formation is correlated to the level of G6PD activity in the inhibited cells (88,89). This again suggests that a reduction in the NADPH level is responsible for the low HMP activity and O_2^- formation, in TPA-stimulated cells treated with DHEA.

According to a recent report (90) DHEA, and some structural analogs more potent inhibitors of G6PD, inhibit stimulation by TPA of prostaglandin E_2 mouse skin content. Again a correlation between G6PD inhibitory activity and potency in blocking the stimulation of prostaglandin E_2 content was seen, suggesting a mechanistic role of G6PD inhibition. These observations, even if preliminary, are interesting because lipoxygenase- and cycloxygenase-dependent metabolism of arachidonic acid, leading to the formation of leukotrienes, prostaglandins, prostacyclins and thromboxanes, has been suggested to be the major source of TPA-stimulated reactive oxygen (91). Inhibitors of lipoxygenase and cycloxygenase activity lead to the blockage of TPA-induced tumor promotion (92).

G6PD deficiency and cancer risk in humans

The Mediterranean variant of G6PD, found primarily in Sephardic Jews, Greeks and Sardinians, and the G6PD defect in people of African origin (G6PD-A^-) are the most common variants of mutant G6PD. The main function of G6PD in red cells, is their protection against oxidative injury, mainly from hydrogen peroxide. G6PD deficiencies are sex-linked hereditary defects relatively frequent in some populations. The Mediterranean deficiency is more severe than the African deficiency and is present in many nucleated cells, such as fibroblasts, lymphocytes, PMN, hepatocytes, etc. (93). DHEA mimics many effects of the G6PD-Med deficiency as concerns carcinogen metabolism (26,27) and O_2^- production from TPA-stimulated PMN (89). However, DHEA-induced deficiency inhibits cell growth _in vivo_ and _in vitro_ (30,35), while inhibition of _in vitro_ growth was not observed in HSF and HL carrying G6PD-Med (26,27). It may be suggested that low amounts of NADPH and R5P are required by the relatively slow growing G6PD-deficient HSF and HL (26,27). However, no difference in average weight has been found between G6PD-deficient and normal newborns, in Sassari (Sardinia, Italy) between 1970 and 1979 (3464 ± 462 g, means ± SD, n = 3000, for normal newborns, and 3437 ± 439 g mean ± SD, n = 982 for G6PD-deficient newborns; T. Meloni, personal communication). This behavior apparently denies the existence of any influence of G6PD deficiency on rapidly growing fetal tissues. Perhaps, variations in enzyme turnover in nucleated cells carrying G6PD-Med, with respect to cell growth rate, could explain the discrepancy between genetically-transmitted and DHEA-induced deficiency.

About 25 years ago Beaconsfield (94) has reported that lower cancer incidence in Israelis of North African and Asian origin, with respect to those of Western European or American Origin, could depend on the higher frequency of G6PD deficiency in the former populations. Naik and Anderson (95) found that the frequency of G6PD-deficiency is significantly lower in Black males and females cancer patients, than in healthy controls. Similar results were obtained by Sulis (96) who found a 13% prevalence of G6PD deficiency in 320 male Sardinians cancer patients, against a 25-30% prevalence in the healthy individuals living in the same geographic area. Mbensa et al. (97) have found, in a case control study on 70 Bantu patients with liver cancer in Zaire, only 13% had G6PD deficiency, against 23% of male controls. These studies suggest a negative correlation between G6PD deficiency and tumor development. A recent case control study (98) on 186 male cancer patients, hospitalized in Cagliari (Sardinia, Italy), compared to 186 patients with other diseases (excepting hemolytic anemia), has given evidence of not reducing cancer risk in G6PD-deficient patients. The above studies largely concern patients with a variety of tumors. The role of carcinogen activation, hormones, viruses, and promoting agents is extremely variable, in human tumors, according to cancer types and individuals, and is not always assessable. Available data on the cancer risk in G6PD-deficient individuals, only permit to conclude that the presence of the mutated G6PD gene does not modify, *per se*, cancer incidence. The same conclusion has been recently reached in some studies on the influence of G6PD-deficiency on hematologic malignancies (99) and breast cancer (100), two cancer types which could be strongly influenced by viral infection and hormonal status, respectively.

Other mechanisms of DHEA anti-tumor effect.

It is known that DHEA treatment may elicit some androgenic (101) and estrogenic effects (102) in mice and rats. This review is not aimed to consider the relationships between DHEA hormonal effects and cancer. However, a mechanism of inhibition by DHEA of methylcholanthrene-induced uterine cervix tumors in mice, based on the estrogenic effect of DHEA, has recently been postulated (37). Indeed, treatment of mice with estrogens inhibits the development of these tumors (103), even if it may stimulate G6PD activity (104). The possibility of anti-tumor effect of DHEA based on a hormonal mechanism is attractive, but needs further experimental support.

Food restriction mimics various DHEA effects on skin carcinogenesis (105). It inhibits the rate of [^{3}H]DMBA binding to mouse skin DNA, abolishes stimulation by TPA of epidermal DNA synthesis and depresses epidermal G6PD activity. DHEA exhibits a well known anti-obesity effect (106). However, a major role of the caloric restriction in the DHEA anti-tumor effect is denied by the fact that topical application of DHEA, while not reducing body weight, inhibits skin tumor development in mice (31).

High G6PD activity, in growing tissues, has been thought to be necessary for the supply of cholesterol, for membrane synthesis, as well as for the formation of mevalonate and some of its metabolites involved in the regulation of DNA synthesis. Recent results in our laboratory (107) have shown that a 15-day feeding a diet containing 0.6% DHEA, started at the end of the selection step in Wistar rats subjected to the RH and triphasic treatments, causes a 81% fall in liver <u>de novo</u> cholesterogenesis (labeled acetate incorporation). This effect was not prevented by DRN administration which, however, as reported above, prevents inhibition of DNA synthesis by DHEA in preneoplastic tissue. Preliminary results have shown that DHEA inhibits cholesterol synthesis in a site before mevalonate synthesis. It thus seems that inhibition by DHEA of preneoplastic tissue development does not depend on decreased stimulation of DNA synthesis by mevalonate (65). Perhaps, constitutively active DNA polymerase in EAF does not need great stimulation by mevalonate or some of its metabolites.

Conclusions and Perspectives

The results of several researches strongly support the existence of an anti-tumor effect of DHEA, largely linked, at least as concerns mouse skin and rat liver, to inhibition of preneoplastic tissue growth, during carcinogenesis promotion. In some instances a DHEA anti-initiating effect, probably dependent on impaired carcinogen activation, has been described. Unfortunately DHEA anti-progression effect has not yet received adequate attention. Initiation and promotion steps of carcinogenesis are hardly recognizable in man and studies on the anti-initiation and anti-promotion effect of DHEA, even if important to assess the anti-tumor action of this hormone, are not always useful to prepare a chemopreventive strategy. The study of the DHEA effect on the progression stage is particularly interesting. Recognizable lesions analogous to PN may be found during carcinogenic transformation of the urinary bladder (papillomas), mammary gland (adenomas), intestinal mucosa (polips), and skin (papillomas). Liver macronodular cirrhosis, positive for HBV antigen, has been suggested to be a precancerous condition (108). Even if there is no conclusive evidence in favor of a close correspondence of rat liver nodules, to those arising during chronic active hepatitis and cirrhosis in man, the possibility of modulating the development of human liver nodules to HCC is attractive. Preliminary results, obtained in our laboratory, suggest an anti-progression effect of

DHEA, which appears to be at least partially reversible.

The treatment of mice and rats with DHEA may cause uterine enlargement in young female rats (101), increase in seminal vesicle weight in castrated male rats (102), liver hypertrophia (6,59,63,80), and increase in liver catalase activity (102). Recently, it has been proposed the use of DHEA analogs, more potent inhibitors of G6PD and tumorigenesis than the parent compound (102,109), and which do not cause major side effects. The elimination of side effects and establishment of a treatment strategy which minimizes reversal of growth inhibition after a cycle of treatment with DHEA, could enable the use of these agents to prevent some human malignancies.

Some evidence strongly indicates that G6PD inhibition is the major mechanism of DHEA anti-tumor effect. This inhibition may limit pentose phosphate supply for DNA synthesis. A similar mechanism has also been suggested to explain the anti-aterogenic effect of DHEA in rabbits (71). Alternative mechanisms of DHEA anti-tumor action have been suggested, but none of them has received adequate experimental support. Perhaps, further studies of DHEA effects at a biochemical and molecular level could give new insights on the mechanism of DHEA action. For instance, a study by _in situ_ hybridization of DHEA effect on the expression of genes, involved in growth regulation and cell differentiation, could permit to identify cell types and preneoplastic lesions sensitive to growth inhibition by DHEA in different stages of the carcinogenesis. These and other studies could be interesting in the aim of possible application of chemopreventive treatments to man.

Acknowledgements.

This research was supported by grants from MURST (programs 40% and 60%).

References

1. Weber, G. 1966. Gann Monograph. 1, 151-178.

2. Koudstaal, J., B. Makkink, S.H. Overdiep. 1975. Eur. J. Cancer 11, 111-115.

3. Heyden, G. 1974. Histochem. 39, 327-334.

4. Evans, A.W., N.W. Johmsin and R.G. Butcher. 1983. Histochem. J. 15, 483-489.

5. Moore, M.A., A. Tsuda and N. Ito. 1986. Carcinogenesis 7, 339-342.

6. Garcea, R., L. Daino, R. Pascale, S. Frassetto, P. Cozzolino, M.E. Ruggiu and F. Feo. 1987. Toxicol. Pathol. 15, 164-169.

7. Columbano, A., S. Dessì, G.M. Ledda-Columbano, C. Chiodino, P. Coni and P. Pani, K.N. Rao. 1987. Toxicol. Pathol. 15, 43-50.

8. Marks, P.A. and J. Banks. 1960. Proc. Natl. Acad. Sci. USA 46, 447-452.

9. Lopez, S. and A. Rene. 1973. Proc. Soc. Exptl. Biol. Med. 142, 258-261.

10. Vande Viele, R. and S. Lieberman. 1960. In: Biological Activities of Steroids in Relation to Cancer (G. Pincus and E. Vollmer, eds.). Academic Press, New York, pp. 93-110.

11. Migeon, C.J., A.R. Keller, B. Lawrence and T.H. Shephard. 1957. J. Clin. Endocrinol. Metab. 17, 1051-1062.

12. Lebeau, M.C. and E.E Baulieu. 1973. In: Metabolic Conjugation and Metabolic Hydrolysis (W.H. Fishman, ed.). Academic Press, New York, vol. 3, pp. 151-187.

13. Bulbrook, R.D., J.L. Hayward, C.C. Spicer and B.S.A. Thomas. 1962. Lancet 2, 1235-1237.

14. Kumaoka, S., N. Sakauchi, O. Abe, M. Kusarna and O. Takatani. 1968. J. Clin. Endocrinol. 28, 667-672.

15. Zumoff, B., J. Levin, R.S. Rosenfeld, M. Markham, G.W. Strain and D.K. Fukushima. 1981. Cancer Res. 41, 3360-3363.

16. Rao, L.G.S. 1972. Nature 235, 220-222.

17. Sonka, J., M. Vitkova, I. Gregorova, Z. Tomosova. J. Hilgertova and J. Stas. 1973. Endokrinologie 62, 61-68.

18. Farwell, V. T., R.D. Bulbrook and J.L. Hayward. 1978. In: Early Diagnosis of Breast Cancer: Methods and Results (E. Grundmann and L. Beck, eds.). Fischer Verlag, Stuttgart, pp. 43-51.

19. McMahon, B. 1973. J. Natl. Cancer Inst. 50, 21-42.

20. Brennan, M.J., R.D. Bulbrook, N. Deshpande, D.Y. Wang, J.L. Hayward. 1973. Lancet 1, 1076-1079.

21. Schwartz, A.G. and A. Perantoni. 1975. Cancer Res. 35, 2482-2487.

22. Henderson, E.A., A. Schwartz, L. Pashko, M. Abou-Gharbia and D. Swern. 1981. Carcinogenesis 2, 683-686.

23. Feo, F., L. Pirisi, R. Pascale, L. Daino, S. Zanetti, V. LaSpina and P. Pani, 1981. In: Recent Trends in Chemical Carcinogenesis (P. Pani, F. Feo and A. Columbano, eds.). E.S.A., Cagliari, pp. 176-196.

24. Feo, F., L. Pirisi, R. Pascale, S. Zanetti, L. Daino and V. LaSpina. 1982. In: Membranes in Tumour Growth (T. Galeotti, A. Cittadini and S. Papa, eds.). Elsevier, Amsterdam, pp. 549-557.

25. Pirisi, L., R. Pascale, L. Daino, S. Frassetto, V. LaSpina, S. Zanetti, L. Gaspa, G.M. Ledda, R. Garcea and F. Feo. 1982. Res. Commun. Chem. Pathol. Pharmacol. 38, 301-311.

26. Feo, F., L. Pirisi, R. Pascale, L. Daino, S. Frassetto, R. Garcea and L. Gaspa. 1984. Cancer Res. 44, 3419-3425.

27. Feo, F., L. Pirisi, R. Pascale, L. Daino, S. Frassetto, S. Zanetti and R. Garcea. 1984. Toxicol. Pathol. 11, 261-268.

28. Schwartz, A.G. 1979. Cancer Res. 39, 1129-1132.

29. Schwartz, A., G. Hard, L. Pashko, M. Abou-Gharbia and D. Swern. 1981. Nutr. Cancer 3, 46-53.

30. Pashko, L.L., R.J. Rovito, J.R, Williams, E.L. Sobel and A.G. Schwartz. 1984. Carcinogensis 5, 463-466.

31. Pashko, L.L., G.C. Hard, R.J. Rovito, J.R. Williams, E.L. Sobel and A.G. Schwartz. 1985. Cancer Res. 45, 164-166.

32. Schwartz, A.G. and R.H. Tannen. 1981. Carcinogenesis 2, 1335-1337.

33. Moore, M.A., W. Thamavit, H. Tsuda, K. Sato, A. Ichihara and N. Ito. 1986. Carcinogensis 7, 311-316.

34. Nyce, J.W., P.N. Magee, G.C. Hard and A.G. Schwartz. 1984. Carcinogenesis 5, 57-62.

35. Hamilton, S.R. and G.B. Gordon. 1988. Proc. Amer. Ass. Cancer. Res. 29, 162. Abstr. No 645.

36. Weber, E., M.A. Moore and P. Bannasch. 1988. Carcinogenesis 9, 1191-1195.

37. Rao, A.R. 1989. Cancer Lett. 45, 1-5.

38. Garcea, R., L. Daino, R. Pascale, S. Frassetto, P. Cozzolino, M. Ruggiu and F. Feo. 1985. Proc. 3rd. Sardinian Int. Meet.: Agents and Processes in Chemical Carcinogenesis. Abstracts, STEF, Cagliari, p. 102.

39. Moore, M.A., W. Thamavit, A. Ichihara, K. Sato and N. Ito. 1986. Carcinogenesis 7, 1059-1063.

40. Garcea, R., L. Daino, S. Frassetto, P. Cozzolino, M.E. Ruggiu, M.G. Vannini, R. Pascale, L. Lenzerini, M.M. Simile, M. Puddu and F. Feo. 1988. Carcinogenesis 98, 931-938.

41. Weber, E., M.A. Moore and P. Bannasch. 1988. Carcinogenesis 9, 1049-1054.

42. Slaga, T.G., S.M. Fisher, K. Nelson and G.L. Gleason. 1980. Proc. Natl. Acad. Sci. Usa 77, 3659-3663.

43. Marks, F., G. Fürstenberger, M.G. Schwendt, M. Rogers, B. Schurich, B. Kaina and G. Bauer. 1988. In: Chemical Carcinogenesis. Models and Mechanisms (F. Feo, P. Pani, A. Columbano and R. Garcea, eds). Plenum Press, New York. pp. 217-234.

44. Kinzel, V., L. Lochrke, L. Goertler, G. Fürstenberger and F. Marks. 1984. Proc. Natl. Acad. Sci. USA 81, 5858-5862.

45. Solt, D.B., A. Medline and E. Farber. 1972. Am. J. Pathol. 88, 595-618.

46. Farber, E. and D.S.R. Sarma. 1987. Lab. Invest. 56, 4-22.

47. Gravela, E., F. Feo, R.A. Canuto, R. Garcea and L. Gabriel. 1975. Cancer Res. 35, 3041-3047.

48. Garcea, R., R. Pascale, L. Daino, S. Frassetto, P. Cozzolino, M.E. Ruggiu, M.G. Vannini, L. Gaspa and F. Feo. 1987. Carcinogenesis 8, 653-658.

49. Lans, M., J. de Gerlache, H.S. Tapier, V. Préat and M.B. Roberfroid. 1983. Carcinogenesis 2, 1283-1287.

50. Garcea, R., L. Daino, R. Pascale, M.M. Simile, M. Puddu, S. Frassetto, P. Cozzolino, M.A. Seddaiu, L. Gaspa and F. Feo. 1989. Cancer Res. 49, 1850-1856.

51. Bannasch, P. 1976. Cancer Res., 30, 2555-2562.

52. Bannasch, P., U. Benner, H. Enzmann and H.J. Hacker. 1986. Carcinogenesis 6, 1641-1648.

53. Pirisi, L., R. Garcea, R. Pascale, M.E. Ruggiu and F. Feo. 1987. Toxicol. Pathol. 15, 115-119.

54. Feo, F., M.E. Ruggiu, L. Lenzerini, R. Garcea, L. Daino, S. Frassetto, V. Addis, L. Gaspa and R. Pascale. 1987. Int. J. Cancer 39, 560-664.

55. Pascale, R., L. Daino, M. Ruggiu, M. Vannini, R. Garcea, D. Frassetto, L. Lenzerini, L. Gaspa, M.M. Simile, M. Puddu and F. Feo. 1988. In: Chemical Carcinogenesis. Models and Mechanisms (F. Feo, P. Pani, A. Columbano and R. Garcea, eds.). Plenum Press, New York, pp. 87-92.

56. Pashko, L.L. and A.G. Schwartz. 1983. J. Gerontol. 38, 8-12.

57. Prasanna, H.R., R.W. Hart and P.N. Magee. 1989. Carcinogenesis 10. 953-955.

58. Thornton, M., M.A. Moore and N, Ito. 1989. Carcinogenesis 10, 407-410.

59. Moore, M.A., E. Weber, M. Thornton and P. Bannasch. 1988. Carcinogenesis 9, 1507-1509.

60. Moore, M.A., W. Thamavit, Y. Hiasa and N. Ito. 1988. Carcinogenesis 9, 1185-1189.

61. Moore, M.A., E. Weber and P. Bannasch. 1988. Virchows Arch. B Cell Pathol. 55, 337-343.

62. Tatematsu, M., G. Lee, M.A. Hayes and E. Farber. 1987. Cancer Res. 47, 4699-4705

63. Cleary, M.P., A. Shepherd, J. Zisk and A. Schwartz. 1983. Nutr. Behav. 1, 127-136.

64. Casazza, J.P., W.T. Schaffer and R.L. Veech. 1986. J. Nutr. 116, 304-310.

65. Chen, M.W. 1984. Fed. Proc. 43, 126-130.

66. Reinke, L.A., F.C. Kauffman, S.A. Belinsky and R.G. Thurman. 1980. J. Pharmacol. Exp. Ther. 213, 70-78.

67. Reinke, L.A., P. McManus, F.C. Kauffman and R.G. Thurman. 1982. Cancer Res. 4, 1681-1685.

68. Sadowski, I.J., J.A. Wright and L.C. Israels. 1989. Int. J. Biochem. 17, 1023-1025.

69. Dworkin, C.R., S.D. Gorman, L.L. Pashko, V.J. Cristofalo and A.G. Schwartz. 1986. Life Sci. 38, 1451-1457.

70. Gordon, J.B., L.M. Shantz and P. Talalay. 1987. Adv. Enzyme Regul. 26, 355-382.

71. Shantz, L.M., P. Talalay and G.B. Gordon. 1989. Proc. Natl. Acad. Sci. USA 86, 3852-3856.

72. Tagliaferro, A.R., J.R. Davis, S. Truchon and N. Van Hamout. 1986. J. Nutr. 116, 1977-1983.

73. Meyer, D., E. Weber, M.A. Moore, I. Letsch, E. Filsinger and P.

Bannasch. 1988. Carcinogenesis 9, 2039-2043.

74. Willmer, J.-S. and T.S. Foster. 1965. Can. J. Biochem. 43, 1375-1377.

75. Tepperman, H.M., S.A. DeLaGarza and J. Tepperman. 1968. Am. J. Physiol. 214, 1126-1132.

76. Hilgertova, J. and J. Sonka. 1973. Horm. Metab. Res. 5, 286-289.

77. Cleary, M.P., S.S. Hood, C. Chando, C.T. Hansen and J.T. Billheimer. 1984. Nutr, Revc. 4, 485-494.

78. Shepherd, A. and M.P. Cleary. 1984. Am. J. Physiol. 246, E123-E128.

79. Setchenska, M.S., E.M. Russanov and J.G. Vassileva-Popova. 1975. FEBS Lett. 49, 297-300.

80. Cleary, M.E., A. Shepherd and B. Jenks. 1984. J. Nutr. 114, 1242-1251.

81. Slaga, T.J., A.J.P. Klein-Szonto, L.L. Triplett, L.P. Yotti and J.E. Frosko. 1981. Science 213, 1023-1025.

82. Kennedy, A.R., W. Troll and J.B. Little. 1984. Carcinogenesis 5, 1213-1218.

83. Emerit, I. and P. Cerutti. 1983. Carcinogenesis, 4, 1313-1316.

84. Hsie, A.W., L. Recio, D.S. Katz, C.Q. Lee, M. Wagner and R.L. Shenley. 1986. Proc. Natl. Acad. Sci. USA 83, 9616-9620.

85. Borek, C. and W. Troll. 1983. Proc. Natl. Acad. Sci. USA 80, 1304-1307.

86. Kensler, T.N., D.M. Busch and W.J. Kozumbo. 1983. Science 221, 75-77.

87. Goldstein, B.D., G. Witz, M. Amoruso, D.S. Stone and W. Troll. 1981. Cancer Lett. 11, 257-262.

88. Whitcomb, J.M. and A.G. Schwartz. 1985. Carcinogenesis 6, 333-335.

89. Pascale, R., R. Garcea, M.E. Ruggiu, L. Daino, S. Frassetto, M.G. Vannini, P. Cozzolino, L. Lenzerini, F. Feo and A.G. Schwartz. 1987. Carcinogenesis 8, 1567-1570.

90. Hastings, L.A., L.L. Pashko, M.J. Lewbart and A.G. Schwartz. 1988. Carcinogenesis 9, 1099-1102.

91. Fisher, S.M. and L.M. Adams. 1985. Cancer Res. 45, 3130-3136.

92. Aizu, E.. T. Nakadate, S. Yamamoto and R. Kato. 1986. Carcinogenesis 7, 1809-1812.

93. Beutler, E. 1983. In: The Metabolic Bases on Inherited Disease (J.B. Stambury, J.B. Wyngaarden, D.S. Fredrickson, J.L. Golstein and M.S. Brown, eds.). McGraw-Hill Book Co., New York, pp. 1629-1653.

94. Beaconsfield. P., R. Rainsbury and G. Kalton. 1965. Oncologia 19, 11-19.

95. Naik, S.N. and D.E. Anderson. 1971. Oncology 25, 356-364.

96. Sulis, E. 1972. Lancet 2, 1985.

97. Mbensa, M., C. Rwakunda and R.L. Verwilinghen. 1978. E. Afr. med. J. 55, 17-19.

98. Cocco, P., S. Dessì, G. Avataneo, G. Picchiri and E. Heinemann. 1989. Carcinogenesis 10, 813-816.

99. Ferraris, A.M., G. Broccia, T. Meloni, G. Forteleoni and G.F. Gaetani. 1988. Am. J. Hum. Genet. 42, 516-520.

100. Forteleoni, G., L. Argiolas, A. Farris, A.M. Ferraris, G.F. Gaetani and T. Meloni. 1988. Tumori 74, 665-667.

101. Knudson, T.T. and V.B. Mahesh. 1975. Endocrinology 97, 458-468.

102. Schwartz, A.G., M.L. Lewbart and L.L. Pashko. 1988. Cancer Res. 48, 4817-1822.

103. Monoharan, K. and A.R. Rao. 1985. Ind. J. Exp. Biol. 23, 566-568.

104. Richards A.H. and R. Hilf. 1972. Cancer Res. 32, 611-616.

105. Schwartz, A.G. and L. L. Pashko. 1986. Anticancer Res. 6, 1279-1281.

106. Cleary, M.P. 1989. In: Hormones, Thermogenesis, and Obesity (H. Lardy and F. Stratman, eds). Elsevier, Amsterdam, pp. 365-376.

107. Feo, F., L. Daino, M.E. Ruggiu, M.M. Simile, P. Cozzolino, R. Pascale, J.A. McKeating, G.P. Davliakos, K.S. Sudol, M.F. Melhem and K.N. Rao. 1989. Proc. 5th Sardinian Int. Meet.: Modulating Factors in Multistage Chemical Carcinogenesis. Abstracts. STEF, Cagliari, p. 72.

108. Blumberg, B.S. and W. Th. London. 1985. J. Natl. Cancer Inst. 71, 267-273.

109. Schwartz, A.G., D.K. Fairman, M. Polansky, M.L. Lewbart and L.L. Pashko. 1989. Carcinogenesis 10, 1809-1813.

MODULATION OF LIVER CARCINOGENESIS BY DEHYDROEPIANDROSTERONE

D. Mayer, E. Weber, P. Bannasch

Abteilung Cytopathologie, Institut für Experimentelle Pathologie, Deutsches Krebsforschungszentrum, 6900 Heidelberg, FRG

Introduction

Dehydroepiandrosterone (DHEA; 3β-hydroxy-5-androstene-17-one) has been found to exert biological effects of remarkable diversity in laboratory rats and mice. These include cancer preventive, antiautoimmune, antiatherogenic and anti-aging effects (1). The molecular mechanisms underlying this extraordinary variety of physiological, pharmacological, and metabolic effects of DHEA are far from being clear. However, one notable property of DHEA is its ability to inhibit mammalian glucose-6-phosphate dehydrogenase (G6PDH), the rate-limiting enzyme in the pentose-phosphate-pathway, a major source of cytosolic NADPH necessary for anabolic processes. This inhibition has been considered to be the central mechanism of at least the cancer preventive action of this steroid, but was also discussed to contribute to the other beneficial effects of DHEA (1,2).

In rodents DHEA inhibits the development of chemically induced tumours in different organs, such as lung, colon, mammary gland, liver and skin (1). However, an increase in the incidence of lung lesions has also been reported (3). Recently, Moore et al. (4,5) and Garcea et al. (6,7) described a reduction by DHEA in number and size of lesions induced in rat liver by chemical carcinogens and a modification of the types of lesions towards increased basophilia. In this contribution, we give an overview on observations of the modulating effect of DHEA on chemically induced rat liver carcinogenesis.

Dehydroepiandrosterone (DHEA)

Materials and Methods

Our own data presented in this review were obtained from rats treated for limited periods (stop experiments) with N-nitrosomorpholine (NNM, 120 mg/litre drinking water for 7 weeks) (8,9), dimethylaminoazobenzene (DAB, 0.05% in pellet diet for 25 weeks) (10) or dihydroxy-di-n-propyl-nitrosamine (DHPN, 2 x 1000 mg/kg body weight, i.p.) (3). The hormone DHEA was administered for limited periods (3,10) or throughout the experiment (3,8,9) in pellet diet in concentrations of 0.25% or 0.6% either alone (11), concomitant with, or subsequent to, treatment with the carcinogen.

Groups of treated animals and concurrent untreated controls were regularly sacrificed without prior starvation between 9 and 11 a.m. in order to avoid diurnal variations of enzyme activities and glycogen content. For enzyme histochemistry the livers were rapidly removed and shock-frozen in liquid isopentane at -140°C to -150°C. Samples were processed for light (8,9) and electron microscopy (12) as described. The activity of the following enzymes was demonstrated by the methods developed or adapted in our laboratory (8,9,11-14): glycogen synthase, glycogen phosphorylase, glucose-6-phosphatase, glucose-6-phosphate dehydrogenase, glyceraldehyde-3-phosphate dehydrogenase, hexokinase, glucokinase, pyruvate kinase, fructose-1,6-bisphosphatase, α-glucosidase, succinate dehydrogenase, catalase, acid phosphatase, γ-glutamyltranspeptidase, and glutathione-S-transferase P-form.

Influence of DHEA on Normal Liver Tissue

Light and electron microscopic findings

Treatment of male Sprague-Dawley rats with 0.25% DHEA in the diet (Altromin^R) for 27 weeks led only to minor cellular alterations in the hepatocytes surrounding the terminal veins. These cells were slightly enlarged and appeared more acidophilic in the H&E-stained sections compared to untreated controls. In the electron microscope an increased number of peroxisomes and mitochondria was evident (Figure 1a). Peroxisomes were often enlarged, irregularly shaped and contained tube- or plate-like structures (Figures 1b,c) which have been described as ma-

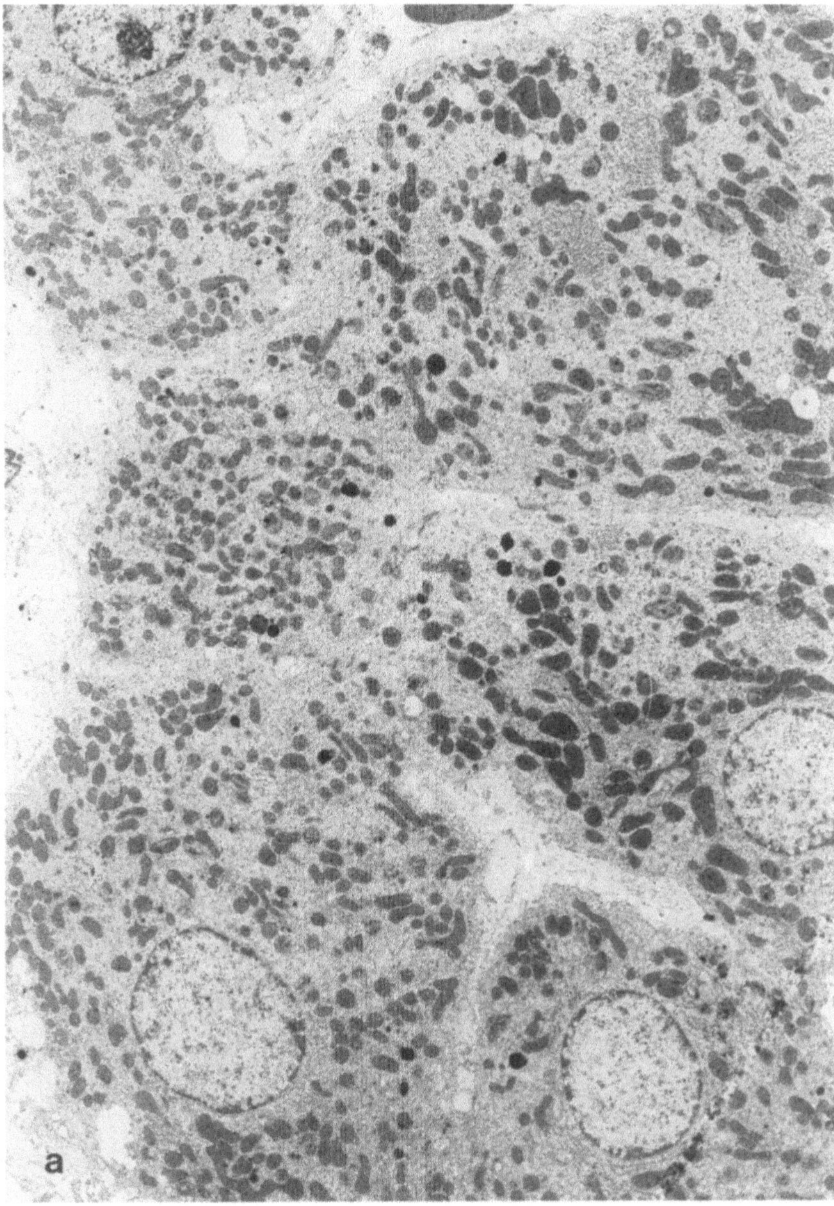

Figure 1 a) Perivenular area in the liver of a DHEA-treated rat. Note high density of cytoplasmic organelles, x 8,320.

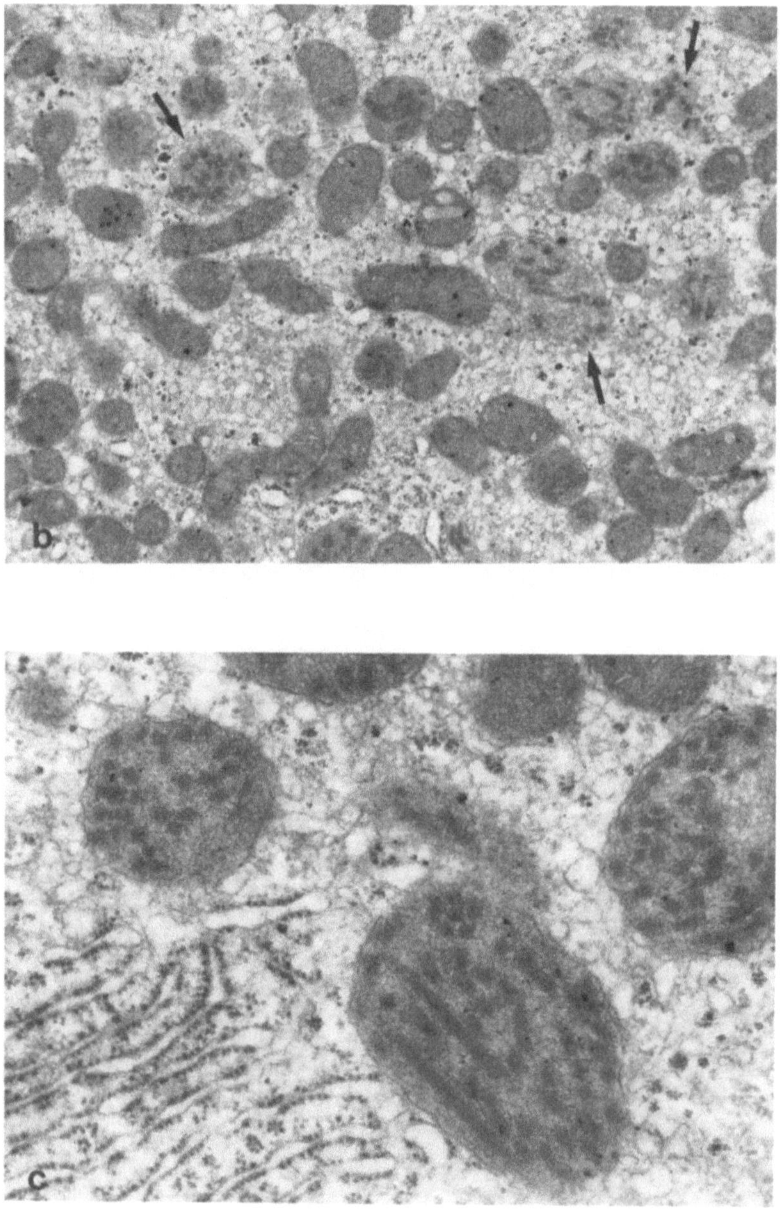

Figure 1 b) Higher magnification of the cytoplasm of a perivenular hepatocyte after DHEA-treatment shows enlarged, irregularly shaped peroxisomes (→) containing tube- or plate-like structures (c). b, x 20,800; c, x 61,600.

Table 1

Glycogen-content and activities of some enzymes of carbohydrate metabolism in DHEA-treated and control rats

Glycogen content (mg/mg protein) and enzyme activities (nmol/mg protein/min)	Control rats (10) mean ± SD	DHEA-rats (10) mean ± SD	Wilcoxon test
Glycogen	0.45 ± 0.07	0.37 ± 0.08	p<0.05
Synthase I	3.5 ± 0.7	4.1 ± 0.9	n.s.
Synthase I + D	23.2 ± 4.2	24.9 ± 4.6	n.s.
Phosphorylase a	166.1 ± 30.5	119.4 ± 11.7	p<0.0005
Phosphorylase a + b	159.3 ± 26.3	114.5 ± 12.3	p<0.0005
Hexokinase	11.1 ± 1.4	6.4 ± 0.9	p<0.0001
Glucokinase	44.8 ± 10.3	21.8 ± 8.4	p<0.0001
Pyruvate kinase	13.1 ± 2.7	9.6 ± 2.1	p<0.005
Glucose-6-Phosphatase	61.8 ± 9.3	75.0 ± 11.2	p<0.005
Fructose-1,6-bisphosphatase	57.8 ± 4.7	39.2 ± 2.5	p<0.00005
Glucose-6-phosphate dehydrogenase	13.9 ± 2.9	9.2 ± 1.8	p<0.001
α-Glucosidase	0.88 ± 0.09	1.09 ± 0.22	p<0.002

n.s.= not significant

trical inclusions after treatment with various other compounds like acetylsalicylic acid, clofibrate and dimethrin (15). DHEA itself did not induce any preneoplastic or neoplastic lesions.

Enzymatic alterations

Biochemical studies.

Biochemical assays of glycogen content and of the activity of a number of key enzymes of carbohydrate metabolism were performed on homogenates or 10^5g supernatants of livers from rats treated with 0.25% DHEA in the diet for 27 weeks and from concurrent untreated controls (11). It was found (Table 1) that the glycogen content was significantly decreased in the livers of DHEA-treated rats compared to the untreated controls, although the activity of glycogen phosphorylase (PHO) was significantly decreased and glycogen synthase (SYN) activity was slightly increased. PHO was present in both groups only in the a-form which is the form being active in vivo. On the other hand, SYN exhibited about 15% of the total activity in the active I-form in both DHEA-treated and control rats. Thus, the proportion of active SYN to active PHO is shifted in favour of glycogen synthesis in DHEA-treated rats. While the phosphorylytic glycogen breakdown may be reduced in DHEA-treated livers, the lysosomal enzyme α-glucosidase which degrades glycogen hydrolytically was significantly increased. This would explain the reduced glycogen content in these livers. Glucose-6-phosphate dehydrogenase (G6PDH), pyruvate kinase and fructose-1,6-bisphosphatase (rate-limiting enzymes of pentose phosphate pathway, glycolysis and gluconeogenesis, respectively) were strongly reduced. Both forms of glucose phosphorylating enzymes, the low K_m hexokinase and high K_m glucokinase showed decreased activity whereas glucose-6-phosphatase (G6Pase) was markedly increased.

Histochemical studies.

The histochemical investigations provided information on the intralobular distribution of enzyme activities and the alterations induced by DHEA-treatment (11). The slight rise in SYN activity observed by biochemical measurements was due to an increase in the periportal part of the liver lobule, and the reduction of PHO and pyruvate kinase activities could be attributed to decreases in the perivenular areas of the liver lobule. The decrease in G6PDH measured in homogenates from DHEA-treated livers could be attributed predominantly to non-parenchymal cells, but some decrease of G6PDH activity could also be observed in hepatocytes. Catalase (CAT) and succinate dehydrogenase (SDH) activities were found to be increased in the perivenular hepatocytes (Table 2).

The enzyme pattern observed in DHEA-treated rat liver suggests a re-

Table 2

Histochemical evaluation of the intralobular hepatic distribution of glycogen and of some enzymes of carbohydrate metabolism in DHEA-treated and untreated rats

Assay	pp/pv*	Controls (10)	DHEA (10)	Change***
PAS	pp	+**	+	n.c.
	pv	++	++	n.c.
Glycogen synthase	pp	+	++	↑
	pv	(+)	(+)	n.c.
Glycogen phosphorylase	pp	+++	+++	n.c.
	pv	+	(+)	↓
Pyruvate kinase	pp	(+)	(+)	n.c.
	pv	+++	+	↓↓
Glucose-6-phosphatase	pp	++	+++	↑
	pv	+	++	↑
Glucose-6-phosphate dehydrogenase	pp	(+)	(+)	n.c.
	pv	(+)	(+)	n.c.

*pp = periportal area, pv = perivenular area;

** intensity of reaction: (+) very low reaction
+ low reaction
++ strong reaction
+++ very strong reaction

*** n.c. = no change; ↑ increase; ↓ decrease

duction of glucose consumption which is manifested in a reduced input of glucose into hepatic metabolism via phosphorylating enzymes and in reduced glycolysis, gluconeogenesis and pentose phosphate pathway which seem to be restrained in these livers. Furthermore, increased glycogen and glucose 6-phosphate hydrolysis as evident from high α-glucosidase and G6Pase activities point to a waste of energy in glucose metabolism due to enhanced energy expenditure via net dephosphorylation of ATP caused by DHEA. A similar mechanism of energy loss has been proposed by Cleary et al. (16) studying DHEA effect on fatty acid metabolism.

Influence of DHEA on Preneoplastic and Neoplastic Lesions

Treatment with hepatocarcinogens

Treatment of rats with hepatocarcinogens like NNM and DAB leads to the development of various types of focal liver lesions. In early stages, the majority of these foci are characterized by an excessive storage of glycogen and appear to be composed of clear and acidophilic cells in the H&E-stained sections (Figures 2 and 3) (12,17,18). These glycogen storage foci gradually develop through mixed cell foci consisting of glycogen-rich and glycogen-poor (basophilic) or intermediate cells to hepatocellular adenomas and finally to hepatocellular carcinomas (17,18). While hepatocellular adenomas may still contain considerable amounts of glycogen, the end stage, the hepatocellular carcinoma, is usually basophilic and free from glycogen (Figure 2). Detailed cytochemical and morphometrical studies supporting this sequence have been reported (19). Enzyme histochemical studies designed to get insight into mechanisms leading to glycogen storage in early stages and to a loss of glycogen in late stages of hepatocarcinogenesis, have shown that glycogen storage foci usually exhibit a typical pattern of alterations of key enzyme activities of carbohydrate metabolism (14,20). As a rule, the activities of PHO, adenylate cyclase and G6Pase were decreased while G6PDH, glyceraldehyde-3-phosphate dehydrogenase (GAPDH) and pyruvate kinase showed increased activity. SYN was unchanged or only slightly increased, and hexokinase was low in these lesions. With progression of the lesions to a more basophilic stage when the glycogen content was gradually reduced, the enzyme alterations persisted or became even more prominent. However, SYN activity was markedly reduced, and hexokinase showed a considerable increase in activity in hepatocellular adenomas and carcinomas. The activity of pyruvate kinase strikingly correlated with the glycogen content of the cells. γ-glutamyltranspeptidase (GGT) and the placental form of glutathione S-transferase (GST-P) were markedly increased in the preneoplastic and neoplastic lesions occurring under these experimental conditions (20). α-Glucosidase activity was reduced in glycogen storage foci, but gra-

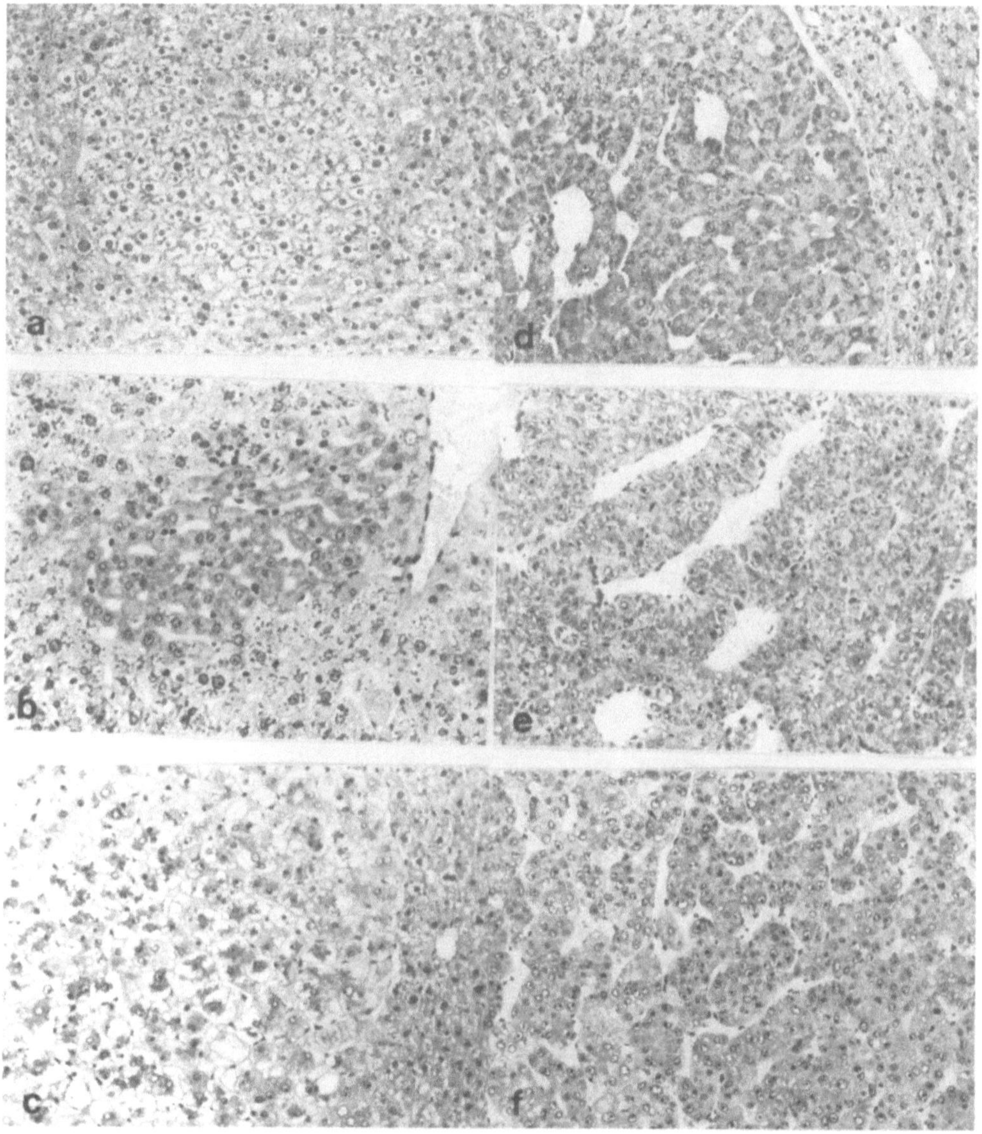

Figure 2 a) Clear cell glycogen storage focus, b) amphophilic focus, c) mixed cell adenoma, d) amphophilic/tigroid cell adenoma, e) dedifferentiated hepatocellular carcinoma, f) differentiated hepatocellular carcinoma, x 140; H&E-staining. a, c and e are lesions observed in rats treated with NNM alone, b, d and f from rats treated with NNM plus DHEA.

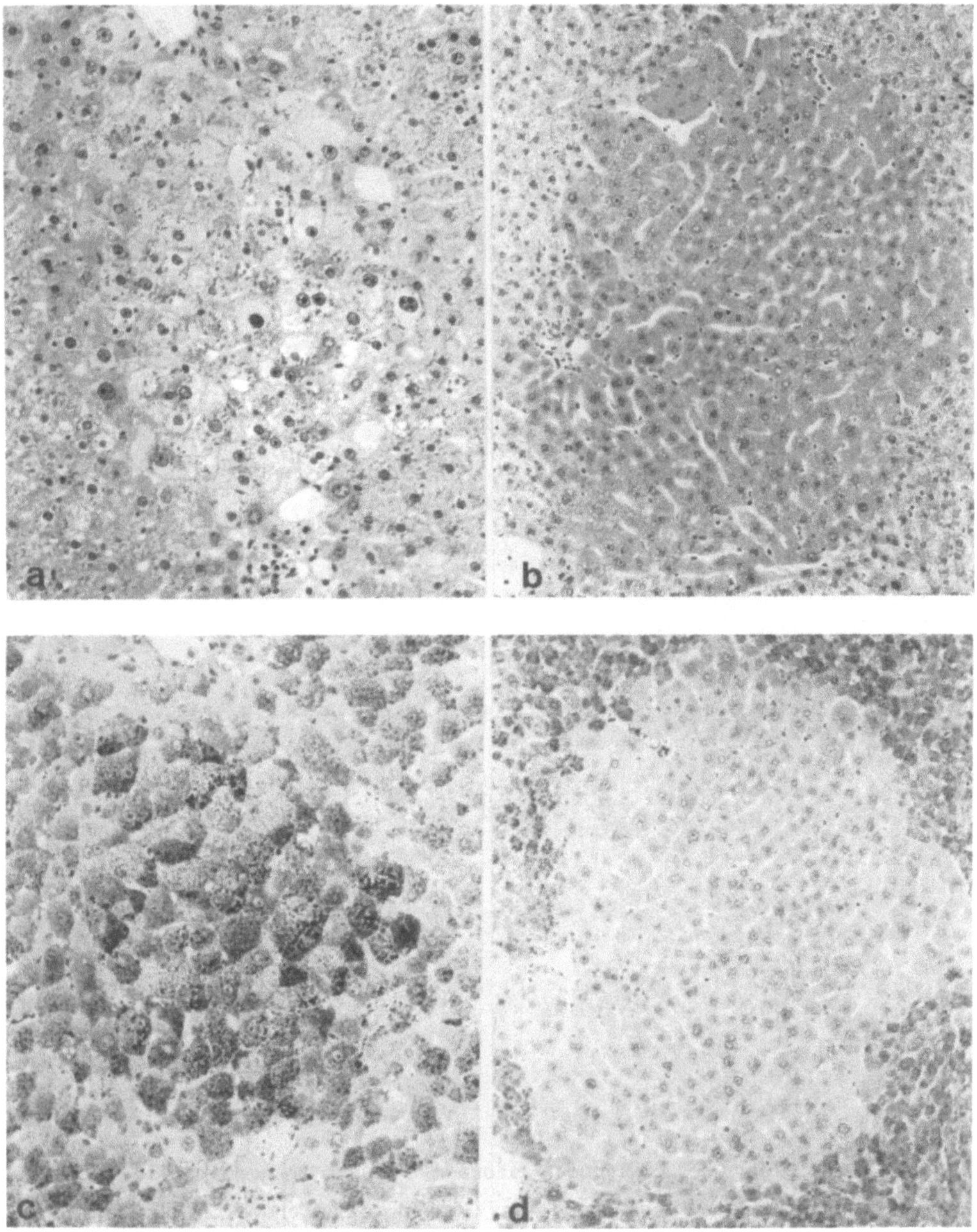

<u>Figure 3</u> a,c) acidophilic glycogen storage focus, x 150; b,d) amphophilic focus, x 100. a,b) H&E-staining, c,d) PAS-staining.

Table 3

Enzyme pattern in glycogen storage foci and amphophilic foci

Parameter	GSF	AMF
Glycogen	↑↑	↓
Synthase	↑ or no change	↓
Phosphorylase	↓	↓
Glucose-6-Pase	↓	↑ or ↓
G6PDH	↑↑	↑
GAPDH	↑	↑
SDH	no change	↑
CAT	no change	↑
Acid phosphatase	↓	↑
GGT	↑	no change or (↓)
GST-P	↑	no change

dually increased during the process of hepatocarcinogenesis being high in the tumours (21).

Treatment with hepatocarcinogens and DHEA

While the administration of the hormone DHEA alone did not result in induction of any preneoplastic lesions in rat liver, the morphological and enzymatic phenotypes of hepatocarcinogen-induced foci and nodules were altered by long-term (3,8,9) and even by a 4 week-DHEA application (10). In addition to the lesions described above another type of focus occurred which in the light microscope became prominent by a low glycogen

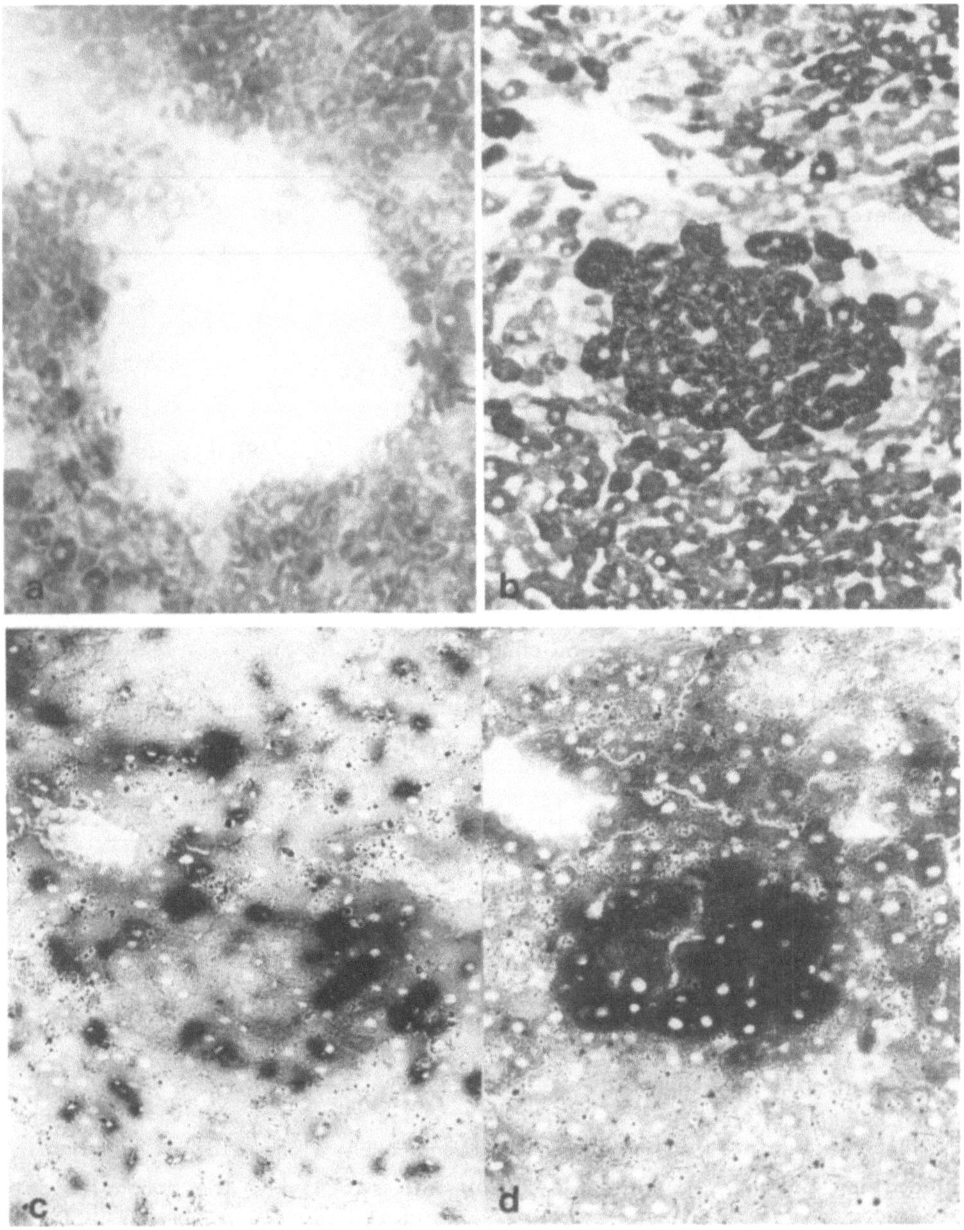

Figure 4 Amphophilic focus, enzyme histochemical demonstration of glycogen synthase (a), glucose-6-phosphatase (b), glucose-6 phosphate dehydrogenase (c), glyceraldehyde-3-phosphate dehydrogenase (d).

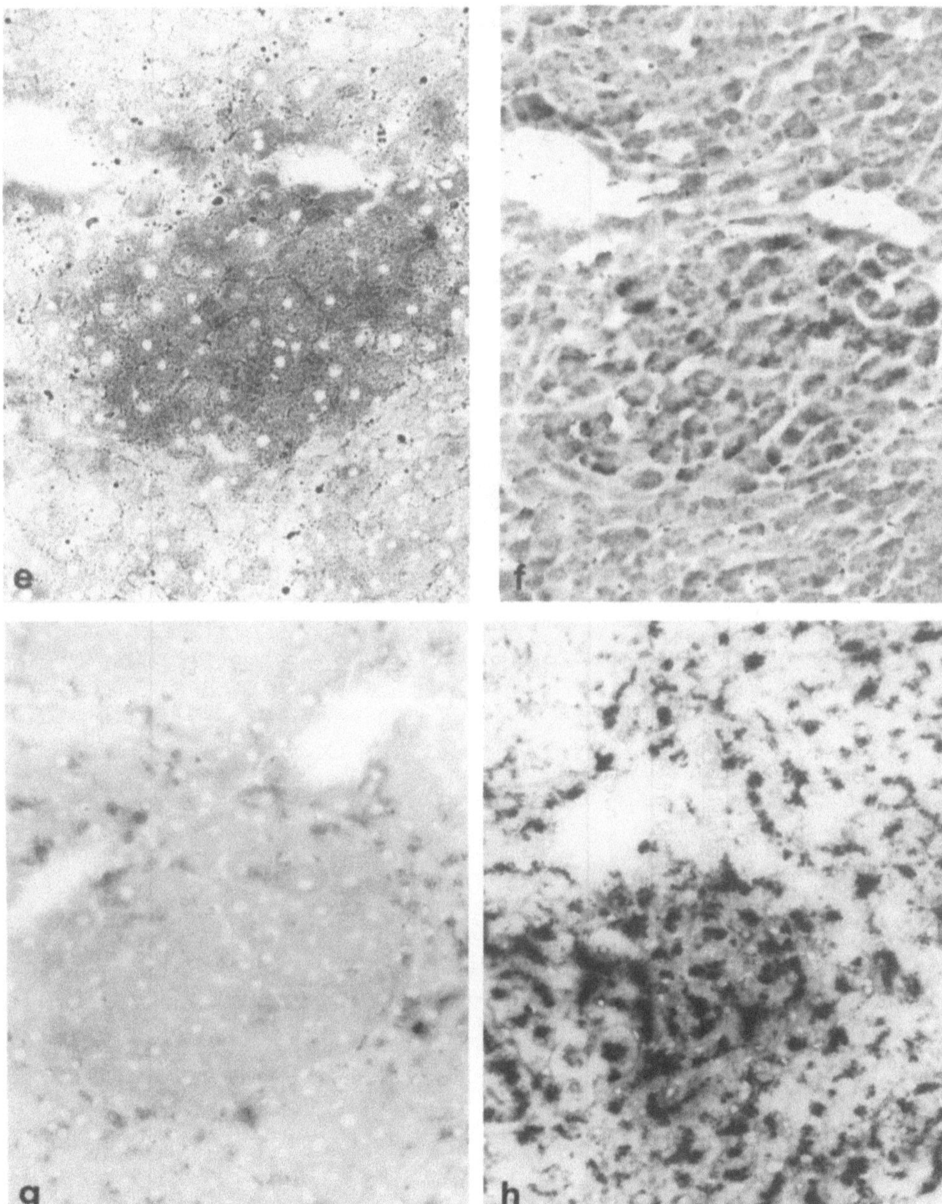

Figure 4 Succinate dehydrogenase (e), catalase (f), γ-glutamyltranspeptidase (g) and acid phosphatase (h). x 145.

Table 4

Percentage of lesions in NNM + DHEA-treated rat liver showing definite enzyme histochemical changes.

	amphophilic/tigroid cell foci				amphophilic/tigroid cell adenomas			
Histochemical 'marker'	no. of investigated foci	Intensity of reaction[a] ↓	n.c.	↑	no. of investigated adenomas	Intensity of reaction[a] ↓	n.c.	↑
Glycogen	40	100	0	0	20	100	0	0
GAPDH	38	3	39	58	20	0	15	85
G6PDH	40	5	20	75	21	0	0	100
G6PASE	39	51	15	34	20	85	10	5
Acid Pase	39	13	44	43	21	19	38	43
GGT	39	0	100	0	21	0	100	0
Catalase	40	0	55	45	20	5	80	15
SDH	40	8	35	57	21	19	19	62

[a] ↓ decrease; n.c. no change; ↑ increase

content and by a homogeneous, granular acidophilic cytoplasm with randomly scattered basophilia (Figures 2 and 3). These amphophilic foci covered about 20% of lesions in rat livers treated with NNM and DHEA (8) and were only very rarely found in rats treated with NNM alone. Adenomas developing from this type of lesions showed an increase in cellular basophilia which was arranged in stripes and classified to be of the tigroid cell type (8,12). Hepatocellular carcinomas originating from amphophilic/tigroid cell adenomas were well differentiated and appeared less malignant than carcinomas induced by NNM alone (Figure 2).

Enzyme histochemically the amphophilic foci and adenomas showed a common phenotype (Table 3) whereby glycogen was decreased, catalase, succinate dehydrogenase and acid phosphatase were increased, suggesting a high number of peroxisomes, mitochondria and lysosomes. GAPDH and G6PDH showed a heterogeneous distribution within these lesions. These enzyme activities were mostly increased but less marked compared to the increase observed in glycogen storage foci and mixed cell foci; G6Pase activity was increased in some foci and decreased in others. PHO was regularly decreased. γ-glutamyltranspeptidase and GST-P showed a striking lack of staining in the vast majority of foci and in all the adenomas investigated (Figure 4, Table 4). Similar results were obtained when DAB or DHPN were used as initiating carcinogens (3,10).

Quantitative Studies on Modulation of Hepatocarcinogenesis by DHEA

Body and liver weights

A number of observations on rats and mice, whether carcinogen-treated (3,6,22,23) or not (22-25) indicated that chronic DHEA-treatment inhibits body weight gain without reducing food consumption. A weight reduction of up to 39% in male and 15% in female rats was observed (Table 5) (3,4). On the other hand, absolute liver weights were unchanged in male or even increased in female rats treated with DHEA (3). This results in a marked increase in the liver/body weight ratio (3,6). Garcea et al. (6) reported an increase in cell size corresponding to an

Table 5

Quantitative data for DHPN-induced liver lesions

	Effective No. Animals	Body Weight (BW)	Liver Weight (% BW ratio)	Liver Lesions: GST-P Positive No./cm²	Liver Lesions: GST-P Positive % Area	Liver Lesions: Amphophilic/Tigroid No./cm²	Liver Lesions: Amphophilic/Tigroid % Area
Male							
Control	10	385 ± 12.2	11.8 ± 0.7 (3.1)	4.6 ± 2.4	<0.1	0.0	0.0
DHEA	10	250 ± 20.9***	9.8 ± 0.6 (3.9)	0.0**	0.0	0.0	0.0
DHPN	14	330 ± 32.1	10.5 ± 1.0 (3.2)	84.1 ± 22.0	2.1 ± 0.8	0.0	0.0
DHEA + DHPN	10	352 ± 27.2	10.8 ± 0.9 (3.1)	24.3 ± 6.9***	1.1 ± 1.4*	0.0	0.0
DHPN + DHEA	15	235 ± 10.8***	9.0 ± 0.7 (3.8)	18.6 ± 6.7***	1.3 ± 1.6*	1.8 ± 2.1**	1.5 ± 2.4
Female							
Control	10	190 ± 8.2	5.9 ± 0.5 (3.1)	1.3 ± 2.1	<0.1	0.0	0.0
DHEA	10	166 ± 8.5*	9.2 ± 0.6 (5.5)**	0.0	0.0	0.0	0.0
DHPN	15	201 ± 11.9	6.4 ± 0.6 (3.2)	16.4 ± 2.0	0.2 ± 0.1	0.0	0.0
DHEA + DHPN	10	204 ± 14.2	6.3 ± 0.7 (3.1)	4.7 ± 1.7***	<0.1	0.0	0.0
DHPN + DHEA	15	163 ± 8.8**	9.9 ± 0.7 (6.1)**	4.1 ± 2.6***	<0.1	0.0	0.0

Significantly different from control or DHPN alone group *p<0.05 **p<0.01 ***p<0.001

enlargement of the cytoplasmic area while the nuclear size remained unchanged.

Morphometric studies on liver lesions

As discussed above, DHEA exerts various effects on the morphology and enzymatic phenotype of focal lesions and tumours induced by chemical carcinogens. In order to obtain quantitative data on the DHEA-effect morphometric studies have been performed on lesions presenting altered morphological (3,8-10) and altered enzymatic phenotype (3,6-10).

Focal liver lesions.

Continuous application of DHEA to rats initiated with NNM leads to the development of amphophilic foci (8) which appear in addition to clear/mixed cell foci predominating in rats treated with NNM alone (Table 6). The numbers/cm² of both types of foci increased with the time course of the experiment, the ratio between amphophilic/tigroid cell foci and clear/mixed cell foci remaining constant (ratio ~0.2). In an experiment using DAB as initiator (10) DHEA-treatment resulted in a reduction of focal lesions of the mixed cell type from 9.2 ± 6.0 foci/cm² to 1.2 ± 0.6 foci/cm² (Table 7). On the other hand, the number of amphophilic/tigroid cell foci rose from 0.2 ± 0.2/cm² in DAB-treated rats to 1.2 ± 0.3/cm² in rats treated with DAB plus DHEA. However, the total liver area occupied by both types of foci was significantly lower in animals treated with DAB plus DHEA compared to those treated with DAB alone. A quantitative evaluation based on enzyme alterations such as G6PDH and GST-P proved to be very difficult in DHEA experiments since the foci in DHEA-treated rat livers were often not homogeneously stained and therefore, could not clearly be distinguished from the surrounding tissue.

Using an "initiation-selection model" in combination with DHEA-treatment Garcea et al. (6,7) observed a significant reduction by DHEA in the number of G6PDH and GGT-positive foci exhibiting a uniform enzyme pattern. Similarly, the surface area and the percentage of liver occupied by foci were significantly decreased. However, also under these experimental conditions, DHEA caused a relative increase in the number of foci showing a non-uniform pattern of histochemical reaction for GGT and irregular boun-

Table 6

Frequency of clear/mixed-cell and amphophilic/tigroid cell lesions in animals treated with NNM + DHEA.

time period (weeks)[a]	no. of animals	no. of clear/ mixed-cell foci per cm^2 [b] (%total lesions)	no. of amphophilic/ tigroid cell foci per cm^2 (%total lesions)	incidence of mixed-cell adenomas (%)	incidence of amphophilic/ tigroid adenomas (%)
7 + 4	20	9.17 ± 7.07 (77)	2.71 ± 2.30 (23)	0 (0)	0 (0)
7 + 20	20	25.01 ± 13.21 (79)	6.69 ± 4.51 (21)	9 (45)	11 (55)

[a] Seven weeks treatment + time period after withdrawal of carcinogen.

[b] Numbers represent mean and standard deviation.

Table 7

Quantitative data for focally altered populations

	Mixed Cell		Amphophilic/Tigroid Cell	
	No./cm^2	% Area	No./cm^2	% Area
DAB:	9.2 ± 6.0**	3.6 ± 3.8*	0.2 ± 0.2**	< 0.1**
DAB + DHEA:	1.2 ± 0.6	0.5 ± 0.5	1.2 ± 0.3	0.2 ± 0.1

Significantly different from DAB + DHEA *p<0.05 **p<0.01

daries with the surrounding liver. These foci were not accessible to a detailed morphometrical analysis.

In an "initiation-selection-promotion model" including phenobarbital (7), DHEA-treatment did not reduce the number of foci, but surface area and percentage of liver occupied by foci were strongly reduced indicating the presence of a large number of small foci (7).

Liver tumours

Liver tumours were induced in rats by application of either NNM alone or NNM plus DHEA (3,8). The tumours included adenomas of the mixed cell type and of the amphophilic/tigroid cell type. While the incidence of mixed cell adenomas was similar in both the NNM plus DHEA and the NNM-alone group, the total number of adenomas was higher in the group treated with NNM alone. On the other hand, incidence and total number of amphophilic/tigroid cell adenomas was much higher in the group treated with NNM plus DHEA than in the group treated with NNM alone (Table 8). Hepatocellular carcinomas of highly differentiated type formed single- or double-cell layered sheets with wide sinusoids. Their cells showed little polymorphism or atypia, but increased tigroid or diffuse basophilia (Figure 2). Less well differentiated hepatomas demonstrated multiple cell layers forming trabecular and adenocarcinomatous patterns with marked nuclear and cytoplasmic polymorphism (Figure 2). In line with these findings indicating decreased malignancy of lesions in hormone-treated animals, the mitotic rate established for

Table 8

Incidence (%age) and frequency of hepatic lesions

Treatment	No. Animals	Hepatocellular Adenomas				Hepatocellular Carcinomas		
		Mixed Cell		Amphophilic/Tigroid Cell		Highly Differentiated	Poorly Differentiated	
		Incidence (%)	Total (No/animal)	Incidence (%)	Total (No/animal)	Incidence (%)	Incidence (%)	Total (No/animal)
NNM	35	32 (91)	204 (5.8)	1 (3)**	1(<0.1)	12 (34)	11 (31)**	26 (0.7)
NNM + DHEA	35	28 (80)	44 (1.3)	32 (91)	154 (4.5)	12 (34)	2 (6)	20 (0.6)
Control	15	0 (0)	0 (0)	0 (0)	0 (0)	0 (0)	0 (0)	0 (0)
Control + DHEA	15	0 (0)	0 (0)	0 (0)	0 (0)	0 (0)	0 (0)	0 (0)

Significantly different from NNM + DHEA *p<0.05 **p<0.01

hepatocellular carcinomas observed after NNM alone (316±207) was markedly higher than for those observed after additional hormone administration (71±52). No significant difference in total number of carcinomas was observed between rats treated with NNM alone or with NNM plus DHEA. Both groups showed the same incidence of highly differentiated hepatocellular carcinomas. However, the incidence of poorly differentiated carcinomas was significantly higher in the NNM-alone group (Table 8). Furthermore, metastases generally composed of trabecular formations, were only observed in animals treated with NNM alone. These findings, together with the observations on alterations in enzyme pattern indicate that amphophilic foci represent a type of lesion which ends up in the development of tumours which are less malignant than those arising from glycogen storage foci. In this context, it may be of interest to note that the number of hemangiosarcomas characterized by marked invasive growth was significantly smaller in the NNM plus DHEA group compared to the NNM-alone group (9).

Some Aspects of the Modulatory Mechanism of DHEA

The inhibitory effect of DHEA on tumour development has initially been solely attributed to the inhibition of G6PDH, the key enzyme of pentose phosphate pathway by this hormone (1). It was argued that a reduced cellular NADPH pool resulting from G6PDH-inhibition might be the limiting factor for lipogenesis and biosynthesis of ribonucleotides and deoxyribonucleotides essential for tumour growth (6,22,24). However, more recently it became evident that the DHEA-effect could not simply be due to G6PDH-inhibition (2). Thus, after a long-term administration of the hormone to rats, G6PDH-activity of the liver is often normal (25,26), and the activity of malic enzyme and isocitrate dehydrogenase which both produce NADPH, are markedly increased (7,16,26) and the $NADP^+$/NADPH ratio is unchanged (7). Recently, Garcea et al. (7) have suggested that not a decrease in NADPH but a relative deficiency of ribo- and deoxyribonucleotides for DNA synthesis, caused by a great fall in pentose phosphate reduction, may be the reason for the inhibitory effect of DHEA on tumour growth. In their experiments the inhibitory DHEA-effect on the development of GGT-positive liver foci could be overcome by the injections of nucleosides.

This agrees with findings of Shantz et al. (27) who observed a block of growth and differentiation to adipocytes of 3T3-L1 fibroblasts by DHEA and a drop of intracellular 6-phosphogluconate which is the product of G6PDH activity. Both, block of growth and differentiation of DHEA-treated 3T3-L1 cells could be overcome by introduction of 6-phosphogluconate into the cells by using liposomes. It may, therefore, be argued that in both DHEA-treated rat liver and 3T3-L1 fibroblasts growth inhibition is not due to a reduction of intracellular NADPH pool but more likely to a fall in intracellular level of sugar phosphates such as 6-phosphogluconate (27) and ribulose 5-phosphate (7).

Other effects of DHEA which may be involved in the modulation of hepatocarcinogenesis have been reported. Enzymatic studies in fatty acid metabolism in rat liver by Cleary et al. (16) revealed an increase in long-chain fatty acyl-CoA hydrolase and a decrease in fatty acid synthetase. This was interpreted as a futile cycle of fatty acid metabolism leading to net ATP consumption, since liberated fatty acids need ATP-dependent reattachment of CoA before further utilization (28).

Increased G6Pase and decreased glucokinase observed in DHEA-treated rat liver (Table 1) may represent such a futile cycle in glucose metabolism (11), since liberated glucose needs rephosphorylation by ATP before entering further metabolism. It may be speculated that enhanced energy expenditure via net dephosphorylation of ATP is a principal mechanism underlying the DHEA-effect.

It has also to be noted that chemical carcinogens induce enzyme alterations which are strikingly opposite to those caused by the "anti-carcinogen" DHEA (Tables 1,3,4; Figure 3). DHEA leads to profound alterations in enzyme activities of liver carbohydrate metabolism in a way that glycolysis and gluconeogenesis are both restrained in DHEA-treated livers and that an increased glucose release via G6Pase may lead to a net output of glucose. An elevated hydrolytic cleavage of glycogen via lysosomal α-glucosidase activity would fit into this conception of glucose output. On the other hand, glycogen storage foci show an increase in activity of glycolytic enzymes and a decrease in G6Pase and α-glucosidase (20,28) which points to a higher intracellular consumption of glucose. Thus, many findings suggest that the mechanism which underlies modulation of liver carcinogenesis by DHEA is closely related to alterations in the regulation of glucose metabolism.

Conclusions

Administration of DHEA to rats previously treated with chemical carcinogens modulates the process of hepatocarcinogenesis. The total number of tumours per animal is reduced compared to animals treated with the carcinogen alone, and these tumours exhibit a less malignant phenotype. The tumours are preceded by preneoplastic lesions of an amphophilic or amphophilic/tigroid appearance which do not or only very rarely occur in livers treated with the carcinogen alone. The amphophilic foci are characterized by alterations in the pattern of enzyme activities of carbohydrate metabolism which are less pronounced or even opposite to those observed in glycogen storage foci which usually represent the prestages of hepatocellular tumours in rat livers treated with the carcinogen alone. The modulation of metabolic pathways such as glycolysis, gluconeogenesis or hexose monophosphate shunt by DHEA may be related to the mechanism underlying the modulation of hepatocarcinogenesis by the hormone.

References

1. Schwartz, A.G., J.M. Whitcomb, J.W. Nyce, M.L. Lewhart and L.L. Pashko. 1988. Advan. Cancer Res. 51, 391-424.

2. Gordon, G.B., L.M. Shantz and P. Talalay. 1987. Advan. Enzyme Regul. 26, 355-382.

3. Moore, M.A., E. Weber, M. Thornton and P. Bannasch. 1988. Carcinogenesis 9, 1507-1509.

4. Moore, M.A., W. Thamavit, A. Ichihara, K. Sato and N. Ito. 1986. Carcinogenesis 7, 1059-1063.

5. Moore, M.A., W. Thamavit, H. Tsuda, K. Sato, A. Ichihara and N. Ito. 1986. Carcinogenesis 7, 311-316.

6. Garcea, R., L. Daino, R. Pascale, S. Frassetto, P. Cozzolino, M.E. Ruggiu and F. Feo. 1987. Toxicol. Pathology 15, 164-169.

7. Garcea, R., L. Daino, S. Frassetto, P. Cozzolino, M.E. Ruggiu, M.G. Vannini, R. Pascale, L. Lenzerini, M.M. Simile, M. Puddu and F. Feo. 1988. Carcinogenesis 9, 931-938.

8. Weber, E., M.A. Moore and P.Bannasch. 1988. Carcinogenesis 9, 1049-1054.

9. Weber, E., M.A. Moore and P. Bannasch. 1988. Carcinogenesis 9, 1191-1195.

10. Moore, M.A., E. Weber and P. Bannasch. 1988. Virchows Arch. B Cell Pathol. 55, 337-343.

11. Mayer, D., E. Weber, M.A. Moore, I. Letsch, E. Filsinger and P. Bannasch. 1988. Carcinogenesis 9, 2039-2043.

12. Bannasch, P., U. Benner, H. Enzmann and H.J. Hacker. 1985. Carcinogenesis 6, 1641-1648.

13. Hacker, H.J.. 1978. Histochemistry 58, 289-296.

14. Hacker, H.J., A.M. Moore, D. Mayer and P. Bannasch. 1982. Carcinogenesis 3, 1265-1272.

15. Hruban, Z., M. Gotoh, A. Slesers and S.-F. Chou. 1974. Lab. Invest. 30, 64-75.

16. Cleary, M.P., S.S. Hood, C. Chando, C.T. Hansen and J.T. Billheimer. 1984. Nutr. Res. 4, 485-494.

17. Bannasch, P., D. Mayer and H.J. Hacker. 1980. Biochim. Biophys. Acta 605, 217-245.

18. Bannasch, P., H.J. Hacker, F. Klimek and D. Mayer. 1984. Advan. Enzyme Regul. 22, 97-121.

19. Bannasch, P., H. Enzmann, H.J. Hacker, E. Weber and H. Zerban. 1989. In: Liver Cell Carcinoma (P. Bannasch, D. Keppler and G. Weber, eds.). Kluwer Academic Publishers, Dordrecht•Boston•London. pp. 55-76.

20. Mayer, D., F. Klimek, H.J. Hacker and P. Bannasch. 1989. In: Liver Cell Carcinoma (P. Bannasch, D. Keppler and G. Weber, eds.). Kluwer Academic Publishers, Dordrecht•Boston•London. pp. 329-345.

21. Klimek, F. and P. Bannasch. 1989. Virchows Arch. B Cell Pathol. 57, 245-250.

22. Schwartz, A.G., G.C. Hard, L.L. Pashko, M. Abou-Gharbia and D. Swern. 1981. Nutrition and Cancer 3, 46-53

23. Schwartz, A.G., L.L. Pashko, E.E. Henderson, R.H. Tannen, M.P. Cleary, M. Abou-Gharbia and D. Swern. 1983. Comment. Res. Breast Dis. 3, 113-130.

24. Yen, T.Y., J.A. Allan, D.V. Pearson and J.M. Acton. 1977. Lipids 12, 409-413.

25. Cleary, M.P., A. Shepherd and B. Jenks. 1984. J. Nutrition 114, 1242-1251.

26. Casazza, J.P., W.T.Schaffer and R.L. Veech. 1986. J. Nutrition 116, 304-310.

27. Shantz, L.M., P. Talalay and G.B. Gordon. 1989. Proc. Natl. Acad Sci. U.S.A. 86, 3852-3856.

28. Shepherd, A. and M.P. Cleary. 1984. Am. J. Physiol. 246, E123-128.

DEHYDROEPIANDROSTERONE ALTERS THE MORPHOLOGY AND PHOSPHOLIPID CONTENT OF CULTURED HUMAN ENDOTHELIAL CELLS

Milton Sholley*, Stephen Gudas*, William Regelson**, Richard Franson***, and Mohammed Kalimi****

Departments of Anatomy*, Medicine**, Biochemistry and Molecular Biophysics***, and Physiology****
Medical College of Virginia, Virginia Commonwealth University
Richmond, Virginia 23298

Introduction

Dehydroepiandrosterone (DHEA), an adrenal corticosteroid produced in large quantities in humans, is known to have anti-obesity effects in some strains of mice (1) and rats (2). In his extensive monograph on DHEA, Szonka (3) reviewed studies suggesting that the anti-obesity effects of this hormone include reduction of serum cholesterol levels and protection against atherosclerosis. Subsequent studies have both challenged and supported these possible beneficial actions of DHEA. In rats made hypercholesterolemic by feeding a high cholesterol diet, DHEA feeding either had no effect on the serum cholesterol level (4) or exacerbated the hypercholesterolemia (5). On the other hand, Nestler et al. (6) reported that oral administration of DHEA to healthy human subjects significantly lowered serum low density lipoprotein (LDL) levels, and Parker et al. (7) demonstrated that the level of LDL in human fetal plasma was inversely related to the level of DHEA. In rabbits made hypercholesterolemic by feeding a high cholesterol diet, DHEA feeding did not significantly reduce the hypercholesterolemia, but it did decrease the severity of atherosclerotic plaques developing both in balloon catheterized aortas (8) and in unperturbed aortas (9). The

Dehydroepiandrosterone (DHEA)

epidemiological study of Barrett-Connor *et al.* (10) demonstrated that the serum level of DHEA is inversely related to deaths from cardiovascular diseases in men over age 50. If DHEA does decrease atherogenesis, this protective function would be a major physiological role of this quantitatively major but little understood hormone; moreover, DHEA or its analogues could be useful as drugs to ameliorate atherosclerotic disease.

All blood vessels are lined by endothelial cells, which provide the interface between the plasma and the deeper layers of the vascular wall or the extravascular compartment. Endothelial cells mediate both active and passive transport of lipids across the vascular wall (11). It has been estimated (12) that endothelial cells provide about half of the total body catabolism of LDL, both by transport of LDL to the extravascular compartment for catabolism and by intraendothelial catabolism. Since endothelial cells limit passage of lipids into the arterial wall, modulation of their lipid metabolism by DHEA could affect development of atherosclerosis. Our laboratory has initiated studies of the effects of DHEA on lipid metabolism in endothelial cells. In this paper we document that DHEA causes changes in morphology and phospholipid composition, particularly an accumulation of sphingomyelin, in cultured human endothelial cells.

Endothelial Cell Culture System

Endothelial cells from human umbilical veins were isolated and cultured using a modification of the technique of Gimbrone *et al.* (13). Following perfusion with phosphate buffered saline (PBS), the lumen was filled with 0.1% collagenase in PBS, incubated for 15 minutes at 37°C, and flushed into a centrifuge tube using Hanks' balanced salt solution (HBSS). The cells were pelleted by centrifugation at 200 x g for 5 minutes and then were resuspended in Medium 199 containing Earle's salts, L-glutamine, 100 U/ml penicillin, 100 µg/ml streptomycin, 0.25 µg/ml amphotericin-B, 20% fetal bovine serum (FBS), 20 µg/ml endothelial cell growth factor (ECGF), and 90 µg/ml heparin. Cells were routinely plated into gelatin-coated polystyrene flasks (14). The endothelial identity of the cultured cells was determined at confluence by the typical "cobblestone" morphology of the monolayer and was confirmed in selected cultures by immunofluorescent staining for factor VIII related antigen (15). Confluent cultures to be harvested were washed with HBSS and cells were released with 0.025% trypsin-0.01% EDTA. Cells were collected in M-199 containing 20% FBS and the diluted cell suspensions were counted using quadruplicate hemacytometer counts. The usual passage density, which yielded confluency within one week, was 1.25 x 10^5 cells/25 sq cm flask. For biochemical procedures, the cultures were washed with PBS and the cells were harvested mechanically by scraping with a rubber policeman.

DHEA was dissolved in dimethylsulfoxide (DMSO) and added to the media on the day of use. Dilutions of steroids were adjusted to standardize the DMSO concentration in the media at 0.1%. In each experiment, control cultures were also run at this 0.1% DMSO concentration without steroid.

Morphological Changes in DHEA-treated Endothelial Cells

Cultured endothelial cells treated with DHEA (1-50 µM) displayed distinctive morphological changes, which are described in detail elsewhere (16). Under low magnification (200x) phase contrast microscopy, cells exposed to DHEA for 24 hours displayed numerous perinuclear cytoplasmic granules. Following *in situ* fixation in glutaraldehyde and osmium and standard dehydration in ethanol, many granules had the ultrastructural

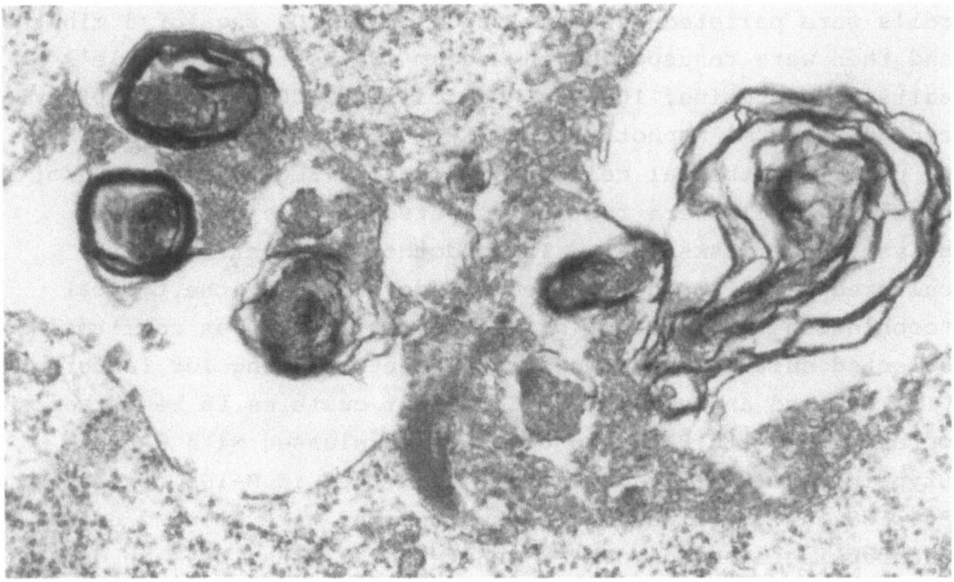

Figure 1. Myelin figures within cytoplasmic vacuoles in an endothelial cell from a confluent culture treated with DHEA (50 µM) for 24 hours. Uranyl acetate and lead citrate, x52,000.

appearance of myelin figures (Fig. 1), which consist of whorls of layered membranes. The ultrastructural morphology of the granules was reminiscent of tubular myelin, a phospholipid-containing surfactant granule found in pneumocytes of anuran amphibians (17) and of sphingolipid granules in lipid storage cells of individuals with inherited sphingolipidoses such as Niemann-Pick disease (18). Less extracted granules corresponded ultrastructurally to multilamellar lipid structures (16), which are pleomorphic lipoprotein-degrading lysosomes that form one type of lipid deposit in foam cells of atherosclerotic plaques (19). Phospholipids comprise 30% of the total lipids in LDL, and these phospholipids primarily include phosphatidylcholine (65%) and sphingomyelin (25%) (20). It is conceivable that the development of phospholipid-containing granules could result from increased degradation of lipoproteins stimulated by DHEA.

Major Phospholipid Content of DHEA-treated Endothelial Cells

To determine whether the cellular content of phospholipids was altered by DHEA, the phospholipid content of control and DHEA-treated endothelial cells was analyzed biochemically. Beginning two days after plating, endothelial cells were grown to confluence over the next 5 days in media containing DHEA (50 µM) or the 0.1% DMSO solvent only. Cells were then harvested mechanically, dispered by pipetting, and counted. Total lipids were harvested from equivalent numbers (1×10^7) of control and DHEA-treated cells using the chloroform-methanol-water method of Bligh and Dyer (21). Phospholipids were separated by two-dimensional thin layer chromatography on silicic acid and quantitated by measuring lipid phosphorus (22). The total amounts of phospholipid were essentially equivalent in control and DHEA-treated cells. However, phospholipid separation and quantitation revealed that the cellular content of sphingomyelin was more than doubled in DHEA-treated cells (Table 1). The amount of phosphatidylserine and phosphatidylinositol was increased by about 19%, whereas phosphatidylcholine and phosphatidylethanolamine were reduced by about 10% and 8%, respectively. The increased sphingomyelin content in DHEA-treated endothelial cells could represent the sphingomyelin component of degraded LDL. If this were the case, one might also expect phosphatidylcholine, the other major phospholipid of LDL, to be increased, but phosphatidylcholine may be more prone to lysosomal degradation than sphingomyelin. It is known that inborn deficits of lysosomal hydrolases lead to accumulation of sphingolipids, as in Niemann-Pick disease, which in some cases is associated with a sphingomyelinase deficiency (23). Thus, it is conceivable that increased uptake of lipoproteins in DHEA-treated endothelial cells could lead to depletion of hydrolases in lipid-overloaded lysosomes and that such depletion could be a mechanism underlying the accumulation of sphingomyelin in multilamellar lipid structures.

Table 1. Effect of DHEA (50 μM) on Phospholipid Composition of Cultured Human Endothelial Cells.

Phospholipid	nmols/10^7 cells		DHEA as % of C
	Control (C)	DHEA	
Phosphatidylserine/ Phosphatidylinositol:	71.4	84.8	118.8%
Sphingomyelin:	29.5	63.7	215.9%
Phosphatidylcholine:	202.4	181.3	89.6%
Phosphatidylethanolamine:	129.4	118.7	91.7%

Concluding Remarks

It has been reported that the sphingomyelin content of grossly normal male aortas increases exponentially with age (24). The cellular or matrix components containing the extra sphingomyelin were not determined (24). It is unknown if this age-linked increase in aortic sphingomyelin bears any relationship to DHEA levels, which, rather than increasing, generally decline with age (3). Thus, it is difficult to speculate whether this *in vivo* increase in aortic sphingomyelin is related to DHEA and/or to endothelial cells.

Biochemical analysis of other lipid species and experiments to trace radiolabeled lipoproteins are necessary to determine whether the increased sphingomyelin content in DHEA-treated endothelial cells is related to altered uptake and degradation of lipoproteins. However, experiments reported elsewhere (16) have shown that the presence of lipoproteins in the culture media is required for induction of the myelin figures and multilamellar lipid structures. Because of the strategic location of endothelial cells between the plasma and the deeper layers of the arterial wall, altered lipid metabolism in these cells could modulate the development of atherosclerosis. It remains to be determined whether or not the DHEA-induced changes in ultrastructure and sphingomyelin content of cultured endothelial cells are related to the anti-atherogenic effects of DHEA *in vivo*.

Acknowledgment

This research was supported by Grant #J100 from the Thomas F. Jeffress and Kate Miller Jeffress Memorial Trust.

References

1. Cleary, M., R. Seidenstate, R. Tannen and A. Schwartz. 1982. Proc. Soc. Exp. Biol. Med. 171, 276-284.

2. McIntosh, M. and C. Berdanier. 1988. Proc. Soc. Exp. Biol. Med. 187, 216-222.

3. Sonka, J. 1976. Acta. Univ. Carol. Med. Monogr. 71, 1-171.

4. Ben-David, B., S. Dikstein, G. Bismuth and F. Sulman. 1967. Proc. Soc. Exp. Biol. Med. 125, 1136-1140.

5. Kritchevsky, D., S. Tepper, D. Klurfield and A. Schwartz. 1983. Pharmacol Res Commun 15, 797-802.

6. Nestler, J., C. Barlascini, J. Clore and W. Blackard. 1988. J. Clin. Endocrinol. Metabol. 66, 57-61.

7. Parker, C., E. Simpson, D. Bilheimer, K. Leveno, B. Carr and P. MacDonald. 1980. Science 208, 512-514.

8. Gordon, G., D. Bush and H. Weisman. 1988. J. Clin. Invest. 82, 712-720.

9. Arad, Y., J. Badimon, L. Badimon, W. Hembree and H. Ginsberg. 1989. 9, 159-166.

10. Barrett-Connor, E., K. Khaw and S. Yen. 1986. New Engl. J. Med. 315, 1519-1524.

11. Fielding, C. and P. Fielding. 1983. In: Biochemical Interactions at the Endothelium (A Cryer, ed.). Elsevier Science Publishers, New York. pp. 275-299.

12. Van Hinsbergh, V., J. Emeis and L. Havekes. 1983. In: The Endothelial Cell - A Pluripotent Control Cell of the Vessel Wall (D. Thilo-Korner and R. Freshney, eds.). Karger, Basel. pp.99-112.

13. Gimbrone, M., R. Cotran and J. Folkman. 1974. J. Cell Biol. 60, 673-684.

14. Thornton, S., S. Mueller and E Levine. 1983. Science 222, 623-625.

15. Maciag, T., G. Hoover, M. Stemerman and R. Weinstein. 1981. J. Cell Biol. 91, 420-426.

16. Sholley, M., S. Gudas, C. Schwartz and M. Kalimi. 1990. Am. J. Pathol. In press.

17. Goniakowska-Witalinska, L. 1984. Europ. J. Cell Biol. 33, 127-133.

18. Takahashi, K., and M. Naito. 1985. Acta Pathol. Jpn. 35, 85-408.

19. Lupu, F., I. Danaricu and N. Simionescu. 1987. Atherosclerosis 67, 127-142.

20. Skipski, V. 1972. In: Blood Lipids and Lipoproteins: Quantitation, Composition, and Metabolism (G. Nelson, ed.). Wiley-Interscience, New York. pp. 471-583.

21. Bligh, E. and W. Dyer. 1959. Can. J. Biochem. Physiol. 37, 911-917.

22. Bartlett, G. 1959. J. Biol. Chem. 234, 466-468.

23. Takada, G., W. Satoh, K. Komatsu, Y. Konn, Y. Miura and Y. Uesaka. 1987. Tohoku J. Exp. Med. 35, 27-36.

24. Rouser, G. and R. Solomon. 1969. Lipids 4, 232-234.

Studies on the Biochemical Action and Mechanism of Dehydroepiandrosterone

Mohammed Kalimi, Justicia Opoku, Ruisheng Lu, Shafagoj Yanal, Mona Khalid, William Regelson, Dastgir Qureshi

Departments of Physiology and Medicine
Medical College of Virginia, Virginia Commonwealth University
Richmond, Virginia 23298

Introduction

Dehydroepiandrosterone (DHEA) is a major adrenal cortical secretory steroid in humans which has been implicated in a wide variety of pathophysiological conditions including cancer, diabetes, aging and obesity (1) . Although it is obvious from various articles presented here that DHEA has a profound effect on various pathophysiological conditions, our knowledge concerning the mechanism (s) by which DHEA exerts its various effects at the cellular and molecular level are largely unknown. It is likely that DHEA responses are mediated by the specific receptors present in the subcellular fractions of target tissues such as cytosol and or nuclei. DHEA may also modulate other steroid hormone receptors such as that of estrogens, androgens, mineralocorticoids, progestins or glucocorticoids. In addition, DHEA responses may be partly mediated by membrane receptors or through its various metabolites in vivo. Therefore, the elucidation of various mechanisms by which DHEA exerts its effects on varieties of tissues and cells is not only of considerable scientific interest but may also lead to the synthesis of improved and more potent DHEA derivatives with higher biological activities. In this chapter, we will review some work carried out in our laboratory regarding the physicochemical characterization of a [^{3}H] DHEA binder in rat liver. We present here some data suggesting that DHEA (i) down regulates rat hepatic glucocorticoid receptors, (ii) increases rat liver tyrosine aminotransferase enzymatic activity and (iii) inhibits [^{3}H] uridine uptake and incorporation in normal and transformed liver cells.

I. Characterization of specific [^{3}H] DHEA binder in rat liver:

A high affinity (Kd 2.3 nM) DHEA binding macromolecule was identified in male Sprague-Dawley rat liver cytosol (2). High specificity of binding for DHEA was shown by competition studies in which [^{3}H] DHEA was displaced 70-80% by unlabelled DHEA, β etiocholanolone and androsterone. Testosterone, progesterone and 17β estradiol were less effective (40-50%). Dexamethasone and aldosterone were ineffective. The specific binding was highest in liver followed by kidney and testis. The specific binding was enhanced (without effecting binding affinity) by addition of0.1 M NaSCN or 0.3 M KCl to the cytosol. The specific binding was almost completely eliminated by trypsin and chymotrypsin. Lipase, deoxyribonuclease and ribonuclease were without any effect. The [^{3}H] DHEA - macromolecular complex was stable upto 2 hours at 35°C. Further stabilization of [^{3}H] DHEA macromolecular complex was obtained by addition of 5 mM dithiothreitol, 10 mM molybdate or 0.3 M KCl. On sucrose density gradients under low and high salt conditions 7-8 S and 3-4 S macromolecular complex were observed. The [^{3}H] macromolecular complex when activated by incubating for 45 minutes at 25°C bound poorly to the DNA-cellulose.

Thus to the best of our knowledge this is the first evidence for the presence of high affinity and specific DHEA binder in rat liver cytosol. It is yet to be determined whether this binder has any physiological relevance that would identify it as a specific receptor. We would like to study the regulation of this binder under various physiological and pharmacological interventions. This will be of considerable interest since there was no correlation between the high affinity in vitro binding observed by us and the large concentrations of in vivo DHEA used by several investigators (1,3). At the present time, the only explanation for this apparent contradiction may be attributable to the lower thermal activation rate of [^{3}H] DHEA complex in terms of its poor binding to the DNA-cellulose.

II. Modulation of rat liver glucocorticoid receptors by DHEA:

Recently, Crudo (4) reported the downregulation of hepatic glucocorticoid receptors in Sprague-Dawley rats given 4 mg/100 gm body wt of DHEA subcutaneously daily for five days. He found no change in the affinity of the receptor for the hormone. He apparently used intact animals and one time point (five days) after subcutaneous doses of DHEA were given. Therefore we have systematically examined the effect of daily im injection of 4 mg/100 gm body weight DHEA on rat hepatic glucocorticoid receptors using adrenalectomized male Sprague-Dawley rats. Adrenalectomized animals are devoid of endogenous glucocorticoids and therefore glucocorticoid receptors are free and can be estimated using binding assays.

Results presented in Figure 1 suggest that there is a gradual decline in rat hepatic glucocorticoid receptors 24 hours after DHEA administration, the level decreases to almost

50% within five days. We have determined the change in the glucocorticoid receptors concentration using a single saturating concentration of [^{3}H] dexamethasone. Therefore we have only measured receptor numbers and not the affinity. Our results showed that beside binding to a specific DHEA binding protein (2), DHEA can also act through regulating other steroid hormone receptors. The amounts of DHEA needed to down regulate glucocorticoid receptors to 50% is very large and requires a number of days as compared to glucocorticoid agonists and antagonists such as RU 486 (5). This indicates that perhaps DHEA metabolite(s) and not DHEA per se is involved in the down regulation of glucocorticoid receptors. Further work is in progress to understand this phenomenon.

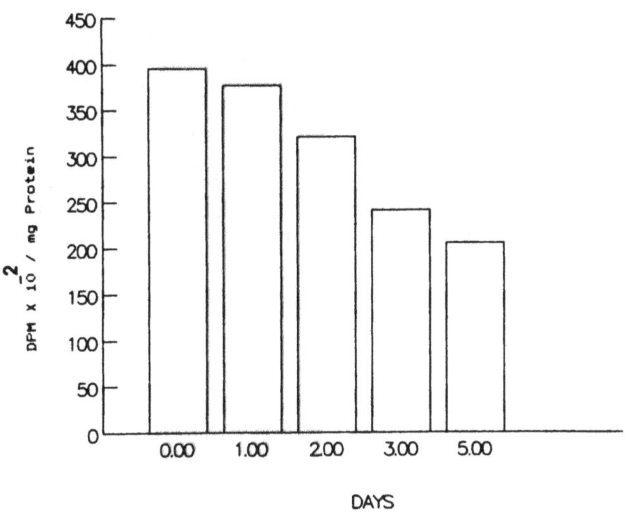

Fig. 1 Effect of DHEA administration on rat hepatic glucocorticoid receptors - 3-5 day-adrenalectomized male Sprague-Dawley rats were injected daily intramuscularly with DHEA (4 mg/ 100 gm body weight) dissolved in 0.2 ml olive oil. Control animals were given 0.2 ml olive oil. The animals were killed on the days indicated. Livers were removed, homogenized (1:2 w/v) in 10 mM Tris-HCl/0.25 M sucrose,pH 7.5 at 4°C. The homogenate was centrifuged at 100,000 x g for 30 minutes. The supernatant (cytosol) was diluted to approx. 10 mg protein/ ml and used for the binding assay. Protein content was determined by the method of Lowry et al. (6). Cytosol was treated with a saturating level (5 x 10^{-8} M) [^{3}H] dexamethasone in the presence and absence of 500-fold excess of unlabelled dexamethasone for 4 hours at 4°C. Receptor- bound steroid was then determined by

the dextran-coated separation method as previously described (7). Results at each time point represent the mean ± S.E. of two experiments performed using two animals at each time point.

III. Regulation of tyrosine amino transferase enzymatic activity by DHEA:

3 days post adrenalectomized male Sprague-Dawley rats were given im injection of 4 mg/100 gm body weight DHEA dissolved in dimethyl sulfoxide or dimethyl sulfoxide alone. Two hours after injection, animals were sacrificed and livers removed and homogenized in 2 volumes of ice-cold 0.9% NaCl solution. The homogenate was centrifuged at 100,000x g for 30 minutes and the resulting supernatant was used for the measurement of tyrosine amino transferase activity according to method of Diamondstone (8). Figure 2 shows that there was about a two fold increase in tyrosine amino transferase activity after DHEA administration. We also detected increased in tyrosine amino transferase activity 24 hours after feeding chow containing 0.4% DHEA to rats (results are not shown). The tyrosine amino transferase activity is known to be induced by glucocorticoids, glucagon,cAMP and other substances(9). DHEA is well known to inhibit glucose- 6- phospho dehydrogenase activity and this inhibition is central to anti proliferative and other

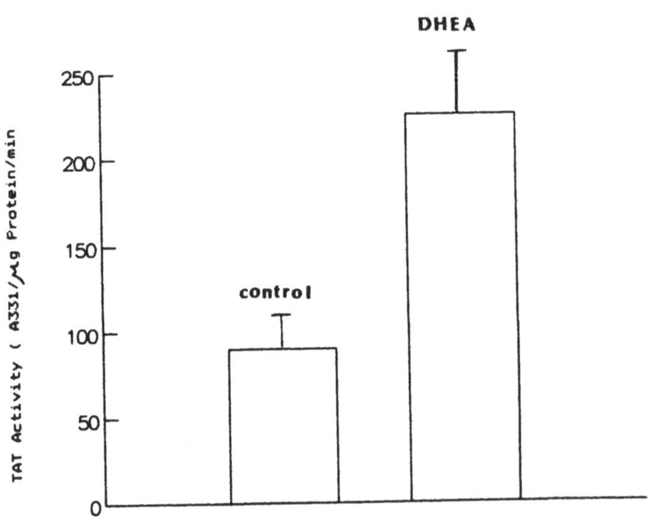

Fig 2. Measurement of tyrosine amino transferase levels after DHEA administration to rats - 3-5-day adrenalectomized male Sprague-Dawley rats were given an intramuscular injection of

(4 mg/ 100 gm body weight) DHEA in 0.2 ml dimethyl sulfoxide. Control animals were given 0.2 ml dimethyl sulfoxide. 2 hours after injection animals were killed, liver removed, homogenized in 2 volumes of ice-cold saline solution. After centrifugation of the homogenate at 100,000 x g for 30 minutes, the resulting supernatant was used for the enzyme assay according to method of Diamondstone (8). The results represent the mean ± S.E. of four experiments using two rats in each groups.

biochemical effects of DHEA (1). It will be of considerable interest to understand the role of this increased enzymatic activity in terms of various biochemical effects of DHEA.

IV. Inhibition of [^{3}H] uridine uptake and incorporation by DHEA into normal and transformed liver cells:

DHEA treatment is known to inhibit growth of various tumors (10,11) including complete inhibition of spontaneous breast cancer formation (12) and lung tumors in certain strains of mice (13). Dworkin et al. (14) have shown that DHEA blocks growth of HeLa cells and WI-38 cells in vitro. In order to understand the mechanism(s) by which DHEA influences cell proliferation, we have studied the effect of DHEA treatment on uptake and incorporation of [^{3}H] uridine into rat hepatoma and hepatocyte cells. Results presented in Figure 3 suggest that DHEA inhibits [^{3}H] uridine uptake in a dose dependent manner with optimum inhibition obtained at 10^{-5} M DHEA. Interestingly

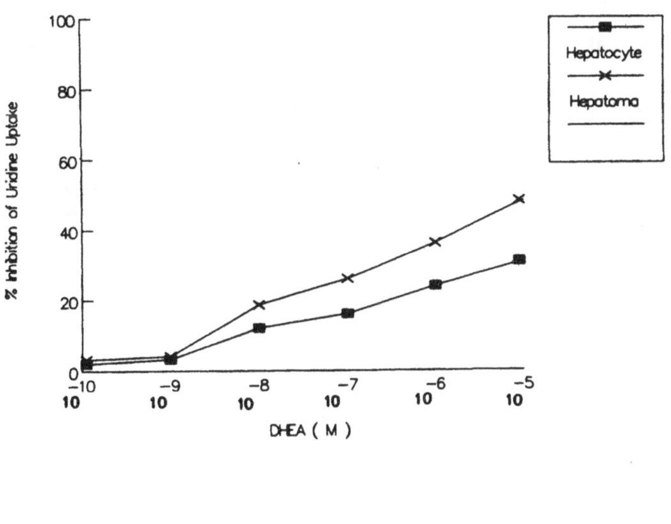

Fig 3. Inhibition of [^{3}H] uridine uptake by DHEA into normal

and transformed liver cells- Hepatocytes were prepared as described previously (15). 2 x 10^6 hepatocytes or hepatoma cells growing in the log phase were plated in triplicate at uniform density in BME supplemented with 5% fetal calf serum, 2 mM L-glutamine and 50 μg gentamicin. After 24 hours the medium was replaced with 2 ml of BME containing 2% fetal calf serum from which endogenous steroids were removed by stirring for 45 minutes at 4°C with dextran coated charcoal. DHEA (dissolved in ethanol) was added in the medium at the indicated concentrations. The final concentration of ethanol was adjusted to 0.1%. After 24 hours the medium was replaced by 2 ml of fresh media containing 5 μCi/ ml [^{3}H] uridine and at the time indicated, the media was decanted and dishes were washed three times with 2 ml of ice-cold PBS. The cells were detached with trypsin-EDTA solution and sonicated in 0.5 ml PBS. The suspension was adjusted to 5% TCA and centrifuged at 3,000 x g for 15 minutes. Radioactivity in the supernatant and pellet (after heating the pellet suspended in 5% TCA for 30 minutes at 90°C) were determined. The viability of the attached cells was judged by trypan blue exclusion test and the cell counts were determined by hemocytometer. The results represent Mean ± S.E. of three experiments done in triplicate.

inhibition was greater in hepatoma cells as compared normal to hepatocytes. Similar results were obtained using [^{3}H] uridine incorporation studies (data are not shown).

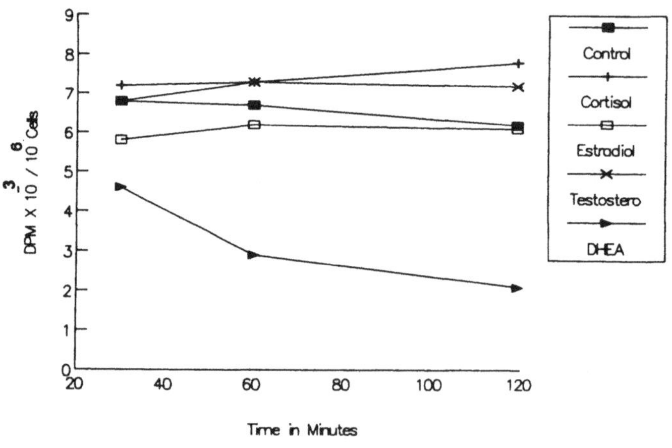

Fig.4. Steroid specificity of DHEA inhibition of [^{3}H] uridine

uptake - 2 x 10^6 hepatocytes were treated for 24 hours in 2 ml of BME supplemented with 5% fetal calf serum. After 24 hours the medium was replaced with 2 ml of BME containing 2% fetal calf serum. Endogenous steroids were removed from the calf serum by previously stirring for 45 minutes at 4°C with dextran coated charcoal. Various steroids dissolved in ethanol were added and final concentration of ethanol was adjusted to 0.1%. 5 μCi/ ml [^{3}H] uridine was added during the last 60 minutes of culture period. The cells were washed, sonicated, adjusted to 5% TCA and the radioactivity in the supernatant was counted as described in legend to figure 3. The results represent Mean ± S.E. of three experiments done in triplicate.

The DHEA effect was found to be specific in that 10^{-5} M hydrocortisone, testosterone or 17β estradiol were without significant effect (Figure 4). Thus it seems that DHEA may influence cell proliferation at least in part through inhibition of RNA synthesis.

Concluding remarks:

DHEA has so many diverse physiological and biochemical actions that it is probable that beside specific receptors, DHEA actions may be mediated by other steroid receptors such as that of estrogen, androgen, and glucocorticoids. We cannot rule out the involvement of membrane receptors and various metabolites of DHEA in the mechanism of DHEA action. In this article we have shown two possible mechanisms by which DHEA can bring out various biological responses. In that regard, there is a distinct possibility that at least some cellular DHEA actions are mediated by a specific DHEA binder protein(s) characterized by us in rat liver cytosol. The antiglucocorticoid properties of DHEA can be explained by the observed modulation of glucocorticoid receptors by DHEA. The inhibition of uridine uptake and incorporation may be related to the antiproliferative effect of DHEA in normal and tumor cells. The preliminary results presented here relates to our long term goal of understanding cellular and molecular mechanism(s) by which DHEA brings out various biochemical responses during cellular growth, differentiation and transformation.

References

1. Regelson, W., Loria, R., Kalimi, M.: Neuroimmunomodulation: Interventions in aging and cancer (Pierpaoli, W., Spector, N. Eds.) Annals of the New York Academy of Sciences 521, 260-273, (1988).

2. Kalimi, M., Regelson, W.,: Biochem.Biophy. Res. Commun. 156, 22-29, (1988).

3. Schwartz, A.,; Molecular Biology of Aging (Woodhead, A. Ed.) Plenum Press New York, 181-191 (1985).

4. Crudo, D.,: Endocrine Society 71st Annual Meeting, Seattle, Washington, Abstract # 1455, (1989).

5. Kalimi, M., Agarwal, M.: Biochem.Biophy.Res.Commun. 153, 365-371 (1988).

6. Lowry, O., Rosenbrough, N., Farr, A., Randall, R.: J. Biol. Chem. 193, 265-275, (1951).

7. Shirwany, T., Hubbard, J., Kalimi, M.: Biochimica et Biophysica Acta 886, 162-168, (1986).

8. Diamondstone, T.: Anal. Biochem. 16, 395-401, (1966).

9. Kalimi, M., Colman.P., Feigelson, P.: J. Biol. Chem. 250, 1080-1086, (1985).

10. Schwartz, A., Tannen,R.: Carcinogenesis. 2, 1335-1337, (1981).

11. Nyce, J., Magee, P., Schwartz, A.: Proc. Am. Assoc. Cancer Res. 23, 91, (1982).

12. Schwartz, A.: Cancer Research, 39, 1129-1132, (1979).

13. Schwartz, A., Hard, G., Pashko, L.: Nutr. Cancer. 3, 46-53.

14. Dworkin, C., Gorman. L., Pashko, L.: Life Sci. 38, 1451-1457, (1986).

15. Diegelman, R., Guzelian, P., Gays, R., Gays, S.: Science. 219, 1341-1343, (1983).

DHEA: SOME THOUGHTS AS TO ITS BIOLOGIC AND CLINICAL ACTION.

William Regelson, Mohammed Kalimi, Roger Loria

Departments of Medicine, Physiology and Microbiology, Medical College of Virginia, Virginia Commonwealth University, Richmond, Virginia 23219

As discussed in our introduction, it is our feeling that DHEA has both intrinsic and extrinsic action acting through its metabolites or directly via its own cellular receptors that will be identified (1). In addition to targeted action, DHEA will be found to exert paraendocrine effects by altering cellular environments through steroid receptor regulation or via enzyme inhibition or activation that will permit the expression of metabolic or cytokine mediated events.

This paper does not purport to integrate all the excellent studies and reviews described by the authors in this text or previously published (2-4). However, we plan to review broad metabolic actions of DHEA, some of which have not been included in this textual review.

Apart from our own work with host resistance (5) (Loria, et al. this text), and hypertension (Kalimi et al, this text) an example of the effective broad potential of DHEA is seen in the detailed studies of DHEA prevention of carcinogenesis, an area of current clinical concern and historically an effective demonstration of the clinical relevance of this steroid. In addition, we will focus on issues relative to tumor growth, radiation, skin, hair, aging, prolactin, pregnancy, thyroid function, bone growth and effects on CNS behavior and response.

Dehydroepiandrosterone (DHEA)

Carcinogenesis

The concepts of DHEA as an anticarcinogen has been reviewed by Feo and Pascale, Mayer, Weber and Pannasch, and Schwartz et al., in this text with contributions by Prough and Wu which also focuses on DHEA's effects on hepatic metabolism.

The key to the prevention of carcinogenesis by DHEA feeding has been thought to relate to its block to glucose-6-PO_4 dehydrogenase (G-6-PD) with inhibition of carcinogen synthesis or activation and/or subsequent inhibition of ribose-5-PO_4 and DNA synthesis.

Van Noorden et al. (6) describe a proportional relationship between G-6-PD activity and hepatocyte ploidy, with mononuclear diploid stem cells having the greatest G-6-PDH activity. It would be of interest to see if DHEA or related sex steroids activate or inhibit this enzyme activity in hepatocytes as controlling factors in proliferation.

Alternatively, DHEA acts through depletion of NADPH which Feo and Pascale speculate may be a factor in the inhibition of free radical formation. This has also been discussed by Schwartz, et al. (4). However, in contrast, Heffner & Milani (6) recently claim that depletion of G-6-PD activity in the isolated rabbit lung enhances pro-oxidant injury to xanthine oxidase.

Moore et al. and Weber et al. (8-10) feel that the anti-carcinogenic effect of DHEA in the liver is mediated by a differentiating effect on pre-neoplastic nodules and that this is associated with peroxisome and catalase induction, an area thoroughly discussed by Prough and Wu and by Mayer et al. (this text). The peroxisomal enhancing effects of DHEA in some way may provide a key to the anticarcinogenic effect of this agent as has been seen for clofibrate (11,12).

In seeking a uniform theory for DHEA's anticarcinogenic action, an alternative hypothesis to those expressed above is that the slowing of hepatic cellular proliferation induced by DHEA could give rise to a cytokine activated differentiation, or alternatively, give rise to cytokine induced cell stasis or death. An example of a cytokine mediated process is seen in the action of tumor necrosis factor (TNF) where TNF blocks lipogenic enzymes in cultured adipocytes (13). In similar fashion to DHEA, TNF is found to block the conversion of fibroblasts to adipocytes by insulin (14). Another example of cytokine activation is seen in the interleukin-1 response to etiocholanolone an inflammatory steroid metabolite of DHEA (15).

In the search for prognostic factors governing the proliferation of breast cancer cathepsin D, an estrogen-induced lysosomal protein has been found associated with enhanced tumor recurrence (16,17). It would be of interest to see if the anticarcinogenic action of DHEA is related to inhibition by DHEA of growth mediating proteases.

DHEA feeding is associated with a reduction in glucose consumption; increased glycogen hydrolysis and increases in long chain fatty acetyl-coA hydrolase with a decrease in fatty acid synthetase ("futile cycle") resulting in a net increase in ATP consumption which could affect tumor proliferative activity (18), (Mayer et al., this text). Berdanier and McIntosh (this text) also discuss the effect of DHEA on energy systems, as does Cleary and Schwartz et al. (this text) which provides us with another anticarcinogenesis concept for DHEA - namely that DHEA works in similar fashion to the anti-carcinogenic action of dietary restriction.

In that regard, we must examine recent work seeking to explain the anti-aging mechanism of dietary restriction which significantly extends the life of rodents (19,20). The anti-aging effect of dietary restriction in mice is now felt to be due to the lowering of body temperature during sleep and there is evidence suggesting that this temperature fall enhances the DNA reparative action via DNA polymerases (20) and decreases the expression of protooncogenes (21). For this reason, it would be of interest to see if DHEA or its metabolites can affect the efficiency of DNA repair to radiation or mutagenesis, which peaks in embryos or neonatal tissue when DHEA levels are highest. Indeed, Sonka (22,23) has reported that depending on the timing of DHEA administration to whole body radiation, it has radiation protective and/or enhancing action. Similar protection is seen for DHEA in stimulating mutagenic repair mechanisms via promotion of sister chromosomal exchange (24).

In further support of anticarcinogenesis as analogous to dietary restriction, Hastings et al. (25) have developed DHEA derivatives which do not have sex steroid action (i.e., 16α fluoro-5-androsten-17 one) but which are three times more potent as anti-obesity agents than DHEA and in addition, block DMBA skin binding and phorbol induced blastogenesis. These observations are supported by the report that DHEA (0.2% in diet) promotes a decline in PGE2 content of mouse skin inhibiting superoxide anion formation and phorbol ester stimulation (25).

The actions of DHEA are dose related in regard to metabolites formed with low dosage resulting in acetylation of DHEA in the mouse liver (26). Whether these metabolites are effective in preventing carcinogenesis or result in other actions for DHEA requires further evaluation.

In conclusion, the search for a unifying answer to the anticarcinogenic action of DHEA has not been found, and there may be no single answer as to its anticarcinogenic effects.

Mammary and other hormonally responsive tumors

DHEA is a sensitive indicator that can distinguish normal women from patients at risk with breast cancer (27-29). This is seen in Bulbrook's studies (27,28) and subnormal plasma and urinary excretion levels of DHEA and DHEA-S have been reported to indicate high risk for breast cancer development, and evidence of a poor prognosis (30). However, breast cancer patients with high DHEA/cortisol plasma ratios respond well to endocrine therapy. O'Higgins et al. (31) feel that this can be used to predict response to metastatic disease. Furthermore, Riley et al. (this volume) and Kalimi (work in progress) have reported that DHEA has an antagonistic effect to stress mediated corticosteroid action which can affect tumor growth or immunoresponses related to tumor/host resistance.

Breast cancer tissue in-vitro can convert DHEA and DHEA-S to hydroxy-DHEA and androstenediol (AED). It is thought that AED can compete with estrogen for estrogen receptors which may provoke proliferative action and that DHEA conversion to AED may play a role in the "western" pattern of breast cancer (32,33).

DHEA-S found in the breast fluid of normal pre-menopausal and post-menopausal women exceeds that of plasma by 50 to 1,000 fold (34). It has been suggested, because of this, that high local levels of DHEA can lead to estrogen synthesis that promotes tumor growth (35). Dogliotti et al. (36) have found a correlate between DHEA-S and cation (K+) concentration in breast cysts. Similar results have been reported by Chasalow et al. (34) with a decrease in DHEA-S in high Na+ cyst fluid which may relate to antibody cross reacting steroids. Of interest to these findings, there is a relationship between DHEA and epidermal growth factor (EGF) in these cystic breast lesions (37) which appears to correlate with the apocrine character of the breast cyst and its sodium content. In this regard, DHEA's effects on tumor growth are varied as it has been found to stimulate mammary tumor growth in rats (38) and block mouse mammary tumor virus induction (39,40) and ductal proliferation (40).

A decrease in urinary DHEA and increase in urinary estriol after castration and adrenalectomy in breast cancer patients suggests that breast cancer tissue may be a source of DHEA (41). Patients whose breast cancer tissue showed no sulfuration of either estradiol or DHEA failed to respond to adrenalectomy (35). When the ratio of DHEA/estradiol was less than one, the prognosis was poor for adrenalectomy (29).

In a study by Zumoff et al. (29), in operable breast cancer, DHEA levels when contrasted with normal controls, did not show an invariant age related DHEA decline. Pre-menopausal operable breast cancer women had subnormal DHEA levels while post-menopausal patients showed a significant increase in DHEA levels. This observation supports a two disease concept of pre and post-menopausal breast cancer based on DHEA/androstenedione ratios which suggests that there may be a channeling of DHEA into sex steroid synthesis as a factor in tumor induction or growth. In regard to the above, enzymatic conversion of DHEA in-vitro can be inhibited by physiological concentration of estradiol and estrone (42).

In ovarian tumors, the circulating levels of DHEA-S are clinically lower that normal, with the lowest values found in ovarian cancer (43). However, Barkan et al. (44) describe a virilizing, hypertension producing ovarian tumor which produced large concentrations of DHEA-S. Paradoxically, DHEA induces granulosa cell ovarian tumors in the SWXJ-9 mouse. In this model, the DHEA systemic effect on vaginal epithelium was estrogenic, and testosterone enhanced tumor incidence, but estradiol inhibited it (8). These paradoxes abound, as an example in another area, Moore et al. (8) have shown that DHEA opposes hepatic carcinogenesis for nitrosamines, while it is permissive for lung cancer induction.

Feo and Pascale (this text and others, (44) discuss DHEA anti-tumor relationships as a function of G-6-PD inhibition with epidemiologic support for this mechanism seen in the decreased incidence of cancer in populations showing G-6-PD deficiencies as a genetic problem. This can relate to both genetic or dietary influences as DHEA levels are significantly lower in Japanese women who have one tenth the western breast cancer incidence. This is discussed by Adams et al. (45), and of interest to this, DHEA and DHEA-S are found in tumor tissue at levels that correlate with age but which also suggests the presence of active sulfatases within the tumor (46). This is in contrast to the situation in ovarian cancer (44) and in Jones & James's (47) immunoassay study, DHEA/DHEA-S was higher in tumor tissue as compared to normal tissue. For this reason, we should be aware that breast and myometrial tissue possess paraendocrine behavior in that they are able to metabolize C_{19}-steroids particularly, DHEA-S and DHEA to androstenediol (AED) and estrone and AED can displace estradiol from breast receptors (48).

Although largely superseded clinically by the use of estrogen antagonists, sex steroids have a long history in suppressing the growth of breast tumors (49). In support of this, DHEA-S can modulate MCF-7 human breast cancer cell growth where it affects the availability of estrone sulfate necessary for tumor proliferation. This action of DHEA-S is a non-competitive inhibition (50) based on inhibition of estrogen metabolizing esterases.

Van Landeghem et al. (51) have reported a decrease in DHEA and AED in the nuclei of malignant breast tissue. The significance of this is not clear, particularly in relation to the possibility that this may have bearing on estrogen levels within the tumor. What is of interest to clinical cancer therapy is that these steroids, along with their proliferative effect, induce progesterone specific binding sites in these estrogen sensitive cell lines suggesting that following DHEA priming, progesterone or progesterone antagonists could then have a clinical modulating influence on breast tumor growth.

Begin et al. (52) have shown that in an androgen responsive mammary tumor (Shionogi Cells) in vitro growth stimulation is obtained with DHEA, DHEA-S, androstenedione and AED which are blocked by flutamide the androgen antagonist. This has relevance to prostatic cancer where Labrie et al. (53) have found DHEA and AED to stimulate growth of the rat ventral prostate, while Huot and Shain (54) found DHEA-S to decrease proliferation of rat prostate cancer cells. Belanger et al. (55) discuss the importance of DHEA-S, DHEA, and AED as factors in promotion of prostatic cancer growth. However, in our clinical Phase I study in 12 of our cancer patients (56), DHEA, given up to 40mg/kg PO for as long as 2 1/2 years, and in our recent multiple sclerosis study in 4 male patients, change in prostatic size was not seen although looked for in those getting DHEA for 3 months or longer.

DHEA effects on tumor induction and growth are blocked by exogenous mixtures of deoxyribonucleosides and ribonucleosides which Gordon, et al, (57) feel is mediated by nucleoside bypass to the decline in ribose-5-phosphate due to the inhibition of the pentose shunt by DHEA. This relates to our rational for our phase I anti-cancer (56) study which was based on DHEA's capacity to inhibit the pentose shunt and DNA synthesis. In animal studies, we did not see significant regression of the P1798 corticosteroid sensitive and resistant tumors in the Balb/c mouse that could not be explained by weight loss. However, in the rat mammary tumor, MTW9, DHEA produced a significant decrease in tumor volume (Vincent Hollander, personal communication). In our long term treated patients, two renal cancers appeared to have an arrest of progressive disease, one for 2 1/2 years, but no tumor regression was seen, although in 6 or 7 dogs and cats significant inhibition of mastosarcomas by DHEA were observed. Combination studies with chemotherapeutic agents that block nucleic acid synthesis are planned with reference to possible anti-tumor synergism because of DHEA's ability to inhibit ribose-5-P synthesis.

Radiation

Sonka and Strakova (23) have reported that DHEA given intraperitoneally to mice when given 0-16 hours before radiation showed a protective effect at 30 days. In contrast, if given 20-52 hours before radiation DHEA showed a radiation sensitizing action. This study was based upon the observation that liver glucose-6-PD was inhibited in mice subjected to sublethal whole body irradiation which affected pentose availability that was felt could be critical to radiation repair.

Of interest to the above, both DHEA and progesterone have similar effects on stimulating sister chromatid exchange following exposure to mutagens in senescent mice (24). This suggests that DHEA may stimulate anti-mutagenic or DNA repair mechanisms which is worthy of serious consideration in view of its anticarcinogenesis action, a point recently made by Wilpart et al. (58).

Skin

Oral DHEA inhibits prostaglandin E2 formation in mouse skin induced by topical application of phorbol esters. As discussed earlier, Hastings et al. (25) feel that this may relate to the mechanism of DHEA's block to cutaneous carcinogenesis and this same mechanism is inherent in Schwartz's structural analogs to DHEA which are anticarcinogenic, but do not have sex steroid activity.

DHEA does not antagonize cortisol induced metabolism of human skin biopsy fibroblasts while androgen does (59). DHEA has site related activity in skin which correlates with lipogenesis and is similar to that seen for AED (60). DHEA excretion by human skin was increased when it was given to acne patients (61) and it is clearly metabolized and secreted by human skin (62) where it has a stimulating effect on sebum secretion. Related to this, dexamethasone and/or oral contraceptives were given to men and women with cystic acne to lower DHEA levels, with significant clinical improvement (63).

Patients with steroid sulfatase deficiency develop ichthyosis with accumulation of cholesterol sulfate. Desulfation of DHEA-S is impaired with a decline in DHEA and androstenedione. One source of desulfation is derived from gastrointestinal flora whose inhibition by ampicillin inhibits desulfating activity (64).

The phylogenetic cutaneous importance of DHEA is seen in the skin of fishes where steroids may act as pheromones with DHEA in the male African catfish being converted to AED as a sex attractant. The skin of fish can synthesize delta-4 water soluble glucuronide conjugates which include androgen related compounds (65).

Psoriasis

DHEA has been found to be decreased in the skin of psoriatics which correlates with high levels of G-6-PD found in the red cells of psoriatics (66). This correlation was also seen in the failure of DHEA in-vitro to accumulate in erythrocytes in psoriatics by as much as 50% as compared to controls (67).

The above has led to a wide range of successful topical studies of DHEA (67) with evidence of up to 50% absorption in 24 hours of a DHEA lauryl-SO_4 preparation on the skin. Clinical studies with a depot enanthate DHEA preparation (68) or an oral DHEA-SO_4 (69) have shown a degree of success in preventing the Koebner phenomenon where cutaneous inflammation or injury induces psoriasis in susceptible individuals.

Hair

The role of DHEA in hair growth has been associated with hirsutism or hair loss in women (70). Pitts (71) has found a correlation between DHEA-S levels and male pattern baldness in young men. Of interest, Jolesz et al. (70) and Faredin et al. (72) have found that hair contains high quantities of DHEA-S associated with the capacity of human skin for DHEA sulfation. More recently, Toth and Faredin (73) report that high concentrations of DHEA-S or testosterone sulfate found in hair is a function of adjacent sweat glands while free steroids found are derived from sebaceous glands. Kasick et al. (74), in an excellent review, propose that DHEA and AED are key factors in androgenic alopecia via inhibition of G-6-PD.

In view of a recent report of the Orentreich Foundation that methionine promotes hair growth, it would be of interest to see if DHEA has affects on methionine or S-adenosyl methionine metabolism.

In patients with hirsutism, suppression of ACTh production with dexamethasone results in decreases in androgen or DHEA production with slowing of hair growth (75). In contrast, Ketoconozole treatment decreases serum testosterone concentration but not DHEA-S, but there is still a decline in hair growth (76). DHEA-SO_4 stimulated hair growth in mice and guinea pigs although it did block induced growth of vibrissae on prolonged administration (77).

The pilosebaceous unit of skin contains the enzymes necessary to convert DHEA to dihydrotestosterone or testosterone. In support of this higher DHEA levels are found in acne patients, and there is a correlation between DHEA and AED levels with acne and hair density in men (61-63). There is also a relationship between male pattern baldness and DHEA-S levels (79). Similar relationships have been found in women with adrenal hyperplasia (80).

Acanthosis nigricans is a form of epidermal hyperplasia associated with malignancy and/or polycystic ovaries. Elevated serum DHEA-S is found in 50% of patients with polycystic ovarian disease along with insulin resistance, hirsutism and obesity (80,81).

Collagenase

The endocrine control of collagen synthesis and degradation may be under DHEA control by itself or via estrogen synthesis (82).

Sakyo et al. (83,84) have described a collagenase inhibitor that is stimulated by both estrogen and progesterone in rabbit uterine cervical fibroblasts. This is thought to be a metalloproteinase inhibitor that is found in many tissues including skin and serum. In contrast, DHEA-S has been reported to stimulate the production of cervical collagenase in the rabbit.

Pertinent to the report of cervical collagenase activation by DHEA-S in pregnancy is the observation by Peh et al. (85) that at term, the human fetal chorion produces DHEA from pregnenolone. In this regard, increases in DHEA-S levels have long been known to parallel gestation (85-87), and DHEA-S administered to pregnant monkeys enhances cervical dilation (88). Similar effects have been reported clinically (89) and Sakyo et al. (83,84) have used DHEA-S in Japan for this purpose. Clinical cervical dilation is explained by DHEA-S activation of cervical collagenases or gelatinolytic metalloproteinases (89) which is mediated by DHEA-S and not DHEA or estradiol.

Estrogen, which can be derived from DHEA, regulates lysyl oxidase an enzyme critically involved in the cross linking and integrity of collagen (82,90) and elastin (91). Lysyl oxidase levels govern the status of elastin (92) and one should determine if there is a relationship between lysyl oxidase activity and cervical dilation. It would be of interest to see if there is an antagonism between lysyl oxidase and DHEA-S levels and if this has pertinence, to collagen metabolism in other tissue sites as seen in diabetes and aging. Ito et al. (93) describe a cytoplasmic cervical DHEA-S binding protein in the rabbit cervix which is remindful of Kalimi and Regelson's (1) observation of DHEA cytoplasmic binding protein in hepatocytes.

Aging

The decline in DHEA and DHEA-S with age is well defined (94-99) in man and monkey (100), but there is also no age related change in DHEA/DHEA-S ratios (101). Of interest to this in young females, the ovaries are responsible for 50% of androstenedione with the remainder coming from peripheral conversion of DHEA. With aging the ovary, despite estrogen loss, is still able to synthesize androstenedione and testosterone (102). However, clinical estrogen substitution does nothing to restore DHEA levels, suggesting a non-estrogen dependent role for ovarian function in regulating DHEA production (94). Cumming et al. (103) have found that ovariectomy has a significant influence on DHEA levels, enhancing the age related decline. However, DHEA decline is probably indicative of changes in the adrenal zona reticularis. The implications of this suggests that we must look at androstenedione, AED and other androgens whose levels are fairly stable despite aging as they may have more physiologic importance to the maintenance of youthful survival which may go beyond their role as sex steroids.

In regard to the place for DHEA decline in aging, Shimomura et al. (104) have shown that adrenalectomy can slow weight gain in young obese db/db mice. In their review, they speak of the value of adrenalectomy in restoring the synthesis of muscle protein to normal along with lowering of plasma glucose in these mice. They discuss aging as related to hyperinsulinemia which may provide insight into DHEA's functional role for preventing obesity, insulin resistance, and hyperphagia in rodents. Nestler et al. (105,106) and in this text, discusses the clinical decline in DHEA-S induced by insulin and "suggests that the age related decline in DHEA/DHEA-S may be a function of increased age related insulin production." However, Buffington et al. (107) suggest there is an opposing action between DHEA and testosterone on insulin resistance. In that regard, Mobbs (108) has postulated that aging is the result of ventral medial hypothalamic damage resulting in hypersecretion of insulin and glucagon. Mobbs (108) feels that hyper-insulinemia, insulin resistance and hyperglycemia play a causative pathologic role in senescence. This is an extension of Sonka's (22) concept of the age related "hyperproductive syndrome" due to the decline in DHEA. Of interest to this, hypophysectomy in aging mice can reverse many of the effects of aging suggesting that the neuroendocrine system can produce a hypophyseal programmed death (109,110). Hypophysectomy blocks collagen cross linking in aging mice (109,111) and as DHEA alters collagen metabolism (82-89), it would be of interest to see the age related effects of hypophysectomy on DHEA values as although DHEA values are low in hypopituitary patients they respond normally to ACTh (111).

In relation to DHEA as an anti-aging hormone, we have to consider its potential as a stress modulating steroid. This is based on the work of Riley et al. reported in this volume and Sulman (113) which suggests that DHEA has an adrenocortical stress mediated blocking action. Crudo (114) has reported the down regulation of liver glucocorticoid receptors, while Rogers et al. (115) in similar fashion to Riley's in vivo experience also report a DHEA in vitro block to dexamethasone thymic involution as well as an in vitro (116) block to blastogenesis. This area has been discussed by Svec & Lopez-S (117) who felt that low concentrations of DHEA to cortisol with age can lead to age related stress induced CNS injury. In that regard, both insulin and dexamethasone induce G-6-PDH in hepatocytes (117), and it would be of interest to see if hippocampal or stress related injury is associated with changes in G-6-PDH that could be blocked by DHEA. We also have Coleman's observations (this text) for the antidiabetogenic action of DHEA which may also relate to DHEA action in blocking corticosteroid responses. In this regard, aging has been likened to a cushingoid response (109), that leads to age related diabetes and degenerative disease which may be impeded by DHEA.

Whether these effects are due to ACTh or other pituitary peptides has not been determined. In macronodular adrenocortical hyperplasia (118) serum DHEA-S may be elevated which has been ascribed a non-ACTh pituitary peptide.

Prolactin

There is evidence that prolonged hyperprolactinemia enhances DHEA-S and DHEA levels (119-154) which may not be true for acute prolactin responses (123). Of interest to this, the baboon adrenal cortex, at mid-gestation, synthesizes DHEA-S under the influence of ACTh, hCG, hCS (Human somatomammotropin) and prolactin (126). Prolactin receptors are found on adrenocortical cells where it can produce a direct in vitro stimulating action on DHEA-S production (127) and the use of agents that decrease prolactin production have a significant effect on decreasing DHEA-S production (128), while prolonged use of sulpiride which induces prolactinemia has the opposite effect (123).

It is important to note that estrogens stimulate dopaminergic and growth hormone release in vitro from pituitary cells of the rat. AED or related C19-delta-5 adrenal steroids have estrogenic function and cause a similar prolactin increase as does DHEA. This effect is time related and similar effects were seen for DHEA-S, while AED was without effect (128).

The prolactin dopaminergic stimulating action of these steroids (129) suggests a clinical place for DHEA and its metabolites in the CNS control of Parkinson's disease where dopaminergic drugs are of value and where prolactin can show dopaminergic effects following striatal injury in rats (130).

In view of L-dopa's action in the treatment of striatal disease, there is controversial but positive evidence of L-dopa affecting DHEA levels (131,132) and DHEA can depress serum luteinizing hormone while increasing FSH and prolactin in rats (133). Of interest to this, on medroxy progesterone treatment in renal and prostatic cancer, serum prolactin is elevated while DHEA is decreased along with testosterone, cortisol and testosterone binding globulin (134).

Clinical

Apart from DHEA acting as a biomarker for aging or pregnancy, DHEA levels are decreased in AIDS patients and in the Alzheimer population (135) (Merrel et al. this text). The latter needs more comparative data related to age, Downs syndrome, and the nutritional and functional state of patients. Utilizing acute physiologic and chronic health evaluation criteria, Wade et al. (136) have also found DHEA-S to be low in intensive care (ICU) patients who also show a reduction in DHEA/cortisol ratios although ACTh responses can be normal.

In this regard, Parker et al. (137) have found that cortisol levels rise in severe illness at the expense of other steroids including DHEA and AED while Lephart et al. (138) have found a DHEA fall in burn patients and in both alcoholic and non-alcoholic livers, where there is significant fall in DHEA-S (139). Similar results were reported in burn & seriously ill patients by Semple et al. (140). Patients with anorexia nervosa have a disassociation between adrenal androgens and cortisol (141) with low levels of DHEA. In exercise, DHEA and DHEA-S rise along with other androgens (142).

Pregnancy

During pregnancy DHEA-S is derived from both fetal and maternal metabolism. During the last weeks of pregnancy, the maternal DHEA-S decreases by one third while fetal DHEA production increases from the fetal adrenal cortex (143,144). Foetal membranes or placenta may play a role in DHEA metabolism (145,146), and DHEA-S loading tests have been used in attempts to delineate fetal growth retardation (147). Walsh (148) suggests that there is a decrease in the supply of DHEA-S to the placenta of women with severe pre-eclampsia which he feels explains the decline in plasma and urinary estrogen levels in women with this problem. The level of DHEA-S is elevated in women with cervical ripening preceding the onset of labor (149).

Moise et al. (150) report an interesting finding that the trophoblast shuts down progesterone secretion in vitro when low density lipoprotein is introduced with DHEA-S. They suggest that this could be a signal for parturition.

Thyroid function

Lardy et al. (151) were the first to speak in detail of parallelisms between the function of thyroid hormones and the action of DHEA. In that regard, the key relationship between thyroid action and DHEA is that both stimulate thermogenesis. Liver malic enzyme are stimulated by both thyroid hormone and DHEA (152). Lardy et al. (151) link the action of thyroxin, catechols and DHEA through a common stimulation of mitochondrial G-3-P dehydrogenase. In similar fashion, diets high in carbohydrate, decrease food consumption in rats and increase these mitochondrial enzymes and Lardy et al. (151) postulate that DHEA works via this route in similar fashion to that seen for an excess of thyroid hormone action.

Pertinent to age related effects of DHEA, it should be noted that thyroid hormones have an accelerating effect on neonatal adrenal maturity and function (152). Oppenheimer et al. (153) showed that T_3 action on induction of the hepatic lipogenic enzymes (malic enzymes) decreased with age independent of insulin action while DHEA induces an increase in hepatic malic enzyme action similar to a thyroid hormonal effect (154-155). In view of DHEA effects on diabetes and insulin sensitivity (Coleman, Nestler, et al., this text) as diabetes is a progeric disease, it is of interest that diabetes is associated with decreased responses to thyroid hormone (155,156) while insulin can decrease the levels of DHEA-S (105,106). Pertinent to this, Lemmen et al. (157) showed in fibroblasts that the lipogenic enzymes acetyl-coA carboxylase and G-6-PDH were synergistically stimulated by insulin and T_3.

We have reviewed the action of thyroid in aging as a state dependent hormone (56), in our discussion of DECO ("decreased oxygen consumption hormone") postulated by Denckla to program aging as a hypothalamic thyroxin blocking factor (110). One must study DHEA's action as its decreased availability, with clinical aging, may influence the trophic actions of thyroid hormone on immunity or cardiac function. Pertinent to this, thyroid hormones have significant developmental effects in younger animals, and ornithine decarboxylase (ODC) levels in the developing rat heart can be dramatically stimulated by T_3 (158). This change in polyamine synthesizing enzymes induced by thyroid action explains thyroid hormonal action on differentiation or growth as a critically timed event. Growth hormone, insulin and hypophysectomy have stimulating or blocking action on ODC hepatic levels (159), and ODC response might constitute a good assay for thyroid interaction and DHEA effects, with particular reference to the report that DHEA can act as an inhibitor of ODC (160) despite the evidence for DHEA's capacity to stimulate hepatic, splenic, liver and testicular organomegaly (Loria et al., this text).

Most importantly, the development of the brain is dependent on thyroid hormone action in neonatal animals (161) and the cerebral cortex also responds promptly to thyroid hormone in adult animals (162). In that regard, brain capillary formation is under thyroid control (163), and Leonard et al. (162) have shown a rapid increase in cerebral cortical iodothyronine 5'-deiodinase activity within 24 hours of rat thyroidectomy. Pertinent to this, Kung et al. (164) have shown that DHEA-S inhibits T_3SO_4 sulfation which relates to deiodinase action as it is enhanced by sulfation of thyroid hormone. This rapid effect in adult animals suggests that this may relate to DHEA's place in CNS physiology. Thyroid hormones also influence the level of acetylcholinesterase activity in the rat brain and pituitary (165) and have permissive effects on nerve and epidermal growth factors in mice (166).

DHEA induces thyromimetic responses on cholesterol and lipoprotein distribution (106,107,167,168) and thyroid hormones maintain the response of androgen dependent mammary tumor cells (169) which may be mediated through modification of cell membrane properties (170). The influence of thyroid hormone on obesity, cholesterol levels and hyperlipemias are well documented and may reflect on sympathetic tone (171). Regardless of mechanism, thyroid hormones can increase high density lipoprotein levels which may be pertinent to atherosclerosis as an age mediated DHEA modulated event (172).

It is of interest that thyroid weight varies as a sigmoid function of systolic blood pressure and is linearly related to renal weight and blood pressure (173). In this regard, the kidneys of both normal and renal hypertensive rats contain a thyroid depressing factor which inhibits both thyroid 131 I uptake and in vitro thyroidal peroxidase activity (174).

Fat Metabolism

In the clinical area of obesity, Cleary et al., and Berdanier and McIntosh and Nestler, et al., MacEwen, et al. and Schwartz et al. (this text) discuss the varied mechanisms wherein DHEA or related metabolites or analogs could contribute to potential clinical value in weight loss.

Of interest conceptually to neuroendocrine control of fat metabolism mediated by DHEA, Kurtz et al. (175) have suggested that DHEA circulating levels could serve as a communicating system with the hypothalamus to determine the size of fat stores. The freely diffusible nature of DHEA into fat stores and the observation of an increased response of DHEA to ACTh in simple obesity (176) are presented as suggestive evidence for such an effect. In support of this, the recent paper by Hardwick et al. (177) showing that corticotrophin releasing factor (CRF) following RU 486 can stimulate the brown fat futile cycle in mice suggests that there might be an interface between CRF and DHEA with reference to DHEA as a hormone which can block corticosteroid responses (113-116) (Riley, et al. this text). In that regard, CRF stimulates the release of ACTh, cortisol and DHEA in pubertal development (178) and angiotensin II and vasopressin enhance the action of CRF (179). Both Lardy et al. (151) and Cleary (this volume) speak about effects of DHEA on the futile cycle as a factor in enhanced energy loss and fat mobilization and it would be of interest to see if CRF can modulate DHEA action at effector sites independently of its stimulation of ACTh.

Clinical Studies of DHEA

These can be summarized under: menopausal or climacteric symptoms; treatment of hypercholesterolemia, hyperlipemia and complications of angina or peripheral vascular insufficiency;

the treatment of gout, cirrhosis, psoriasis, cervical dilation, and cancer and its use as an anabolic steroid. Unfortunately, space limitations forbid our discussing these areas in detail. There is also an extensive clinical literature as DHEA-S is available in Europe as prasterone or astenile where given alone or with estrogen systematically it is thought to be useful in the treatment of menopausal symptoms or depression.

Schinasi (this volume) has reported on a direct in vitro HIV anti-viral action for DHEA. In our own clinical observations, one AIDS patient had 2 years of symptomatic improvement, but no improvement in immune response. In one ARC patient, prompt reduction of lymphadenopathy and fever was seen at 40 mg/kg/day of DHEA, but the patient died of a lymphoblastic leukemia one month later. Clinical phase I studies of DHEA in advanced HIV patients are underway. (Elan Corp., Athlone, Ireland).

We feel it is of interest to comment briefly on the action of DHEA in arteriosclerosis which has been discussed by Cleary, Nestler and MacEwen, et al. in this volume and Arad et al. (180) and Gordon et al. (181,182)

The anecdotal experience of Sonka (21), and Lopez-S et al. (183), Kask (184) and others has been given reenforcement by the epidemiologic studies of Barrett-Connor et al. (185,186), this volume) and by Gordon and Arad et al.'s experimental work in rabbits (180-182) and Nestler et al. in man (106,107 this volume). Most recently, Eich et al. (187) have shown DHEA to be protective to the rabbit heart aortic graft where it delays the arteriosclerotic vascular injury seen on cyclosporine utilization to prevent host graft rejection. In regard to these observations, one has to determine if the clinical effects seen are pharmacologic or physiologic. In the latter case, replacement therapy to reproduce youthful blood levels would be in order.

The observations of Sholley et al. in this text, describing the increase in sphingomyelin following in vitro exposure of endothelial cells to DHEA suggests that DHEA may control the phospholipid content of cell membranes. In view of the fact that LDL is the major DHEA serum transport system, it is appropriate to speculate that DHEA might control cholesterol transport or cell membrane fluidity through an effect on the ratio of phostidylcholine to sphingomyelin transported to cell membranes by LDL. Alternatively, DHEA may work metabolically to affect phospholipid cholesterol ratios in the cell matrix or on the endothelial surface. As sphingomyelin is vulnerable to phospholipase C the lysosomal lipase, it would be of interest to examine the action of DHEA and its metabolites on this lipolytic system.

Of interest to Sholley et al. and Loria et al. (this text), Milewich et al. (188) describe the in-vitro conversion of DHEA to androstenedione and AED in pulmonary endothelial cells, which suggests that endothelium may have a profound metabolic effect on the conversion of DHEA to distinctive intermediates of pertinence to vascular physiology.

In regard to the effects on fat mobilization induced by DHEA and its effects as a factor in the prevention of atherosclerosis, an inverse correlation has been found between low-density lipoprotein cholesterol and DHEA-S in human fetal plasma (189). This is felt to reflect on the utilization of cholesterol to make fetal estrogen by the fetal adrenal. In this regard, Boizel et al.(190) have found a decrease in DHEA-S with homozygous familial hypercholesterolemia during ACTh infusion.

Bone

DHEA clinical physiologic replacement in delayed puberty showed no effect on bone maturation or adrenarche induction (191), although DHEA has bone growth promoting responses in immature mice that was similar to what has been seen for testosterone. There was no effect of ovariectomy on bone responses to DHEA or testosterone. DHEA was more active than the sulfated DHEA-S (192) and Drucker (193) states that DHEA may play a role in bone maintenance. Katz et al. (194) have shown DHEA-S values to correlate with bone age, body fat and adolescence. Similar results have been reported by Bing et al. (98) in Chinese girls.

In an old study, Puche and Romano (195) found DHEA to have differentiating effects on chick embryo frontal bones, and endogenous androgens have an anabolic effect on bone density where serum DHEA-S and AED levels show a decrease with age that correlate with bone density (196-198).

With aging, osteoporosis is an accompanying problem, particularly in women. However, there is no major correlation between osteoporosis and testosterone levels which suggests that we must examine DHEA intermediates like androstenedione or AED on the pathway to testosterone or dihydrotestosterone synthesis for their potential primary role in controlling skeletal matrix integrity, a point made by Buchanan et al. (196). Brody et al. (199) have shown there to be a significant correlation between trabecular bone mineralization and DHEA-S levels. The importance of DHEA on normal bone metabolism may be reflected in the abnormal as in Prader-Willi syndrome there is a patient subset with unusual sulfoconjugates of DHEA alongside the normal DHEA-S. This disease is associated with growth failure; delayed skeletal growth and low somatomedin C levels (200).

The CNS and Behavior

The physiologic role of DHEA in the brain and nervous system has not been fully ascertained and is the subject of Robert et al's chapter in our text. DHEA-S has been used as early as 1952 in the treatment of neurasthenia and in selected schizophrenics (201-204) where DHEA values were reported to be depressed (201,202).

Based on the work of Robert's group on DHEA stimulation of neurite formation and our own observations of the anti-viral activity of DHEA, we were stimulated to clinically study DHEA administration in advanced multiple sclerosis patients as was Roberts' group as reported in this text. In both of our studies at 90 mg/day and 40 mg/kg/day improvement in mood and energy in one third of patients were seen in non-randomized studies. This effect might relate to the high testosterone levels resulting from DHEA metabolism as we saw hirsutism and increased libido in several of our female patients. This improvement in mood is not surprising in view of DHEA-S marketing in Europe alone or combined with estrogen, for post menopausal depression.

In that regard, Recordati, in Italy (204), has studied DHEA-S in depression or neurasthenic fatigue syndromes with significant clinical results. Fava et al. (205) feel that there is a subset of "panic disorder" in patients who show an increased DHEA-S/cortisol ratio and that this is a measure of increased adrenocortical stimulation. Panic disorders are now frequently identified with the "mitral valve prolapse syndrome" which accompanies aging. Panic is a common, but serious impediment to normal function in the elderly and it would be of interest to see if increasing DHEA levels have value in this syndrome.

DHEA or DHEA-S levels may be inversely correlated with type A high coronary risk populations (206), and Fava et al. (207) review the literature and present preliminary papers and summary data of an inverse correlation between aggressive, high catechol stress personalities and low DHEA-S levels (207). DHEA-S levels have been negatively correlated in adolescent girls with an interest in sports (208). There is also a stress related increase in DHEA in hockey players (209) along with increases in cortisol and prolactin and similar changes are seen in female marathon runners (210).

Robel et al. (211) have described the accumulation of pregnenolone and dehydroepiandrosterone in the rat brain. In the monkey, DHEA and pregnenolone are not affected by dexamethasone and apparently shows a regular circadian variation. In similar fashion to what Paroli et al. reported (212), DHEA given to male castrated rats reduces olfactory and visual aggressive stimulation. In sexual stimulation, pregnenolone is decreased in the olfactory bulb but DHEA levels are increased (213).

Pertinent to clinical problems phynantoin and carbamazepine in the control of temporal lobe epilepsy lowers DHEA-S levels (214).

In a clinical autopsy study, DHEA and related metabolites were found in all regions of the brain with no specific regional concentrations for any steroid (215). However, in the study by Farmer & Harrington (216) in the rat the highest exogenous concentration of DHEA uptake was found in the pineal within one hour of its administration. Significant uptake was also found in amygdala, hippocampus, thalamus, midbrain, and frontal cortex and Corpechot et al. (217) found similar results with a rise in hypothalamic DHEA.

The rapid concentration of DHEA in the pineal is of particular interest in view of its role in controlling circadian and seasonal rhythms that affect sexual maturity, mating behavior and sleep patterns (218). DHEA has evident clinical sexual cycling patterns in the endometrium in premenopausal women (219), and DHEA production shown an autumn-winter increase in post menopausal women (220) and an age related circadian decline in elderly men (221). Melatonin is the key neuro-chemical mediator of the pineal, as well as TRH, and both agents stimulate thymic mediated immune responses (222) which may be related in some way to our own observations of an immuno up regulating action of DHEA (Loria et al. (163), this volume). In addition, the pineal is under sympathetic or dopaminergic neural control (222) which can directly or indirectly govern hormonal functional, and behavioral responses directly or indirectly through prolactin secretion, an area of physiologic response under DHEA influence.

The studies of Roberts et al. (this text) and the observations of Sunderland et al. (135) (this text) of a DHEA related decline in alzheimers that may be unrelated to that of an age related decrease is of particular interest in view of Bonnet & Brown's (this text) experience in achieving a clinical improvement in traumatic induced memory loss in one patient.

Summary and Conclusion

DHEA or DHEA-S have had a surprisingly wide past clinical experience and apart from having a dose dependent, mild androgenic action are remarkably free of side effects. Space limits our discussion of past application and it is increasingly difficult to keep up with the growing interest and applications for this hormone and its metabolites or synthetic analogs. Currently, clinical activity is focused on prevention or treatment of cancer and atherosclerosis for DHEA or its congeners. However it has broader potential: As an example, DHEA has been found by Koo et al. (223) to be useful in the treatment of patients with hereditary angioneurotic edema, a disorder of C esterase complement inhibition. The digitalis like action of DHEA is another area of major interest which is discussed in our text by Chesalow et al. and there is a growing interest in its potential role in hypertension (Kalimi, et al. this text). The ornithine decarboxylase inhibiting action, akin to retinoic acid, has been described by Merriman (160), and there is a potential for its considered use in porphyria (224), cirrhosis (22,23,213), and as discussed previously, psoriasis (66-69).

DHEA in vitro clearly inhibits G-6-PD activity at physiologic concentration (225), while the G-6-PD inhibiting activity of DHEA must be considered as contributing to a potential hemolytic problem. This has never been reported although looked for in many clinical studies including our own. Heffner et al.'s (7) recent observation of in vitro pro-oxidant lung damage due to G-6-PD inhibition is a caveat that must be kept in mind in any clinical studies. In another area, Pashko et al. (226) describe a DHEA induced decrease in proteinuria in aging rats, an observation independent of weight loss, which may be pertinent to prevention or treatment of nephrosclerosis.

Our recent observation of the immunostimulating anti-infectious action of DHEA (5), Loria et al. (this text) or its AED metabolite contrast with DHEA's action in blocking the lupus-like syndrome in the NZB mouse (4,227). DHEA effects may vary, depending on its route of administration and where, in the future, the role for this precursor steroid and its metabolites will lead is anyone's guess, but the potential for clinical application is before us in multiple systems and disease states. It is our feeling that DHEA and its congeners will rival the glucocorticoids in their physiologic versatility and DHEA or its metabolites will be used for both physiologic replacement, i.e., to replace the age related decline or for pharmacologic effects. In summary, DHEA, as a precursor steroid, can be likened to a "mother steroid" whose idealized role is to provide useful progeny and to help these varied steroids provide for a state dependent or hormone modulating existence.

References

1. Kalimi, M., W. Regelson. 1988. Biochem.Biophys.Res. Commun. 156, 22-29.

2. Lardy, H., F. Stratman. 1988. Hormones, Thermogenesis and Obesity. Elsevier, New York. pp. 337-452.

3. Parker, L. 1988. Adrenal Androgens in Clinical Medicine. Academic Press, New York.

4. Schwartz, A.G., J. Whitcomb, J.W. Nyce, et al. 1988. Adv. Cancer Res. 51, 391-424.

5. Loria, R.M., T.H. Inge, S.H. Cook, et al. 1988. J. Med. Virol. 26, 301-314.

6. Van Noorden, C.J., I.C.M. Vogels, J.M. Houtkooper, et al.. 1984. Eur.J.Cell Biol. 33, 157-162.

7. Heffner, J.E., M. Milam. 1990. Am.J.Respir.Cell Mol. Biol. 2, 257-261.

8. Moore, M.A., E. Weber, M. Thornton, P. Bannasch. 1988. Carcinogenesis 9, 1507-1509.

9. Weber, E., M.A. Moore, P. Bannasch. 1988. Carcinogenesis 9, 1191-1195.

10. Weber, E., M.A. Moore, P. Bannasch. 1988. Carcinogenesis 9, 1049-1054.

11. Bendayan, M., J.K. Reddy. 1982. Lab. Invest. 47, 364-369.

12. Perera, M.I.R., H. Shinozuka. 1984. Carcinogenesis 5, 1193-1198.

13. Torti, F. M., B. Dieckmann, B. Beutier. 1985. Science 221, 867-869.

14. Garcea, R., L. Daino, S. Frassetio, et al. Carcinogenesis 9, 931-938, 1988;

15. Watters, J.M., P.Q. Besey, C.A. Dinarello, et al.. 1985. Surgery. 98, 298-306.

16. Tandon, A.K., G.M. Clark, G.C. Chamness, et al.. 1990. N.Eng.J.Med. 322, 297-302.

17. Ingle, J.N.. 1990. N.Eng.J.Med. 322, 329-331.

18. Mayer, D., E. Weber, M.A. Moore. 1988. Carcinogenesis 9, 2039-2043.

19. Duffy, P.H., R.J. Fevers, J.A. Leakey, et al.. 1989. Mech.Age Develop. 48, 117-133.

20. Lipman, J.M., A. Turturro, R.W. Hart. 1989. Mech.Age Develop. 48, 135-143.

21. Nakamura, K.D., P.H. Duffy, M.-H. Cu, et al.. 1989. Mech.Age Develop. 48, 199-205.

22. Sonka, J. 1976. ACTA Univ. Carol 71, 1-137, 146-71.

23. Sonka, J., M. Stravkova. 1970. Aggressologie 5, 421-426.

24. Reff, M.E. and E.L. Schneider. 1980. Biological Markers in Aging. Bethesda, NIH. Pub.#82-2221. Personal Communications.

25. Hastings, L.A., L.L. Pashko, M.L. Lenbart, A.G. Schwartz. 1988. Carcinogenesis 9, 1099-1102.

26. Casey, M.L., C.H.K. Shackleton, P.C. MacDonald. 1988. J.Steroid Biochem. 30, 1-6.

27. Bulbrook, R.D., J.L. Hayward, C.C.F. Spicer. 1971. Lancet 2, 395-398.

28. Wang, D.Y., R.D. Bulbrook, M. Herian, et al. 1974. Eur. J.Cancer 10, 477-482.

29. Zumoff, B., J. Levin, S. Rosenfeld, et al. 1981. Cancer Res. 41, 3360-3363.

30. Thomas, B.S., P. Kirby, E.E. Symes, D. Y. Wang. 1976. Europ.J.Cancer 12, 405-409.

31. O'Higgins, N. J., P. Carson, I. Mitchell, N. Deshpande. 1976. Clin. Oncol. 2, 267-275.

32. Li, K., T. Foo, and J.B. Adams. 1978. Steroids 31, 113-127.

33. Li, K., D.P. Chandra and T. Foo. 1976. Steroids 28, 561-574.

34. Chasalow, F.I., S.L. Blethen and H.L. Bradlow. Steroids. 52, 205-215.

35. Forrest, A.P.M. 1971. Proc. R. Soc. Med. 64, 509-516.

36. Dogliotti, L., F. Orlandi, M. Torta, et al.. 1986. Br.J.Cancer Res. 22, 1301-1307.

37. Boccardo, F., G. Valent, S. Zanardi, et al. 1988. Cancer Res. 48, 5860-5863.

38. Adams, J.B. 1977. Cancer 40, 325-332.

39. Schwartz, A.G. 1979. Cancer Res. 39, 1129-1132.

40. Pashko, L.L., A.G. Schwartz, M. Abou-Gharbia. 1981. Carcinogenesis 2, 717-721.

41. Adams, J.B., M.S.F. Wong. 1968. J.Endocrinol. 41, 41-52.

42. Deshpande, N., J. Yates. 1973. J.Endocrinol. 59, 35-

43. Heinonen, P.K., T. Koivlila, P. Pystynen. 1987. Gynecol. Obstet. Invest. 23, 271-274.

44. Barkan, A., F. Cassorla, P.L. Loriaux, et al.. 1984. Obs.Gynecol. 64, 287-295.

45. Adams, J.B., L. Archibald, K. Seymour-Munn. 1980. Cancer Res. 48, 3815-3820.

46. Maynard, P.V., M. Bird, P.K. Basu. 1978. Brit.J.Cancer 14, 549-553.

47. Jones, D.L., V.H.T. James. 1987. J.Steroid Biochem. 26, 151-159.

48. Maynard, P.V., A.W. Pike, A. Weston, K. Griffiths. 1975. Eur.J.Cancer 13, 971-975.

49. Huggins, C., K. Mainzer. 1957. J.Exp.Med. 105, 485-499.

50. MacIndoe, J.H. 1988. Endocrinol. 123, 1281-1287.

51. Van Landeghem, A.A.J., J. Poortman, M. Nabuurs, J.H.H. Thijssen. 1985. Cancer Res. 45, 2907-2912.

52. Begin, D., I.A. Luthy, F. Labrie. 1988. Molec. & Cellular Endocrinol. 58, 213-219.

53. Labrie, C., A. Belanger, F. Labrie. 1980. Endocrinol. 123, 1412-1417.

54. Huot, R.I. and S.A. Shain. 1988. J.Steroid Biochem. 29, 617-621.

55. Belanger, B., A. Belanger, F. Labrie. 1989. J.Steroid Biochem. 32, 695-698.

56. Regelson, W., R. Loria, M.K. Kalimi, 1988. Ann. N.Y. Acad. Sci., 521, 260-273.

57. Gordon, G.B., L.M. Shantz, P. Talalay. 1989. In: Hormones, Thermogenesis and Obesity (H.A. Lardy, and F. Stratman, eds.). Elsevier, New York. pp. 339-354.

58. Wilpart, M., A. Speder, P. Ninane, and M. Roberfroid. 1986. Teratogenesis, Carcinogenesis, Mutagenesis. 6, 265-273.

59. Harvey, W., R. Grahme, G.S. Panay, 1971. Ann. Rheum. Dis. 32, 272-278.

60. Chakravorty, J. 1970, Br.J.Dermatol. 83, 477-481.

61. Brehm, G., H. Lachner, G.W. Oertel, P. Menzel. 1969. Arch. Clin. Exp. Derm. 234, 117-124.

62. Oertel, G.W., L. Treiber. 1969. Eur.J.Biochem. 7, 234-238.

63. Marynick, S. P., Z. H. Chakmakjian, D. L. McCaffree, J. H. Herndon, Jr. 1983. N.Engl. J. Med. 308, 981-986.

64. Bergner, E.A. and L.J. Shapiro. 1988. J.Inher.Metab.Dis. 11, 403-415.

65. Ali, S.A., E.J. Schoonen, J.G.D. Lambert. et al. 1987. Gen. Comp. Endocr. 66, 415-424.

66. Dennenbaum, R., G. Hoffman, G.W. Oertel. 1972. Horm. Metab. Res. 4, 383-385.

67. Hoffman, G., B. Modsches, U. Dohler. 1972. Rach. Derm. Forsch. 243, 18-30.

68. Holzman, H., R. Krapp, B. Morsches, et al. 1971. Aerptliche Forsch. 25, 345-353.

69. Holzman, H., B. Morsches, R. Knapp, G. Hoffman. 1973. Arch. Derm. Forsch. 247, 23-28.

70. Jolesz, M., I. Faredin, I. Toth. 1966. Acta Med.Hung. 22, 49-52.

71. Pitts, R.L.. 1987. J.Am.Acad.Dermatol. 16, 571-573.

72. Faredin, I., I. Toth, A.G. Fazekas. 1968. J. Endocr. 41, 295-296.

73. Toth, I. and I. Faredin. 1983. Acta Med.Hung. 40, 139-145.

74. Kasic, J.M., W.F. Bergfeld, W.D. Steck, and M.K. Gupta. 1983. Cleve.Clin.Q. 50, 111-122.

75. Shapiro, M.S., S. Zelefsky, R. Chayen, M.M. Werber. 1989. J. Clin. Chem. Clin. Biochem. 27, 27-31.

76. Martikainen, S., J. Heikkinen, A. Ruokonen, A. Kapuppila. 1988. J. Clin. Endocrinol. Metab. 66, 987-991.

77. Bosse, I.K., R. Fehr, G. Franz, H.J. Wiontzek. 1972. Arch. Derm. Forsch. 243, 177-187.

78. Lookingbill, D.P., N. Egan, R.J. Santen, L.M. Demers. 1988. J.Clin.Endocrinol.Metab. 67, 986-991.

79. Pitts, R.L. 1987. J.Am.Acad.Dermatol. 16, 571-573.

80. Kasick, J.M., W.F. Bergfeld, W.D. Steck, M.K. Gupta. 1983. Cleve Clin. Q. 50, 11-22.

81. Alper, M.M., M. B. Garner. 1987. Fertil. and Sterility 47, 255-258.

82. Ozasa, H., T. Tominaga, T. Takeda. 1986. ACTA Obstet. Gynecol. Scand. 65, 543-545.

83. Sakyo, K., A. Ito, Y. Mori. 1986. J. Pharmacobiodyn. 9, 276-286.

84. Sakyo, K., A. Ito, C. Ogawa, Y. Mori. 1986. Biophys. Acta. 883, 517-522.

85. Peh, K.L., R. Gunasegaram, A. Loganath. 1987. Horm. Metabol. Res. 19, 231-232.

86. Parker, G.R., Jr., K. Leveno, B. R. Carr. 1982. J. Clin. Endocrinol. Metab. 54, 1216-1220.

87. Yuen, B-H, E.K. Mincey. 1987. Am.J.Obstet Gynecol. 156, 396-400.

88. Ohshima, K., M. Hayashi, N. Nose. 1979. Acta Obstet. Gynaecol. Jpn. 31, 1853-1861.

89. Mochizuki, M., S. Tojo. 1980. In: Dilation of the Uterine Cervix, (F. Naftolin, and P.G. Stubblefield, eds.) Raven Press, New York, pp. 267-286.

90. Sakyo, K., A. Ito, Y. Mori. 1987. Biol.Repro. 36, 277-281.

91. Shikata, J., H. Sanada, T. Yamamuro, T. Takeda. 1979. Connect. Tissue Res. 7, 21-27.

92. Ozasa, H., T. Tominaga, T. Nishimura, J. Takeda. 1981. Endocrinol. 109, 618-621.

93. Ito, A., S. Katsufumi, H. Sano, et al. 1986. Chem. Pharm. Bull. 34, 2118-2125.

94. Pavlov, G. P., S. M. Harman, G. P. Chrousos, et al. 1986. J. Clin. Endocrinol. Metab. 62, 767-772.

95. Shikata, J., H.,Sanada, T. Yamamuro, T. Takeda. 1979. Connect. Tissue Res. 7, 21.

96. Orentreich, N., J.L. Brind, R.L. Rizer, J.H. Vogelman. 1984. J. Clin. Endocrinol. Metab. 59, 551-555.

97. Ruutianen, K., R. Erkkola, M.A. Gronroos, and H-L. Kaihola. 1988. Fertl. and Sterility 50, 255-259.

98. Bing, C., S.E. Xu, G.D. Zhang, and W.Y. Wang. 1988. Ann. Human Biol. 15, 421-429.

99. Smith, C.P., D.B. Dunger, A.J.K. Williams, et al.. 1989. J. Clin. Endocrinol. Metab. 68, 932-938.

100. Mann, D. R., V. D. Castracane, F. McLaughlin, et al. Biol. Repro. 28, 279-284.

101. Carlstrom, K., S. Brody, N.O. Lunell. 1988. Maturitas 10, 297-306.

102. Purifoy, F.E., L.H. Koopmans, R.W. Tatum. 1980. Human Biol. 52, 181-191.

103. Cumming, D.C., R.W. Rebar, B.R. Hopper, S.S. Yen. 1982. J. Clin. Endocrinol. Metab. 54, 1069-1071.

104. Shimomura, Y., G.A. Bray, M. Kee. Horm. Metabol. Res. 19, 295-299.

105. Nestler, J.E., J.N. Clore, J.F. Strauss, III, W.G. Blackard. 1987. J. Clin. Endocrinol. Metab. 64, 180-184.

106. Nestler, J. E., K. S. Lisiskin, C. O. Barclascini, et al. 1989. J. Clin. Endocrinol. Metab. 69, 1040-1046.

107. Buffington, C. K., J. R. Givens, A. E. Kitbachi. 1988. Clin. Res. 36, 41.

108. Mobbs, C.V. 1989. Neurohumoral hysterisis as a mechanism for aging in humans and other species: comparative aspects development, maturation and senescence of neuroendocrine systems: a comparative approach. (M.P. Schreibman and C.G. Scanes, eds.). Academic Press, New York, pp. 223-252.

109. Regelson, W. 1983. In: Intervention in the Aging Process, Part 2 (W. Regelson and M. Sinex, eds.) Alan R. Liss, New York, pp. 3-52.

110. Denckla, W.D.. 1977. Mech. Aging & Develop. 6, 143-152.

111. Harrison, D.E., J.R. Archer, C.M. Astle. 1982. J. Immunol. 129. 2673-2678.

112. Ranke, M. B., W. Rosendahl, D. Gupta. 1982. Hormone Res. 16, 32-41.

113. Sulman, F.G.. 1980. American Lecture Series, Thomas, Springfield, Ill..

114. Crudo, D. F., 1989. Proc. Am. Soc. Endocrinol. Abs. 1455. pp. 386.

115. Rogers, W. M., K. L. Blaver, W. Beanton. 1989. Proc. Am. Soc. Endocrinol. Abs. 668, pp. 189.

116. Blaver, K. L., W. M. Rogers, E. W. Bernton. 1989. Proc. Am. Soc. Endocrinol. Abs. 25. pp. 29.

117. Svec, F., A. Lopez-S., 1989. Lancet II: 1335-1336.

118. Jackson, J.A., JK.D. Fachnie, R.C. Mellinger. 1989. J. Endocrinol. Invest. 12, 269-272.

119. Kurtz, J. W., W. Wells. 1981. J. Biol. Chem. 256, 10870-10875.

120. Schiebinger, R., G.P. Chrousos, G.B. Cutler, Jr., and D.C. Loriaux. 1986. J.Clin.Endocrinol.Metab. 62, 202-209.

121. Evans, W.S., R.J. Schiebinger, D.L. Kaiser, et al. 1982. Endocrinol. 101, 235-241.

122. Vermeulen, A. and S. Ando. 1978. Clin.Endocrinol. 8, 295-303.

123. Wiebe, R. H. S. Handwerger. 1983. J. Repro. Med. 28, 206-208.

124. Oseko, F., K. Morikawa, A. Nakano, et al. 1986. Andrologia 18, 523-528.

125. Yamaji, T., M. Ishibashi, F. Takaku, et al.. 1989. Acta Endocrinol. (Copenh). 120, 655-660.

126. Pepe, G.J., B. Waddell, and E.D. Albrecht. 1988. Endocrinol. 122, 646-650.

127. Higuchi, K. et al., 1985. Horm. Metab. Res. 17, 451-453.

128. Simard, J., A. Vincent, R. Duchesne, and F. Labrie. 1988. Mol. Cell Endocr. 55, 233-242.

129. Fukieda, K., C. Faiman, F.I. Reyes, et al.. 1981. J.Clin.Endocrinol.Metab. 53, 690-693.

130. Josephs, J.A., G.S. Roth, and J.R. Whitaker. 1983. In: Intervention in the Aging Process, Part 2 (W. Regelson and M. Sinex, eds.) Alan R. Liss, New York, pp. 187-202.

131. Adams, J.B.. 1985. Molec.& Cell Endocrinol. 41, 1-17.

132. Devesa, J., R. Perez-Fernandez, L. Bokser, et al.. 1988. Horm.Metabol.Res. 20, 57-60.

133. Ward, R.C., A. Costoff, and V.B. Mahesh. 1978. Biol.Reprod. 18, 614-623.

134. Tomic, R., B. Ljungberg, and J.E. Damber. 1988. Scand.J.Urol.Nephrol. 72, 15-18.

135. Sunderland, T., C.R. Merril, M.G. Harrington. 1989. Lancet 2, 570.

136. Wade, C.E., J.S. Lindberg, J.L. Cockrell, et al.. 1988. J.Clin.Endocrinol.Metab. 67, 223-227.

137. Parker, L.N., E.R. Levin, and E.T. Lifrak. 1985. J. Endocrinol.Metab. 60, 947-951.

138. Lephart, E.D., C.R. Baxter, E.R., Parker. 1987. J. Clin. Endocrinol. Metab. 64, 842-848.

139. Bannister, P., J. Oakes, P. Sheridan, and M.S. Losowsky. 1987. Quart. J. Med. 63, 305-313.

140. Semple, C.G., C.E. Gray, G.H. Bearstall. 1987. Acta Endocrinol. (Copenhag). 116, 155-160.

141. Zumoff, B., T. Walsh, J.L. Katz, et al. 1983. J. Clin. Endocrinol. Metab. 56, 6768-672.

142. Diamond, P., G.R. Brisson, B. Carudas, F. Peronnet. 1989. Eur. J. Appl. Physiol. 58, 699-704.

143. Herecz, P., L. Ungar, P. Siklos, and R.G. Farquharson. 1988. Europ.J Obs.& Gyn.& Repro.Biol. 29, 1-5.

144. Parker, C. R. Jr., K. Leveno, B.R. Carr, et al. 1982. J. Clin. Endocrinol. Metab. 54, 1216-1219.

145. Sulcova, J., L. Starka, J.E. Jirasek. 1982. Endocrinol. Exp. 16, 9-17.

146. LaPlante C., C. Branchaud, Gates, C. Goodyear, S. Kipowski, 1983. J. Clin. Endocrinol. Metab. 56, 761-766.

147. Keller, P.J., A. Lyskienicz, R. Huch. 1983. Gynecol. Obstet. Invest. 15, 365-371.

148. Walsh, S.W.. 1988. Obstet.Gynecol. 71, 222-226.

149. Zuidema, L.J., F. Khan-Dawood, M.Y. Dawood, B.A. Work, Jr. 1986. Am. J. Obstet. Gynecol. 155, 1252-1254.

150. Moise, J., J. Ilekis, A. Scommegna, R. Benveniste. 1986. Am. J. Obstet. Gynecol. 154, 1080-1085.

151. Lardy, H., S. Ching-Yuan, N. Kneer, and S. Wieglus. 1988. In: Thermogenesis & Obesity (H. Lardy and F. Stratman, eds.). Elsevier, New York. pp. 415-426.

152. Martin, S.M. and G.P. Moberg. 1981. Life Sci. 29, 1683-1686.

153. Oppenheimer J.H., C.N. Mariash, A. Towle, et al.. 1981. Life Sci. 28, 1693-1697.

154. Casazza, J.P., W.T. Schaeffer, and R.L. Veech. 1986. J.Nutr. 116, 304-310.

155. Gonzalez, C., E. Montoya, T. Jolin, and M. Gonzalez. 1982. Endocrinol. 107, 2099-3004.

156. Bagchi, N., T.R. Brown, B. Shieris, et al.. 1981. Endocrinol. 109, 1428-1432.

157. Lemmen, C. C.A. Barth, G. Wolfram, N. Zollner. 1983. Biochim. Biophys. ACTA. 755, 137-143.

158. Lau, C.L. and T.A. Slotkin. 1980. Molec.Pharmacol. 18, 247-252.

159. Siegler, J.M., M.N. Kazalrinoff. 1983. Nutr.& Cancer. 4, 176-180.

160. Merriman, R.L., L.R. Tanzir, N.B. Stamm, et al.. 1981. 13th Int.Cancer.Cong., Seattle, Wash.

161. Seiger, A. and A.C. Granholm. 1981. Cell Tissue Res. 220, 1-15.

162. Leonard, J.L., M.M. Kaplan, T.J. Visser, et al.. 1981. Science. 214, 571-573.

163. David, S. and E.J.H. Nathaniel. 1981. Exp.Neurol. 73, 243-253.

164. Kung, M.-P., S.W. Spaulding, and J.A. Roth. 1988. Endocrinol. 122, 1195-1200.

165. Owasoyo, J.O., G.N. Bunike, C.A. Iramain. 1981. Endokrinologie. 77, 242-246.

166. Watson, A.Y., K. Radie, M. McCarthy, et al.. 1982. Endocrinol. 110, 1392-1401.

167. Ben-David, B., S. Dikstein, G. Bismuth, and F.G. Sulman. 1967. Proc.Soc.Exp.Med.Biol. 125, 1136-1140.

168. Furman, R.H. and R.P. Howard. 1962. Metabolism. 11, 76-93.

169. Natoli, C., G. Sica, V. Natoli, et al.. 1981. J.Steroid Biochem. 15, 409-413.

170. Sica, G., C. Natoli, P. Marchetti, and S. Iacobelli. 1981. J.Steroid Biochem. 15, 415-419.

171. Levin, B.E., J. Triscari, A.C. Sullivan. 1982. Am.J.Physiol. 24, R170-R178.

172. Engelken, S.F. and R.P. Eaton. 1981. Atherosclerosis. 38, 177-188.

173. Fregly, M.J. and R.M. Threatte. 1982. Life Sci. 30, 589-599.

174. Davis, P.J., F.B. Davis, and S.D. Blas. 1982. Life Sci. 30, 675-682.

175. Kurtz, B.R., J.R. Givens, S. Komidr, et al.. 1987. J.Clin.Endocrinol.Metab. 64, 1261-1267.

176. Givens, J.R., R.N. Anderson, J.B. Ragland, et al.. 1975. J.Clin.Endocrinol.Metab. 40, 988-1000.

177. Hardwick, A.J., E.A. Linton, and N.J. Rothwell. 1985. Enocrinol. 124, 1684-1688.

178. Attanasio, A., R. Robkamp, S. Bernasconi, et al.. 1987. Ped.Res. 22, 41-44.

179. Gaillard, R.C., A.M. Riondel, N. Ling, and A.F. Muller. 1988. Life Sci. 43, 1935-1944.

180. Arad, Y., J.J. Badimon, L Badimon, et al. 1989. Arteriosclerosis. 9, 159-166.

181. Gordon, G.B., L.M. Shantz, and P. Talazy. 1988. In: Hormones, Thermogenesis, & Obesity (H. Lardy and F. Stratman, eds.). Elsevier, New York. pp. 339-354.

182. Gordon, G.B., D.E. Bush, H.F. Weisman. 1988. J. Clin. Invest. 82, 712-720.

183. Lopez S.A., C. Wingo, J.A. Herbert, et al. 1976. Atherosclerosis. 24, 471-481.

184. Kask, E.. 1959. Angiology. 10, 358-68.

185. Barrett-Connor, E. and K.-T. Khaw.. 1987. N.Eng.J.Med. 317, 711.

186. Barrett-Connor, E., K.-T. Khaw, and S.S.C. Yen. 1986. N.Eng.J.Med. 315, 1519-1524.

187. Eich, D.M., D.E. Johnson, J.E. Nestler. 1990. Clin. Res. In Press.

188. Milewich, L., T. Hendricks, A. Largelius, et al. 1982. Maturits, 4, 113-118.

189. Parker, C.R., Jr., E.R. Simpson, D.W. Bilheimer, et al.. 1980. Science 206, 572-514.

190. Boizel, R., E. dePeretti, A.M. Cathiard, et al. 1986. Clin. Endocrinol.25, 363-371.

191. Forest, M.G., E. dePeretti, M. David, M. Sempe. 1982. Annales d'Endocrinol. (Paris) 43, 465-495.

192. dePeretti, G. and M.G. Forest. 1978. J.Clin.Endocr.Metab. 47, 572-577.

193. Drucker, W.D.. 1977. Bull. N.Y. Acad. Med. 53, 347-357.

194. Katz, S.H., M.L. Hediger, B.S. Zemel, J.S. Parks, 19085. Human Biol. 57, 401-413.

195. Puche, R.C., and M.D. Romano. 1968. Tiss.Res. 2, 133-144.

196. Buchanan, J.R., P. Hospodar, C. Meyers, et al. 1988. J.Clin.End.Metab. 67, 937-943.

197. Deutsch, S., F. Benjamin, V. Seltzer, et al.. 1987. J.Gynecol.Obstet. 25, 217-222.

198. Wild, R.A., J.R. Buchanan, C. Meyers, et al.. 1987. Proc.Soc.Exp.Biol.Med. 186, 355-360.

199. Brody, S., K. Carlstrom, A. Largelius, et al. 1982. Maturitas 4, 113-118.

200. Chasalow, F.I., S.L. Blethen, and J.G. Tobash. 1987. Am.J.Med.Genetics. 28, 857-864.

201. Tourney, G. and L.M. Hatfield. 1972. Arch.Gen Psychiat. 27, 753-755.

202. Tourney, G. and L.M. Hatfield. 1973. Biol.Psych. 6, 23-36.

203. Strauss, E.B.. 1955. Psychotherapie U.Med.Psychologie. 5, 2-10.

204. Recordati, Informational Pachage, Dehydroepiandrosterone industria chimica. I. Farmaceutica . s.p.a. Milano

205. Fava, M., J.F. Rosenbaum, R.A. MacLaughlin, et al.. 1988. Psych.Res. 28, 345-350.

206. Fava, M., A. Littman, and P. Halperin. 1987. Intl. J. Psychiatry in Med. 17, 289-307.

207. Fava, M., P. Halperin, and A. Littman, et al.. 1987. Neuro-Endocrinology Letters. 9, 189.

208. Warren, M.P. and J. Brooks-Gunn. 1989. J. Clin. Endocrinol. Metab. 68, 77-83.

209. Tegelman, R., K. Carlstrom, and A. Pousette. 1988. Int.J.Androl. 11, 361-368.

210. Keizer, H.A., O.P. Van Soest, H. Kuipers, and E. Beckers. 1984. Med.Sci.in Sports and Exercise. 16, 118-119.

211. Robel, P., E. Bourreau, C. Corpechot, et al.. 1987. Steroid Biochem. 27, 649-655.

212. Paroli, G.C., P. Pantalioni, P. Valer, et al.. 1978. Arch.Int. Pharm.Ther. 196, 147-150.

213. Lanthier, A. and V.V. Patwardhan. 1987. J.Steroid Biochem. 28, 697-701.

214. Levesque, W.A., A.G. Herzog, M.M. Seibel. 1986. J. Clin. Endocrinol. Metab. 63, 243-245.

215. Lacroix, C., J. Fiet, J.-P. Benais, et al.. 1987. J.Steroid Biochem. 28, 317-375.

216. Farmer, R.W. and C.A. Harrington. 1974. Brain Res. 66, 552-556.

217. Corpechot, C., P. LeClerc, E.E. Baulieu, and P. Brazeau. 1985. 45: 229-234.

218. Regelson, W. and W. Pierpaoli. 1987. Cancer Investigation. 5, 379-385.

219. Bonney, T.C., M.J. Scanlow, D.L. Jones. 1984, J. Endocr. 101, 181-188.

220. Deslypere, J.P., G. DeBiscop, A. Vermeulen. 1983. Clin. Endocrinol. 18, 25-30.

221. Montanini, V., M. Simoni, G. Chiossi, et al. 1988. Horm. Res. 19, 1-6.

222. Pierpaoli, W. and Y. Changxian. J. Neuroimmunol. 1990. In Press.

223. Koo, E., K.G. Feher, T. Feher, and G. Fust. 1983. Klin Woehenschr. 61, 715-717.

224. Bradlow, H.L., P.N. Gillette. and K.A. Gallagher. 1973. J.Exp.Med. 130, 754-763.

225. Niort, G., G. Boccuzzi, E. Brignardello, et al. 1985. J. Steroid Biochem. 23, 657-661.

226. Pashko, L.L., D.K. Fairman, A.G. Schwartz. 1986. J. Gerontol. 41, 433-438.

227. Lucas, J.A., S.A. Ahmed, M.L. Casey, P.C. MacDonald. 1985. J. Clin. Invest. 75, 2091-2093.

Author Index

Subject Index

Bei Fragen zur Produktsicherheit wenden Sie sich bitte an:
If you have any questions regarding product safety,
please contact:

Walter de Gruyter GmbH
Genthiner Straße 13
10785 Berlin
productsafety@degruyterbrill.com